The *Higher Calculus:*
A History of Real and Complex Analysis
from Euler to Weierstrass

Umberto Bottazzini

The *Higher Calculus:* A History of Real and Complex Analysis from Euler to Weierstrass

Translated by Warren Van Egmond

Springer-Verlag
New York Berlin Heidelberg
London Paris Tokyo

Umberto Bottazzini
Università degli Studi di Bologna
Dipartimento di Matematica
40127 Bologna
Italy

Warren Van Egmond *(Translator)*
Arizona Center for Medieval
  and Renaissance Studies
Arizona State University
Tempe, Arizona 85287
U.S.A

With 8 Illustrations

AMS Classification: 01A50, 01A55, 26-03, 30-03

Library of Congress Cataloging in Publication Data
Bottazzini, U. (Umberto)
  The higher calculus.
  Bibliography: p.
  Includes index.
  1. Mathematical analysis—History.   2. Mathematics—
History—19th century.   I. Title.
QA300.B67   1986   515'.09   86-6623

Il Calcolo sublime: storia dell'analisi matematica da Euler a Weierstrass.© 1981 Editore
Boringhieri società per azioni Torino, corso Vittorio Emanuele 86.

Printed and bound by R.R. Donnelley and Sons, Harrisonburg, Virginia.
Printed in the United States of America.

9 8 7 6 5 4 3 2 1

ISBN 0-387-96302-2 Springer-Verlag New York Berlin Heidelberg
ISBN 3-540-96302-2 Springer-Verlag Berlin Heidelberg New York

To Vieri

CONTENTS

INTRODUCTION

"The true method of foreseeing the future of mathematics is to study its history and its actual state."

With these words Henri Poincaré began his presentation to the Fourth International Congress of Mathematicians at Rome in 1908. Although Poincaré himself never actively pursued the history of mathematics, his remarks have given both historians of mathematics and working mathematicians a valuable methodological guideline, not so much for indulging in improbable prophecies about the future state of mathematics, as for finding in history the origins and motivations of contemporary theories, and for finding in the present the most fruitful statements of these theories.

At the time Poincaré spoke, at the beginning of this century, historical research in the various branches of mathematics was emerging with distinctive autonomy. In Germany the last volume of Cantor's monumental *Vorlesungen über die Geschichte der Mathematik* had just appeared, and many new specialized journals were appearing to complement those already in existence, from Eneström's *Bibliotheca mathematica* to Loria's *Bollettino di bibliografia e di storia delle scienze matematiche*. The annual *Jahresberichte* of the German Mathematical Society included noteworthy papers of a historical nature, as did the *Enzyklopädie der mathematischen Wissenschaften*, an imposing work constructed according to the plan of Felix Klein.

Towards the end of his life, during the war years from 1914 to 1919, Klein himself held seminars on the history of recent mathematics at his home for a select number of participants. These lectures were collected by Courant and Neugebauer and published posthumously as the *Vorlesungen über die Entwicklung der Mathematik im neunzehnten Jahrhundert*. Even today they remain the best available comprehensive history of mathematics in the last century.

Klein had originally conceived of his lectures as part of a grand project to synthesize "the culture of the present day," and it was this spirit that animated his conception of history. Klein proceeded by themes, and although he did not omit biographical information about their creators, the real object of his research was the mathematical arguments themselves, the connections between them, and the motives that lay behind their development.

On this point the views of Klein and Poincaré seem to have been much alike. Notwithstanding their differing points of view and the arguments that often separated them, they both conceived of a history of mathematics "for the present."

Although schematic, this view is opposed to that which sees history as a collection of more or less curious notes gathered by antiquarian specialists in mathematics, to which we often find attached the fairly widespread idea that a research mathematician has no real need to know the history of his field "because his job is to refute it," as Cavailles has written (1938, p. 8). This is indeed a regrettable attitude when it leads students to include Euler with Eudoxus and Euclid among the mathematicians of classical antiquity, or causes a young research assistant to discover by chance, in the course of a conversation I once overheard, that Lie is not a contemporary Chinese mathematician who invented homonymous algebra, but a Norwegian of the last century.

But, even if we treat the history of mathematics "for the present," there are still different points of view. One fairly widespread view is to see the development of mathematics as a kind of "comedy of errors" which finally redeemed itself by taking on a definitive and rigorous form only in the most recent past. By following this approach we arrive at a *formal reconstruction* of the development of mathematics, one that reveals the errors, misunderstandings, and "lucky intuitions" of the mathematicians of the past *from our point of view*. This effort is perhaps useful, but history cannot limit itself to this. In fact, if we want to understand the real dynamics of the development of mathematics, it is essential to emphasize the uniqueness of the contexts and motives that surround each development, the changing points of view, the contradictions, the generalizations, and the juxtapositions of various theories.

"Science advances by a series of combinations, in which chance does not play a minor role," Galois wrote more than a century ago. "Its path is rough and resembles those of minerals that grow by juxtaposition" (1962, p. 15). In a note found among his papers we read this further, penetrating observation: "It is generally believed that mathematics is a series of deductions" (1962, p. 19). It is an observation that merits rethinking today, in the face of the repeated statement that mathematics is a "hypothetico-deductive system."

Certainly, Galois writes, if

the whole of mathematical truth ... could be deduced regularly and machine-like from a few principles in combination with a uniform method, then no more obstacles, no more of those difficulties that the scholar finds in his explorations, which are often imaginary, but also no more place for the scholar.   It is not so; if the task of the scholar is more laborious and therefore more attractive, the march of science is also less regular (1962, pp. 13, 15).

It is thus possible to understand history "for the present" by accentuating the "less regular" paths of science and, by reversing Lakatos' famous joke (1976, p. 395), to return "real" history to the center of the investigation with his "rational reconstruction" confined to the notes, abandoning with it the idea of an uninterrupted, continuous development towards ever more rigorous theories and rejecting the determinism implicit in this thesis.

In fact, mathematical rigor is itself a "historical" concept and consequently in process. "Personally we do not believe that absolute rigor will ever be attained and if a time arrives when this is thought to be the case, it will be a sign that the race of mathematicians has declined," the mathematician and historian James Pierpont asserted sixty years ago at the annual meeting of the American Mathematical Society (Pierpont, 1928, p. 23).   His views were anticipated by Picard, Lie, and Klein, among others.

Indeed, calling upon the needs of rigor to explain the development of mathematics constitutes a circular argument. In actual fact, new standards of rigor are formed when the old criteria no longer permit an adequate response to questions that arise in mathematical practice or to problems that are in a certain sense external to mathematics. When these are treated mathematically, they compel changes in the theoretical framework of mathematics. It is thus not by chance that mathematical physics and applied mathematics have generally been formidable stimuli to the development of pure mathematics.

Nor is it purely accidental that new criteria of rigor have more often appeared in the formulation of *definitions* than in *demonstrations*. Definitions in fact appear within the complex structure of a "mature" theory and are subsequent to true mathematical discovery.

Moreover, different notions of rigor often correspond to radically different ways of understanding mathematics. The case of Riemann and Weierstrass is famous. The first expresses a "geometrical" and "physical" style of analysis, the other an "arithmetical" one. It is well known that Weierstrass, together with Kronecker, did not consider Riemann to be particularly rigorous.

Therefore, if our conception of mathematical rigor has changed radically with respect to that of Euler or of Cauchy or of Riemann, the most instructive thing, particularly when thinking of history "for the present", is not so much to present historical theories and results with contemporary formalism and rigor, but to understand the ideas thay lay at the origins of these theories together with the motives and problems that led to their generalizations and mutations, the field of rigor included. These ideas often have complex connections and causes both within the theory itself and within the wider context of science, culture, and society at a particular historical moment.

In this connection there are two ways of understanding the development of mathematics, which are usually and somewhat schematically called "internalist" and "externalist" history. This artificial distinction has been the cause of numerous debates. But it is clear that maintaining a rigid separation between the "externalists" and the "internalists" by positioning oneself on one or the other side of the theoretical divide that separates them is a sterile approach. The *real* development of mathematics, in fact, cannot be understood if we limit ourselves to one of these two aspects.

"To deal with conceptions seems to me the chief present function of history," Jourdain wrote at the beginning of this century (1913, p. 663) using the term "conceptions" in a psychological sense. It seems to me that we can accept his opinion, provided we give a much wider meaning to these "conceptions." They must include not only elements deriving from specific relationships within a particular scientific discipline, but aspects and influences arising in the social and cultural environment in which mathematicians work. Jourdain continues by saying,

> Some are of opinion that history is an end in itself, and some think that the only good in history is its heuristic value. It seems clear to me that history provides an enormously valuable -- perhaps the only -- means of attaining a just idea of our knowledge by stimulus to criticize. Then, it gives a stimulus to original discovery; and then, again to criticism (*ibid.*).

It is an opinion that I hope the following pages can serve to confirm.

This book originated with a series of lectures on nineteenth-century analysis which I gave at the University of Calabria in 1977-79. The study of nineteenth-century mathematics presents an extraordinarily rich and fertile field for historical research, particularly when we realize that the production of mathematics in the last century was greater than in all the preceding centuries put together, and that many contemporary theories have their roots in this period.

Adopting the words with which Volterra characterized the past century at the International Congress of Mathematicians held in Paris in 1900, we can speak of the "century of the theory of functions" and attempt to give an idea of nineteenth-century mathematics by tracing the history of real and complex analysis. In reality, the name "theory of functions" at that time included much more than it does today and the term "analyst" had a connotation as vast as that which "geometer" had had before it. Thus, for example, a fundamental part of analysis in the nineteenth century consisted of the theories of elliptic and Abelian functions, which today are generally taken to be parts of the history of algebraic geometry and its origins (see, for example, Dieudonné, 1974).   It was the same for the theory of invariants and the theory of algebraic functions.

There are two themes that cut across the history of mathematical analysis in the nineteenth century and form a kind of guiding line. One is the theory of series, particularly Fourier and power series; the other is the theory of real and complex integration. Attached to these are the problem of the foundations of the infinitesimal calculus, the theory of sets, the changes in the concept of function, and Cauchy's, Riemann's, and Weierstrass' theories of functions of a complex variable. In addition, the related problems of mathematical physics (potential theory and harmonic functions) are naturally connected with these. To maintain a separation between these topics, as I have done here, is required only by the needs of exposition.

In the historical literature about nineteenth-century analysis, an exceedingly large amount of attention has been devoted to real analysis and the foundations of the calculus with respect to complex analysis. If one were to judge from the number of volumes devoted to the first two subjects, one would conclude that the mathematicians of the nineteenth century were entirely preoccupied by questions of rigor and the foundations of real analysis. But in reality these problems, although very important, became primary only in certain moments. In my view we can fully understand these problems only if we are aware of the close ties that connected the various parts of analysis and the great importance that complex analysis eventually acquired, both in research and in teaching. It is in this respect that Weierstrass' work acquires its value as a paradigm. These considerations explain the substantial differences between the English and Italian editions of this book.

I hope that this volume can be of some use not only to students of the history of mathematics, but more generally to university students taking their first course in mathematical analysis. The level of knowledge required for reading this book is not above that acquired in the first years of university study and perhaps an historical presentation can provide motives and reasons for understanding the ideas and theorems that usually appear in the textbooks of analysis.

Following the appearance of the Italian edition of this book, many friends and colleagues came forward with suggestions and critical remarks that have contributed to this new English edition in more or less direct ways.

I would like to thank Clifford A.   Truesdell who initially suggested making a translation of this book, and called my attention to the need for clarifying Euler's numerous contributions to the topics discussed in Chapter I. A. P. Yushkevich's suggestions on these and other topics were also extremely useful. I would further like to thank Ivor Grattan-Guinness for his careful reading of a preliminary version of this book and the wealth of suggestions he gave me.   I am also indebted to him for a great deal of information about the French mathematicians which I acquired by reading preliminary drafts of his imposing work on this subject which is currently under preparation, as well as from numerous private communications. Many reflections on aspects of complex analysis were stimulated by the fruitful collaboration and the long and pleasant conversations I had with Jeremy Gray.

Equally stimulating were the exchanges of ideas on various issues with Jean Dhombres, Christian Houzel, Pierre Dugac, Jesper Lützen, and Henk Bos. I would also like to thank Hubert C. Kennedy, who pointed out various misprints in the Italian version of this book.

In addition, I am indebted to my many colleagues for the leads they gave me during the periodic meetings on the history of mathematics at Oberwolfach. For this I must above all thank Christoph J. Scriba, who gave me this opportunity by inviting me to these annual conferences.

I am particularly indebted to Springer-Verlag, who were willing to publish the English edition of this book and for their competent help in the realization of this plan. I am also grateful to Warren Van Egmond for making the English translation and for the care he took in including the many revisions and changes with respect to the Italian edition, as well as for the accurate revision of the bibliography.

Finally, there is one person who has contributed more to this work than any other, although indirectly and with complete indifference to its content. This is my son, Vieri, who by growing and making me play with him, has made me understand many things better, and by tolerating the many times that I was away from him, taught me, among other things, to fully value the importance of time.

Chapter 1
THE ELEMENTS OF ANALYSIS IN THE EIGHTEENTH CENTURY

1.1. Euler's Concept of a Function

In the "Summary of Results" to the first book of Bourbaki's *Elements of Mathematics*, which is dedicated to the fundamental structures of analysis, we find this definition of functional dependence:

> Let E and F be two sets, which may or may not be distinct. A relationship between a variable element x of E and a variable element y of F is called a *functional relation in y* if, *for all $x \in E$, there exists a unique $y \in F$ which is in the given relation with x.*
> We give the name of *function* to the operation which in this way associates with every element $x \in E$ the element $y \in F$ which is in the given relation with x; y is said to be the *value* of the function at the element x, and the function is said to be *determined* by the given functional relation. Two *equivalent* functional relations determine the *same* function (Bourbaki, 1968, p. 351).

The concept of function here appears to be firmly based on the theory of sets, in terms of which the idea of a functional relation between two sets is expressed as a subset of the Cartesian product $E \times F$. This has today become the standard definition.

The concept of a function as a mapping between abstracts sets thus appears to be the outcome of a debate that has accompanied the history of analysis since the origins of the calculus in the second half of the seventeenth century.

To be sure, according to some historians the idea of functional dependence, like so many of the fundamental concepts of mathematics, already appears in older texts. Bell has even spoken of the "instinct for functionality" found in the Babylonians (1940, p. 32). Certainly some idea of function was already present in antiquity in particular cases. Empirically tabulated functions were commonly used in ancient astronomy, as were tables for finding squares, cubes, square and cube roots, and the like. And yet, as Yushkevich has observed, "the mathematical thought of antiquity created no general notion of either a variable quantity or of a function" (1976, p. 44).

While Arabic science does not seem to have contributed any substantially new elements to the development of the concept of functionality, the same cannot be said for medieval scientists like Bradwardine and Oresme or for the "schools" of Oxford and Paris. The study of the "latitude of forms," even if it did not present any substantive innovations with respect to the techniques of the algebraic calculus, made familiar the idea of magnitudes and quantities being dependent on other magnitudes and quantities, in particular on time, and introduced the first rudiments of graphical representation. Nevertheless, it was not until the end of the sixteenth century and the beginning of the seventeenth that advances in algebraic symbolism made it possible to translate the mathematical study of variable magnitudes into analytic terms. The primary motivation for this lay in the study of kinematics.[1]

The interaction with problems of a physical nature is a recurrent phenomenon in the development of mathematics, both in ancient and modern times, and not only in the vast field encompassing the physical applications of mathematical results. In fact, there are many mathematical theories, sometimes even those that seem the most abstract at first sight, which have their origins and motives in physical research.

The study of variable quantities is the step that clearly separates classical mathematics and medieval science from modern mathematics. According to the German mathematician Hankel (1839-1873), "modern mathematics dates from the moment when Descartes went beyond the purely algebraic treatment of equations to study the variation of magnitudes that an algebraic expression undergoes when one of its generally denoted magnitudes passes through a continuous series of values" (Hankel, 1870, p. 63).

This is a decisive step whose ultimate result was the infinitesimal calculus. It is the calculus that accounts for the radical difference between modern analysis and the geometric techniques of the ancients, as well as between the kinematics of Galileo (1564-1642) and the dynamics of Newton (1643-1727).

Likewise, it is with Newton that there emerges, on the basis of physical motivations, a strict relationship between the concepts of function and those of variation and the fluxional calculus. For Newton the motion of bodies is the center of research; the method of fluxions offers a mathematical instrument to describe variations in fluent magnitudes (i.e. functions), things that "have a true place in nature."[2]

Leibniz (1646-1716), on the other hand, was primarily interested in the study of curved lines, the problems of tangents and of inverse tangents, and arrived at an elaboration of the fundamental concepts of the differential calculus by this means (see Hofmann, 1949). Both the idea of function and the distinction between algebraic and transcendent curves occurred to Leibniz when he faced problems of a geometrical nature connected with the new calculus.[3]

The term "function" appears for the first time in print in Leibniz' articles of 1692 and 1694,[4] but in these years the discussion of the concept of function and the symbols used for representing functions (the characteristic, in Leibniz' terminology) arose repeatedly in the correspondence between Leibniz and Johann I Bernoulli (1667-1748). For example, in a letter dated September 2, 1694, where Bernoulli sends Leibniz the expansion of the integral $\int n\ dz$ in an infinite series

$$nz - \frac{z^2}{2!}\frac{dn}{dz} + \frac{z^3}{3!}\frac{d^2n}{dz^2} - \cdots$$

(which was already known to Leibniz), he adds, "by n I understand a quantity somehow formed from indeterminate and constant [quantities]."[5]

Bernoulli later published the following definition of a function in an article on the problem of isoperimeters which appeared in the Mémoires of the Paris Academy in 1718: "I call a function of a variable magnitude a quantity composed in any manner whatsoever from this variable magnitude and from constants." (*Opera* 2, p. 241).

This definition was used repeatedly throughout the eighteenth century, reflecting the greater success of Leibniz' approach and that of his followers with respect to Newton's.[6]

The definition of function given by Leonhard Euler (1707-1783), who was surely the most original mathematician of the century and the most productive of all times,[7] was entirely consistent with this. He defined it as follows: "A function of a variable quantity is an analytic expression composed in any way from this variable quantity and from numbers or constant quantities" (1748, I, p. 18). This is the definition that we read at the beginning of the first volume of the *Introductio in analysin infinitorum* (1748), the "standard" treatise of eighteenth-century analysis.

For Euler the term "analytic expression" meant an expression composed of symbolic magnitudes and numbers by means of *algebraic operations* (which include addition, subtraction, multiplication, division, raising to powers, and the taking of roots, "to which the resolution of equations is also to be added"), *or transcendental operations*, which include the exponential and the logarithmic "and innumerable others which the integral calculus supplies in abundance" (1748, I, p. 19).

This distinction correlates with the distinction between algebraic and transcendental functions: the first are obtained by means of a finite number of elementary operations (it was Euler's opinion that algebraic equations could in principle be resolved algebraically), while the second could certainly be expanded in infinite series (or at least by means of an infinite number of elementary operations) without much difficulty.   Neither the demonstration nor the legitimacy· of such expansions posed much difficulty, or so Euler maintained.

Thus, in Chapter 4 of the *Introductio*, he considered the most general means of expressing a function to be an infinite series of the form

(1.1.1) $A + Bz + Cz^2 + Dz^3 + \cdots$.

"If anyone doubts this," Euler writes, "this doubt will be removed by the expansion of every function" (1748, I, p. 74).

However, since Euler could not actually *demonstrate* that any function $f(z)$ can be expanded in a series of ascending powers of z, he left open the possibility of using any exponents of z in the expansion (1.1.1). He expressed this possibility in the following terms:

> In order that this explanation may be made even more extensive, in addition to the powers of z having positive exponents, any power whatsoever ought to be admitted. Then there will be no doubt that every function of z can be changed into an infinite expression of this kind,

(1.1.2) $Az^\alpha + Bz^\beta + Cz^\gamma + Dz^\delta + \cdots$,

> the exponents α, β, γ, δ, etc., denoting any numbers whatsoever."
> (*ibid.*)

Therefore any function of z can be expressed as the finite or infinite sum (1.1.2). It is precisely this property that Euler has in mind when he expands functions in infinite series or products or into continuous fractions, which are so common in the *Introductio*.

This is the case, for example, for the elementary transcendental function $\log x$, for which Euler gives the expansion

$$\log(1 + x) = x - \frac{x^2}{2} + \frac{x^3}{3} - \cdots$$

by using Newton's binomial series,

$$(1 + x)^m = 1 + mx + \frac{m(m - 1)}{1 \cdot 2} x^2 + \cdots ,$$

with which he expands the expression

$$(1 + x)^{1/i} ,$$

where i is a "number larger than any preestablished quantity" and x

is sufficiently small.

Besides, as Yushkevich has noted (1976, p. 63), the functions used in analysis at Euler's time were, for the most part, analytic in the sense common today, except at isolated points of the defining domain. In particular cases fractional or negative exponents could be present. Euler observes that the expression (1.1.2) can fail in "exceptional" isolated points, but he substantially limits himself to considering algebraic functions and extends their properties to transcendental functions in general.

> Since one could directly derive the expansion in series of algebraic functions according to the powers of an increment, the derivatives, and the integral, one not only held that it was possible to assume the existence of such a series, derivative, and integral for all functions in general, but one never even had the idea that herein lay an assertion, whether it now be an axiom or a theorem -- so self-evident did the transfer of the properties of algebraic functions to transcendental ones seem in the light of the geometrical view of curves representing functions. And examples in which purely analytic functions displayed singularities that were clearly different from those of algebraic functions remained entirely unnoticed (Hankel, 1870, pp. 64-5).

Hankel supports his argument with examples of functions like

$$\sin(1/x)e^{1/x}, \quad 1/(1 + e^{1/x}), \quad \int^x dx/(1 + e^{1/x}), \quad \text{etc.,}$$

taken at the point $x = 0$.

It is certainly not difficult, from our point of view, to show the inadequacy of Euler's classification of algebraic and transcendental functions. In fact, the algebraic or transcendental character of a function cannot be revealed by the particular kind of "analytic expression" used in its definition. Thus, infinite series of increasing powers of x can define *algebraic* functions, as is the case for

$$y = 1 + \frac{x^2}{2} - \frac{x^4}{2 \cdot 4} + \frac{1 \cdot 3}{2 \cdot 4} \frac{x^6}{6} - \cdots = +\sqrt{1 + x^2}, \quad \text{for } |x| \leqslant 1,$$

as well as *transcendental* ones, as with

$$y = \frac{x^2}{1} - \frac{x^4}{2} + \frac{x^6}{3} - \cdots = \log(1 + x^2), \quad \text{for } |x| \leqslant 1$$

(Pringsheim, 1899, pp. 5-6). Pringsheim himself stresses that only rational functions can be expressed by means of a finite number of elementary operations.

Euler's method of reasoning, in which the supposed analogy between the finite and the infinite plays an essential role, is typical of the epoch. In addition, some procedures that are today held to be illegitimate were not unrigorous for Euler; they are so only with respect to our modern criteria of rigor, which have passed through the filter of two hundred years of the development of analysis. This is an important point, which it is essential to understand if we want to comprehend the *real* development of mathematics and not see it deformed by the "rational" lens of contemporary criticism.

1.2. Working with "Imaginary" Quantities

In the very first pages of the *Introductio*, Euler had asserted that a variable can assume any value whatsoever, including an "imaginary" one. For example, in order to clarify that "a function of a variable quantity is itself a variable quantity," he asserted that there is no value which the function is not capable of assuming, "since a variable quantity also includes imaginary values" (1748, I, p. 18).

He did not give an explicit definition of "imaginary" numbers in the *Introductio*, as he would later do in the *Vollständige Einleitung zur Algebra* of 1770,[8] but he did demonstrate a great mastery in the manipulation of such quantities.

Complex variables made their official appearance in §30 of Volume I of the *Introductio*, where Euler demonstrates the theorem of factorization of a polynomial Z in $\mathbb{C}$. He had previously asserted that the linear factors into which Z is split can be found by determining all the roots of $Z = 0$, including the "imaginary" ones. He now adds that these simple factors are real or imaginary, according to the nature of the root, and that, in the latter case, they are even in number.

In §§100-102 Euler then defines the function $\exp(z)$ for $z \in \mathbb{C}$ and its inverse, $z = \log y$. He defines the latter, however, only for real positive numbers. As for the logarithms of negative numbers, Euler here limits himself to asserting that these "are not real, but imaginary" (1748, I, p. 107).

The definition of $\exp(z)$ enables him to obtain the celebrated "Euler formulas,"

$$\cos v = \frac{e^{iv} + e^{-iv}}{2} \ , \quad \sin v = \frac{e^{iv} - e^{-iv}}{2i} \ ,$$

and their companions for the other trigonometric functions, as well as the exponential form of complex numbers, $e^{iv} = \cos v + i \sin v$ (*ibid.*, §138).

The delicate question of the logarithms of negative numbers was presented in the second volume of the *Introductio* (§§515-517). Here Euler sought to justify his assertions concerning the nature of such numbers with arguments that he himself did not hestitate to call paradoxical.

Thus, in studying the logarithmic curve

$$y = ae^{x/b},$$

Euler asserted that it is formed from two parts which are similar with respect to the asymptote; one is continuous while the other is formed from an infinite number of points corresponding to the values of the exponent x/b given by irreducible fractions of the form $2k + 1/2h$, with h and k integer. "It follows," he says, "that the logarithmic [curve] will have an infinite number of discrete points below the asymptote which do not form a continuous curve, even though with their infinitely small intervals (*par leur rapprochement infiniment grand*) they seem like a continuous curve. This is a paradox which never occurs with algebraic curves" (1748, II, p. 293).

Similarly, he adds, (§517) the points corresponding to the equation $y = (-1)^x$ form two infinitely dense sets of points distributed on the lines $y = 1$, $y = -1$.

But this is not the only paradoxical fact that accompanies these transcendentals. For example, Euler asserts that $\log(-1) = -\log(-1)$, even though $\log(-1) \neq 0$, because $-1 = +1/-1$. He thus concludes that "every number has an infinite number of logarithms, among which no more than one is real" (1748, II, p. 294). This assertion is less surprising that it might seem at first sight, Euler adds, since from $x = \log a$ it follows that $a = e^x$, and hence, by expanding in series,

$$a = \sum_{n=0}^{\infty} \frac{x^n}{n!}$$

which is an equation of infinite degree. It is therefore not particularly strange that it has an infinite number of roots!

The question about the logarithms of negative numbers became a major issue in 1745, the year after the *Introductio* was written, when Cramer (1704-1752) published the correspondence between Leibniz and Johann I Bernoulli (Cramer, 1745). In fact, at the beginning of the eighteenth century there had been a long controversy between these two mathematicians. While Leibniz had held the view that the logarithms of negative numbers are imaginary, Bernoulli concluded from the equivalence of $d \log x = d \log(-x) = dx/x$ that $\log x = \log(-x)$ and therefore that these logarithms were real.

Euler himself, about twenty years earlier, had had the opportunity to discuss this issue with Bernoulli, and had explained to him the difficulty one faced in studying the function $y = (-1)^x$. In a brief exchange of letters around the end of 1727 and the beginning of 1728, Bernoulli had reaffirmed his convictions, but Euler had remained unconvinced.

The publication of the correspondence between Bernoulli and Leibniz thus gave Euler an opportunity to reopen the entire matter. He criticized the arguments of both men and presented the solution to the entire matter in a letter written to D'Alembert (1717-1783) on December 29, 1746.[9]

"The logarithm of every number," he wrote, "has an infinite number of different values, among which there is only one that is real when the number is positive. But when the number is negative, all the values are imaginary" (*Opera* (4a) 5, p. 252).

Thus, for any number $a > 0$, $\log a = \log a + \pi(0 \pm 2n)\sqrt{-1}$, "where in the second member $\log a$ indicates the ordinary logarithm of a" and $\log(- a) = \log a + \pi(1 \pm 2n)\sqrt{-1}$ (*ibid.*).   "All of this," Euler concludes, "follows from the formula $\log(\cos \theta + i \sin \theta)^k = (k\theta + 2mk\pi \pm 2n\pi)\sqrt{-1}$, where m and n indicate any whole numbers.   The truth of this is easy to demonstate" (*ibid.*, pp. 252-3).   The next year, in fact, Euler presented two different demonstrations to the Berlin academy.[10]

D'Alembert, however, replied to Euler, "Although your reasons are very formidable and very learned, I admit, sir, that I am not yet completely convinced" (*ibid.*, pp. 257-8).   In fact, the French mathematician, without apparently having been aware of it at first,[11] had adopted an opinion about the logarithms of negative numbers which was similar to that of Johann Bernoulli.   Euler's letter marked the beginning of a long dispute with D'Alembert, who subsequently reasserted his own ideas 15 years later (D'Alembert, 1761a).   For his part, Euler, after allowing a little time to pass, abandoned   the   arguments   being   continually   reasserted   by D'Alembert, seeing the impossibility of convincing him.

"The dispute over logarithms is only one example of discussions sometime scientific, sometimes personal, that are pursued in the correspondence between D'Alembert and Euler," Yushkevich and Taton note (*Opera* (4a) 5, p. 19).

Their discussions in fact ranged over a variety of topics, from the theory of the perturbations of the orbits of Saturn and Jupiter to the extraordinary proposal advanced by Clairaut (1713-1765) (which he later retracted) of introducing a corrective factor into Newton's law to account for the irregularities observed in the motion of the moon, and from concrete problems of hydrodynamics to the study of the vibrations of a homogeneous string fixed at its end points (see §1.3 below).

In the correspondence between Euler and D'Alembert mechanical arguments intertwine naturally with those of geometry or of analysis.   A good example of this is the study of singularities of algebraic curves or the problems posed by the theory of integration.

The latter was the subject of a paper that D'Alembert had presented to the Berlin academy (D'Alembert, 1748).   In order to reduce the integral of a rational function $\int [P(x)/Q(x)]\ dx$ to a sum of trigonometric and logarithmic functions, D'Alembert was forced to discuss the problem of factoring the polynomial $Q(x)$ into factors of the first and second degree.   His *Recherches sur le calcul integral* (1748) thus began with a demonstration of the fundamental theorem of algebra, that "every rational polynomial with no divisor which is composed of a variable x and of constants, when it is of an even degree, can always be divided into trinomial factors $xx + fx + g$, $xx + hx + i$, all of whose coefficients f, g, h, i etc. are real" (D'Alembert, 1748, p. 182).

Fundamentally, this idea was not new. More than one hundred years earlier the French mathematician, A. Girard (1595-1632) had expressed it more or less clearly in his *Introduction nouvelle en l'algebre* (1629), as had Descartes (1596-1650) in his *Geometrie* (1637). What was completely original was the path taken by D'Alembert to demonstrate it.

He began by establishing the following theorem:

Let *TM* be any curve whose coordinates $TP = z$, $PM = y$ and in which $y = 0$ or ∞ when $z = 0$. If one takes z positive or negative, but infinitely small, the value of y in z can always be expressed by a real quantity when z is positive, and when z is negative, by a real quantity or by a quantity $p + q\sqrt{-1}$, in which p and q are both real (1748, p. 183).

In the demonstration he notes (and this is the decisive point) that in a sufficiently small neighborhood of the origin, y is expandable in a "very convergent" series,

$$(1.2.1) \qquad y = \sum_n a_n z^{r_n} ,$$

where the exponents r_n are given by a sequence of increasing, positive rational numbers. D'Alembert then asserts that, if one takes positive values of z close to zero, then the corresponding values of y are real, while when z is negative, y will have real values if there are no fractions with even denominators in the sequence r_n. In the opposite case, by setting

$$z^{m/2h} = e + f\sqrt{-1}, \qquad (m \text{ odd}),$$

where e and f are real quantities, the ordinate y corresponding to (1.2.1) will be a complex value $p + q\sqrt{-1}$. This case is verified even if, for negative z, parts of the terms of the series are real and parts are complex. In other words, D'Alembert concludes that for (positive or negative) values of z in the neighborhood of zero, (1.2.1) will give real or complex values of the form $p + q\sqrt{-1}$ for y.

In the corollaries that follow the demonstration, D'Alembert tacitly abandons the hypothesis that the curve has an asymptote at the origin. If one increases the abscissa by an infinitesimal quantity, the ordinate would have a corresponding infinitesimal (real or complex) increment. Similarly, he adds, we can conclude that for a finite increment, the corresponding ordinate can be expressed as $p + q\sqrt{-1}$, "at least up to a certain term."

"It is thus clear," D'Alembert concludes, "that since the value of y in z is infinitely convergent when z is infinitely small, one can assume a finite value for z, such that the corresponding value of y is also expressed by a very convergent series. And if one imagines that this series, which is wholly composed of an infinite number of terms, is substituted into the equation of the curve in place of y, the result of this substitution will be infinitely small or zero, whether for the case of z positive, or for the case of z negative" (1748, p. 187). (In the latter case, for complex values of y, it is necessary to consider the modulo of the corresponding value of y.)

D'Alembert then shows that, "whatever may be the finite quantity CQ by which one increases the abscissa AC, the corresponding imaginary ordinate can always be supposed equal to $p + q\sqrt{-1}$" (ibid., p. 187).

D'Alembert reasons by reduction to the absurd. He supposes that α is the smallest value for which the corresponding ordinate of the curves is $p + q\sqrt{-1}$. But, from what was said above, by varying α by an infinitesimal quantity, the corresponding value of p can be supposed equal to $t + i\sqrt{-1}$ and that of $q = \beta + \delta\sqrt{-1}$, where t, i, β, δ are real. In fact, D'Alembert notes, the value of p and q in α will be given by two equations obtained by substituting $p + q\sqrt{-1}$ in place of y in the equation of the curve (which, however, he does not write) and equating both the real part and the coefficient of the imaginary part to zero. Transforming "by the known methods" these two equations in z, p, and q into two equations in z and p, z and q respectively, these latter can be interpreted geometrically as curves and one can apply the theorem given above. Then, by varying α by an infinitesimal quantity (and hence also by a finite quantity) the corresponding value for every curve will be given by $t + i\sqrt{-1}$ and $\beta + \delta\sqrt{-1}$ respectively, and finally by $t - \delta + (i + \beta)\sqrt{-1}$. This contradicts the hypothesis that $\alpha \neq 0$ is the minimum. Consequently, there exist real or complex values of y for which $z = 0$.[12]

D'Alembert can then conclude,

Let $x^m + ax^{m-1} + \cdots + fx + g$ be a polynomial such that there is no real quantity which, when substituted in place of x, makes all the terms disappear. I say that there will always be a quantity $p + q\sqrt{-1}$ to substitute in place of x which renders this polynomial equal to zero (1748, p. 189).

As an immediate consequence of this theorem, D'Alembert notes that the polynomial is divisible by $x \pm (p + q - \sqrt{-1})$ and therefore that it can be split into factors of the second degree with real coefficients.

He then shows that any expression with imaginary quantities can always be rewritten in the form $p + q\sqrt{-1}$. He finally applies the results thus obtained to the calculus of integrals.

On December 29, 1746, Euler wrote to D'Alembert,

The way in which you prove that every expression $x^n + Ax^{n-1}$ + etc. $= 0$ which has no real roots must have one of the form $p + q\sqrt{-1}$, and that consequently it must have a factor of this form, $xx + ax + b$, completely satisfies me. But since it proceeds by

expanding the value of x in an infinite series, I am not certain
that everyone will be convinced (Euler, *Opera* (4a) **5**, p. 252).

Euler here isolates one of the weak points of D'Alembert's
demonstration, that of the existence of the local uniformization
parameter (1.2.1), which would be be demonstrated more than 100
years later by Puiseux (see §4.6 below).

For his demonstration of this same theorem Euler had in fact used
an algebraic proof, which he announced to D'Alembert in the
following manner: "I have recently read a note on this same subject
at a meeting of our Academy, where I demonstrated in a manner
that must be within the reach of everyone that every expression x^n +
Ax^{n-1} + Bx^{n-2} + etc., if n is a binary power, is solvable in real
factors of the form $xx + ax + b$, and from this the same thing
clearly follows for equations of every degree" (*ibid.*).

In his (1751), Euler began by demonstrating his claim for equations
of degree four, discussing among other things the example of the
equation

$$x^4 + 2x^3 + 4x^2 + 2x + 1 = 0,$$

which had been proposed to him some time before by "a very
learned Geometer," who was in all probability Nicolas I Bernoulli
(1687-1759).

He then stated the theorem that every algebraic equation of odd
degree has at least one real root. In order to demonstrate this Euler
utilized the evident geometrical interpretation of the intermediate
value theorem, (see §3.3) that is, that the curve of the equation,

$$y = x^{2m+1} + Ax^{2m} + \cdots + N,$$

is continuous and necessarily cuts the x axis in at least one point.

After having shown that every algebraic equation of the 8th and
16th degree can be split into two factors of the 4th and 8th degrees,
respectively, Euler states the theorem: "Every equation of a degree
whose exponent is a binary power, such as 2^n (n being a whole
number larger than 1) is resolvable into two real factors of degree
2^{n-1}" (1751, p. 105).

In Euler's view, this theorem is the key, since every equation of
any degree whatever can be reduced to the case in question by
multiplying by the factor ax^h with h a suitable integer, and, by then
repeating the factorization, it can finally be reduced to a product of
real polynomials of the second degree.

In order to demonstrate this Euler considers (what is always
possible without loss of generality)

(1.2.3) $x^{2^n} + Bx^{2^n-2} + Cx^{2^n-3} + \cdots = 0,$

supposing that it is split into the two factors sought,

(1.2.4) $x^{2^{n-1}} - ux^{2^{n-1}-1} + \alpha x^{2^{n-1}-2} + \cdots = 0,$

$x^{2^{n-1}} + ux^{2^{n-1}-1} + \lambda x^{2^{n-1}-2} + \cdots = 0,$

where the number of unknown coefficients u, α, λ, ... is $2^n - 1$.

Multiplying equations (1.2.4) by themselves and equating them to the coefficients of the equation obtained with the coefficients B, C,... of (1.2.3), we can find the coefficients α, λ, ... as rational functions (*reellement sans extraction de racines*) of u and of B, C, ... By eliminating α, λ, ..., we finally obtain an equation in u of degree

$$\binom{2^n}{2^{n-1}},$$

in which the known term (as Euler shows) is negative. Therefore, by the intermediate value theorem, there exists at least one root of the equation.

The process of elimination on which Euler's reasoning is based is not without criticism, however, and cannot be considered general, as Gauss would later observe (see §4.1). Nevertheless, in (1816a) he himself uses Euler's idea of demonstrating the theorem by induction on the power of 2 dividing the degee of the equation. For his part, Euler added that, "as far as the solidity of the demonstration is concerned, ... I believe that there is nothing to criticize" (1751, p. 107). In any case, if "one wants to make difficulties in recognizing the correctness of these demonstrations, I am going to add several propositions relative to this subject that are not dependent on the preceding, and whose truth will serve to raise all the doubts that one could still have" (*ibid.*, p. 107). As a special case he shows that equations of degree 6, $4n + 2$, $8n + 4$, $16n + 8$, etc., always contain at least one factor of the second degree.

Like D'Alembert, Euler states the theorem in the form, "If an algebraic equation of any degree has imaginary roots, each one will be included in the general formula $M + N\sqrt{-1}$, the letters M and N indicating real quantities" (*ibid.*, p. 112). As an "important consequence" of this he deduces "that any imaginary quantity, no matter how complicated it is, is always reducible to the formula $M + N\sqrt{-1}$" (*ibid.*, p. 121). He illustrates his assertion by showing how general expressions of the form $(a + b\sqrt{-1})^{c+d\sqrt{-1}}$ can be written in the requisite form, and similarly for the elementary transcendental functions log, sin, cos, etc., when these are calculated for an imaginary value.

In the course of these demonstrations Euler was led to consider $x + y\sqrt{-1}$ as a variable and then take the differential $dx + dy\sqrt{-1}$, a method that had first been introduced by D'Alembert (1747b, p. 141). Some time later D'Alembert himself used it systematically in a work on the resistance of fluids (D'Alembert, 1752a), which he submitted for the prize on this topic offered by the Berlin Academy in 1750.[13]

D'Alembert's *Essai* (1752a) is a paper of great importance in several respects. As Truesdell has noted, "despite its many defects, the *Essay* is a turning point in mathematical physics. For the first time, a theory is put (however obscurely) in terms of a *field* satisfying partial differential equations. ... In addition there are certain definite new results" (in: Euler, *Opera* (2) **12**, lvii). Among these are special cases of the continuity equation or the condition for circulation-preserving motion. Of particular interest is the solution of partial differential equations in terms of complex functions.

The so-called 'Cauchy-Riemann equations' (see §§4.3 and 6.2) appear here for the first time when D'Alembert faces the problem of finding a "method for determining the velocity of a fluid at any point whatever" (1752a, p. 60). This involves finding the components p and q of the velocity. D'Alembert supposes that the following hypothesis, "which is the simplest," is valid:

$$dq = M \, dx + N \, dz,$$

$$dp = N \, dx - M \, dx.$$

The physical problem is thence transformed into the following mathematical problem: "Let $M \, dx + N \, dz$ and $N \, dx - M \, dz$ be complete differentials. It is proposed to find the quantities M and N" (*ibid.*, p. 61).

Since the linear combination of complete differentials is still a complete differential, he obtains

$$(M + \sqrt{-1} \, N)(dx + dz/\sqrt{-1}),$$

$$(M - \sqrt{-1} \, N)(dx - dz/\sqrt{-1}).$$

Then, by setting

$$dx + dz/\sqrt{-1} = du, \quad dx - dz/\sqrt{-1} = dt,$$

he obtains

$$u = F + x + z/\sqrt{-1}, \quad t = G + x - z/\sqrt{-1},$$

where F and G are constants.

Moreover, if we set

$$M + \sqrt{-1} \, N = \alpha,$$

$$M - \sqrt{-1} \, N = \delta,$$

then $\alpha \, du$, $\delta \, dt$ are also complete differentials, D'Alembert observes, and therefore $\alpha = \xi(u)$, $\delta = \zeta(t)$. Then

$$M + \sqrt{-1}\, N = \xi(F + x + z/\sqrt{-1}),$$

$$M - \sqrt{-1}\, N = \zeta(G + x - z/\sqrt{-1}).$$

These are sufficient to determine M and N.

Immediately after this D'Alembert presents "the following method, which is simpler" (1752, p. 62).

"Since

$$(1.2.5) \qquad \frac{\partial p}{\partial z} = -\frac{\partial q}{\partial x}, \qquad \frac{\partial p}{\partial x} = \frac{\partial q}{\partial z},$$

therefore

$$q\, dx + p\, dz \quad \text{and} \quad p\, dx - q\, dz$$

will be complete differentials. Consequently

$$q + \sqrt{-1}\, p = \xi(F + x + z/\sqrt{-1}),$$

$$q - \sqrt{-1}\, p = \zeta(G + x - z/\sqrt{-1}),"$$

which are sufficient to obtain p and q by immediate calculation.

D'Alembert then illustrates this second method with an example where ξ and ζ are real polynomials of the 3rd degree.[14]

Euler also arrived at equations (1.2.5) in a series of articles in which he showed how one can use complex functions to evaluate real integrals. Written between 1776 and the year of his death, these were published posthumously from 1788 on (*Opera* (1) **19**).

The fundamental idea Euler used in these works is that, if $M + \sqrt{-1}N$ is a function of $z = x + \sqrt{-1}y$, for $z = x - \sqrt{-1}y$ it takes the form $M - \sqrt{-1}N$.

In order to calculate the integral

$$(1.2.6) \qquad V = \int Z(z)dz, \qquad z \in \mathbb{R},$$

we let $z = x + \sqrt{-1}y$. Then $Z = M + \sqrt{-1}N$, $V = P + \sqrt{-1}Q$, and, by the preceding observation, we obtain from (1.2.6) the following two equations,

$$P + \sqrt{-1}\, Q = \int (M + \sqrt{-1}\, N)(dx + \sqrt{-1}\, dy),$$

$$P - \sqrt{-1}\, Q = \int (M - \sqrt{-1}\, N)(dx - \sqrt{-1}\, dy),$$

from which we can obtain

$$(1.2.7) \qquad P = \int M\, dx - N\, dy, \qquad Q = \int N\, dx + M\, dy.$$

$M\, dx - N\, dy$ and $N\, dx + M\, dy$ are complete differentials, and hence, Euler observes, M and N "possess the remarkable property"

$$(1.2.8) \quad \frac{\partial M}{\partial y} = -\frac{\partial N}{\partial x}, \quad \frac{\partial N}{\partial y} = \frac{\partial M}{\partial x}.$$

The real and imaginary parts of a complex function thus verify equation (1.2.8), Euler declares. But the most interesting thing for him is the use of (1.2.7) to calculate the integral (1.2.6). Euler attains this end by setting $z = r(\cos\theta + \sqrt{-1}\ \sin\theta)$ and keeping θ constant.

D'Alembert's and Euler's work with complex variables led to a considerable extension of the techniques of analysis and important developments in the concept of a function. Nevertheless, neither for D'Alembert nor for Euler did the consideration of complex variables lead to an investigation of the properties of complex functions; it instead led to the study of the real and imaginary parts of such functions. As we will see in Chapter 4, this attitude still dominated mathematicians in the early decades of the nineteenth century.

1.3. The Debate over Vibrating Strings

Towards the middle of the eighteenth century the discussion over the concept of function became a central issue in a problem of a physico-mathematical character, that of studying the vibration of a string in a plane. The solution of this problem, which was to have an important impact on "pure" mathematics as well, fueled a long and spirited debate among the greatest mathematicians of the century, beginning with D'Alembert and Euler, and later including Daniel Bernoulli and J. L. Lagrange.[15] Echoes of the affair could still be heard at the beginning of the nineteenth century in the work of J. B. Fourier (see §2.3).

The discussion began with a work which was written by D'Alembert in 1747 and published two years later (1749a), which presented the first successful attempt to integrate the partial differential equation that describes the infinite shapes assumed by a homogeneous string held in tension and placed in vibration in a plane. Johann I Bernoulli had partially solved the problem in 1727, but he had limited himself to considering a loaded string, studying the behavior of n equal masses placed at equal distances and joined by a string imagined to be weightless, flexible, and inextensible. For the displacement y_k of the kth mass Bernoulli had obtained the finite difference equation

$$(1.3.1) \quad \frac{d^2 y_k}{dt^2} = a^2(y_{k+1} - 2y_k + y_{k-1}),$$

where a^2 depends on the tension of the loaded string (which is taken to be constant during the vibration), on the total mass of the n bodies, and on their distances from each other. Bernoulli solved this equation and then went on to treat the case of a continuous string, showing that at every instant t the string assumes a sinusoidal form whose equation he obtained by integrating the differential equation

$d^2y/dx^2 = -ky$ (J. Bernoulli, 1727 and 1728).   This result had already been given some time before by Taylor (1685-1731).

When D'Alembert began to interest himself in the problem, his objective was to show that a vibrating string assumes an infinite number of shapes other than the sinusoidal ones. By introducing a Cartesian reference frame and considering in place of y_k a continuous function $y = y(t,x)$ that varies continuously with x from 0 to ℓ, the length of the string, D'Alembert obtained, in place of the finite difference equation (1.3.1), the following equation,

(1.3.2) $$\frac{\partial^2 y}{\partial t^2} = a^2 \frac{\partial^2 y}{\partial x^2},$$

which he integrated for the case of $a^2 = 1$.

Since by differentiating one has the identity

$$d\left(\frac{\partial y}{\partial x} \pm \frac{\partial y}{\partial t}\right) = \frac{\partial^2 y}{\partial x^2}\,dx + \frac{\partial^2 y}{\partial x\,\partial t}(dt \pm dx) \pm \frac{\partial^2 y}{\partial t^2}\,dt$$

for (1.3.2), it follows that

$$d\left(\frac{\partial y}{\partial x} \pm \frac{\partial y}{\partial t}\right) = \left[\frac{\partial^2 y}{\partial t^2} \pm \frac{\partial^2 y}{\partial x\,\partial t}\right](dt \pm dx)$$

"whence it follows

(1)   That $\partial^2 y/\partial t^2 + \partial^2 y/\partial x\partial t$ is a function of $t + x$, and that $\partial^2 y/\partial t^2 - \partial^2 y/\partial x\partial t$ is a function of $t - x$;

(2)   That consequently we have ...

$$\frac{\partial y}{\partial t} = \phi(t + x) + \Delta(t - x), \qquad \frac{\partial y}{\partial x} = \phi(t - x) - \Delta(t - x) \ ...$$

therefore ... $y = \int((\partial y/\partial t)dt + (\partial y/\partial x)dx)$, or

(1.3.3) $y = \Psi(t + x) + \Gamma(t - x)$."

"But it is easy," D'Alembert continues, "to see that this equation includes an infinite number of curves. To show this, consider here only a special case, namely $y = 0$ when $t = 0$; that is, let us suppose that the string, when it starts to vibrate, is stretched out in a straight line" (D'Alembert, 1749a, pp. 216-7; trans. Truesdell, 1960, pp. 238-9).

From (1.3.3), it follows that $\Psi(x) + \Gamma(-x) = 0$. In addition, the conditions $y(t,0) = y(t,\ell) = 0$ that express the fact that the string is fixed at its ends, give, for every t,

(1.3.4) $\Psi(t) + \Gamma(t) = 0$,

and

(1.3.5) $\Psi(t + \ell) + \Gamma(t - \ell) = 0$,

respectively. The first equation, together with $\Psi(x) + \Gamma(-x) = 0$, indicates that $\Psi(x)$ is an even function, while the second equation expresses the fact that Ψ is periodic with period 2ℓ. This reduces the solution (1.3.3) to $y = \Psi(t + x) - \Psi(t - x)$.

Since the initial velocity $(\partial y/\partial t)|_{t=0}$ of the string is given by

$$V = \Psi'(x) - \Psi'(-x),$$

and since Ψ is even, Ψ' will be odd and hence, D'Alembert concludes, "the expression for the initial velocity ... must be such that when reduced to a series it includes only odd powers of x. Otherwise ... the problem would be impossible, that is, one could not assign a function of t and x such as to represent in general the value of the ordinate of the curve for any abscissa x and any time t" (1749a, pp. 218-9; trans. Truesdell, 1960, p. 239).

This paper was immediately followed by a second one on the same subject (D'Alembert, 1749b). Here, "in order to render the solution that we will give more extensive and more general" (*ibid.*, p. 230), D'Alembert supposed the initial condition for the string to be $y(0,x) = f(x)$, and the initial velocity to be $v(0,x) = g(x)$.

From these conditions he obtained the equations

$$\Psi(x) - \Psi(-x) = f(x),$$

$$\Psi'(x) - \Psi'(-x) = g(x),$$

or

(1.3.6) $$\Psi(x) + \Psi(-x) = \int g(x)dx,$$

"and thus the problem is impossible unless $f(x)$ and $g(x)$ are odd functions of x, that is, functions where only odd powers of x enter" (*ibid.*, p. 231).

If this condition is satisfied, we obtain from (1.3.6),

$$\Psi(x) = \frac{1}{2}\int g(x)dx + \frac{1}{2}f(x), \quad \Psi(-x) = \frac{1}{2}\int g(x)dx - \frac{1}{2}f(x),$$

which completely solves the problem.

In conclusion, D'Alembert says, "the general solution of the problem of the vibrating string is reduced to two things: (1) to determine the generating curve [i.e. the curve $z = \Psi(u)$] in the most general way, (2) to find in any particular case the curve from the values of $f(x)$ and $g(x)$" (*ibid.*, p. 235; trans. Truesdell, 1960, p. 240).

After discussing several cases of generating curves, D'Alembert asserts that "one must take heed that f and g may not be given at will, since they must satisfy certain other conditions, as has been seen above in this memoir" (1749b, p. 239; trans. Truesdell, 1960, p. 240).

He then lists a long series of conditions, but, as Truesdell has observed (1960, p. 241), "the essential is that D'Alembert *restricts the initial shape and initial velocity* of the string to curves whose 'equations' are odd functions of period 2ℓ (1960, p. 241). Finally, D'Alembert emphasizes the fact that $f(x)$ is subject to the law of continuity, meaning by this that $f(x)$ is given by a unique analytic expression, i.e., it is a 'continuous' function. Moreover, it should be twice differentiable, otherwise it could not satisfy (1.3.2).

The following year Euler entered the debate with a paper entitled *Sur la vibrations des cordes*[16] (Euler, 1750). Technically Euler's solution is not very different from D'Alembert's; Euler says simply that he wishes to add "a few fairly interesting observations on the application of the general formulas" (1749, p. 51; 1750, p. 64). He in fact considers the same equation studied by D'Alembert (1.3.2) and emphasizes the intention of seeking the maximum possible generality in the solution "so that the initial shape of the string can then be set arbitrarily" (1749, p. 55; 1750, p. 69). This can in fact be given by a curve that is "either regular and contained in a certain equation, or irregular and mechanical" (1749, p. 58; 1750, p. 72). He then shows how, beginning from such a curve, one can construct the solution at time t geometrically. He writes this as

$$(1.3.7) \qquad y = \frac{1}{2} f(t + x) + \frac{1}{2} f(t - x).$$

However, it is precisely over the character of $f(x)$, that is, the function that describes the initial shape of the string, that D'Alembert and Euler's positions substantially differ.

The difference becomes clear in D'Alembert's reply to Euler's paper (D'Alembert, 1752b). The object of contention is essentially the concept of function, which one immediately faces whenever one tries to specify the nature of the mathematical objects that are the "solutions" of the differential equation (1.3.2).

> In effect, it seems to me that one cannot express y analytically in a more general manner than by supposing it to be a function of t and of $s[-x]$. But under this supposition one can only find the solution of the problem for the case where the different forms of the vibrating string can be written in a single equation [i.e. it forms a "continuous" curve]. In all the other cases it seems to me impossible to give y a more general form (1752b, p. 358).

The idea of the continuity of curves to which D'Alembert refers was at that time dominant in analysis. Euler himself had given the authoritative statement of it some years before in his *Introductio*. In the second volume of this treatise, after having introduced a Cartesian reference system in the plane, he wrote,

> Although several curved lines can be described by the continuous mechanical motion of a point, which presents the entire curved line to the eye at one time, nevertheless here we will chiefly consider the origin of curved lines from functions, since this is more analytic, more widely accessible, and better

suited to calculation. Thus any function of x will give a line, whether straight or curved, whence it will be possible to recover curved lines from functions in turn. Consequently the nature of any curved line can be expressed by some function of x ...

From this idea of curved lines immediately follows their division into *continuous* and *discontinuous* or *mixed*. A *continuous* curved line is so defined, that its nature is expressed *by a single definite function of x* [my emphasis]. But if a curved line is defined in such a way that its different parts BM, MD, DM, etc., are expressed by different functions of x, so that, when a part BM is defined by one function, a part MD is described by another function, we call curved lines of this kind *discontinuous* or *mixed and irregular*, because they are not formed according to any constant law and are composed from parts of different continuous curves.

But in geometry the concern is chiefly with continuous curves and it will be shown below that curves described by uniform mechanical motion according to any constant rule are expressed by a single function and are consequently continuous (Euler, 1748, vol. 9, pp. 11-12).

Euler illustrates his argument thus. (Figure 1)

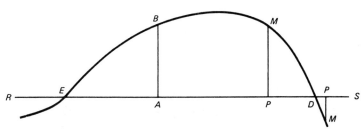

Figure 1

The classification of curves that Euler gave remained standard for a long time and was still found at the beginning of the nineteenth century (see §2.3b).

In the vague terminology of the epoch, in addition to that just given, mathematicians also spoke of "totally discontinuous" curves, curves "traced with a free movement of the hand," "absolutely arbitrary curves," curves that are "irregular or traced at will," and so on.

The debate over vibrating strings thus furnished the occasion for a discussion over the concepts of curve and function and consequently over the functions admissible in analysis, on which depends the generality of the solutions found. As Burkhardt has observed (1901-8, p. 14), D'Alembert and Euler used the same word, "function," but meant different things by it. The first came to the conclusion that the solution is valid when y is a "continuous" function, analytically given by a unique "equation" in x and t. "In every other case the problem cannot be resolved, even by my method, and I do not know but that it might surpass the capacity of known analysis" (D'Alembert, 1752b, p. 358).

Euler, on the other hand, went beyond this, motivated by the physical nature of the problem. As he saw it, it is completely reasonable to suppose that the string, when placed in vibration, initially assumes an arbitrary shape. Thus, in stating the problem Euler wrote,

> the first vibration depends on our pleasure, since we can, before letting the string go, give it any shape whatsoever. This means that the vibratory movement of the same string can vary infinitely, according to whether we give the string such and such a shape at the beginning of the movement (1749, p. 51; 1750, p. 64).

Euler himself suggested considering as initial shapes any polygonal figures given by possible "discontinuous" curves. Consistent with this, he wrote in a later paper that

> the various similar parts of the curve are therefore not connected with each other by any law of continuity, and it is only by the description that they are joined together. For this reason it is impossible that all of this curve should be included in any equation (1755a, p. 250).

The way in which Euler obtained the solution, beginning from the initial shape of the string, showed that "the simple consideration of the trace of the curve suffices to let us know the movement of the string, without subjecting it to calculation" (*ibid.*).

On the other hand, Euler's repeated assertions over the course of the long debate that solutions with corners are also admissible and hence that (1.3.7) represents the complete solution even if the functions $f(x + t)$ and $f(x - t)$ are "discontinuous" -- i.e. functions whose derivative is in general discontinuous in the modern sense -- were open to D'Alembert's objections that the solution must be given by a twice differentiable function.

Euler sought to justify his ideas by making an analogy with the definite integral.[17] A more satisfactory explanation requires more refined mathematical instruments than those which were at the disposition of the eighteenth century, which are today common in the method of continuum physics. Euler consequently reasoned in a different way. As Truesdell notes, "Once he has (1.3.7), *he discards the differential equation altogether*. That is, Euler takes the functional equation (1.3.7) rather than the differential equation (1.3.2) as *the complete mathematical statement of the physical principle of plane wave propagation*, apart from boundary and initial conditions" (1956, p. xliii). The behavior (and the eventual singularities) of the solution at the boundary points was a delicate question, one which neither Euler nor D'Alembert had any clear ideas about and whose study occupied mathematicians for many years, even in the nineteenth century.

The difference of opinion between D'Alembert and Euler over the nature of the solutions of equation (1.3.2) corresponds to two different ways of integrating it. While D'Alembert obtains the solution that is today called "classic," Euler seems to move in the direction of considering "weak" or "generalized" solutions.[18]

Euler's approach to the problem of vibrating strings had important consequences for the development of analysis. In fact, the functions that satisfy the functional equation (1.3.7) can only be piecewise smooth, and "it is clear from every one of the many examples and discussions given by Euler that for him a 'function' is what we now call a *continuous function with piecewise continuous slope and curvature*" (Truesdell, 1960, p. 247).

Aware of the implications connected with his approach, Euler wrote to D'Alembert on December 20, 1763, that "the consideration of those functions which are not subject to any law of continuity opens to us an entirely new field of analysis, of which we already see a very remarkable example in the propagation of sound"[19] (*Opera* (4a) 5, p. 327).

The difference of opinion between D'Alembert and Euler also implied a deeper philosophical difference over the general nature of physical laws. In fact, according to Leibniz, the law of continuity governs natural phenomena and is translated mathematically in terms of "continuous" functions, that is, in modern terms, analytic functions. These have the property of being completely determinate in their entire range by their value in an arbitrarily small domain.

But Euler's interpretation of (1.3.7) shows that the differential equation can admit solutions given by functions that do not obey the law of continuity, and which are possibly nonanalytic.

As Speiser first noted, this constitutes "one of the most important discoveries of Euler, shaking the physical system of Leibniz to its foundations" (1952, p. xxiv).

Agreeing with Speiser's view, Truesdell wrote that, "Euler's refutation of Leibniz's law was *the greatest advance in scientific methodology* in the entire century" (1960, p. 248). According to him, the truest and deepest motive of the dispute lay in this: "Both Euler and D'Alembert realized immediately what was at issue in the otherwise rather tedious problem of the vibrating string. This is the only scientific reason for the sharpness of the controversy that Euler and D'Alembert were to carry on until their death at the end of the century" (*ibid.*).

If this is true, it is no less true that this "rather tedious problem" forced mathematicians to confront their various conceptions of analysis and of its fundamental concepts on the relatively new field of differential equations. Among these fundamental concepts were those of function, continuity, and discontinuity.

In 1753 Johann Bernoulli's son, Daniel (1700-1782), entered the debate with a paper that was published as (1755a). His approach was shaped by his previous research into acoustics. Precisely on the basis of physical considerations (the superposition and composition of waves), he asserted that the movement of a vibrating string can *in general* be represented by an equation such as

$$(1.3.8) \quad \begin{aligned} y = &\alpha \sin \frac{\pi x}{\ell} \cos \frac{\pi ct}{\ell} + \beta \sin \frac{2\pi x}{\ell} \cos \frac{2\pi ct}{\ell} \\ &+ \gamma \sin \frac{3\pi x}{\ell} \cos \frac{3\pi ct}{\ell} + \cdots, \end{aligned}$$

which, however, he never actually wrote down in his paper. Euler had discussed the special case when $t = 0$ at the end of (1750), but Bernoulli was convinced that one could set D'Alembert's and Euler's general solutions into this form. In a second paper which he wrote immediately afterwards (1755b), he took up his father's argument by considering the vibrations of n weights ranged along a weightless string and then letting n increase without limit. In treating the special case of $n = 2$ he confirmed his opinion by determining the two fundamental vibrations.

Daniel Bernoulli's attention to the physical significance of the problem (and how little he estimated D'Alembert as a physicist) can also be seen in a letter he sent to Euler in January of 1750. "I hold Mr. D'Alembert to be a great mathematician *in abstractis*," he wrote, "but when he makes incursions into applied mathematics, I lose all of my esteem ... and it would be better for the true physics if there were no mathematics on the earth" (Fuss, 1843, p. 650).

However, Bernoulli did not argue his ideas in a mathematically precise way, and Euler's and D'Alembert's criticisms of his work rested precisely on this fact.

In (1755a) Euler emphasized the importance of Bernoulli's observations on the physical nature of the problem, but declared as mathematically unacceptable the conclusion that (1.3.8) can represent an arbitrary function (such as the required integration of (1.3.2)). In addition, he observed, a function given by a trigonometric series is periodic, and for the case where y does not have this property, something that happens more often than not, we certainly cannot represent it with a series of this type.

> But perhaps [Euler writes] one will reply that equation (1.3.8), because of the infinite number of undetermined coefficients, is so general that it includes all possible curves; and it must be acknowledged that if this were true, Mr. Bernoulli's method would furnish a complete solution.[20] But, apart from the fact that this great geometer had not made that objection, all the curves contained in this equation, even when we increase the number of terms to infinity, have certain characteristics which distinguish them from all other curves (1755a, pp. 236-7).

Among these characteristics Euler strongly emphasizes their periodicity, but, as Grattan-Guinness has written, "Euler's preference for the periodicity argument is seen to be completely mistaken when we remember that the analysis has reference only to the portion AB of the x-axis over which the string is stretched; what happens outside AB is irrelevant to the vibrating string, and therefore to the mathematics used to describe it" (1970b, p. 10).

We are here facing a misunderstanding that reveals one aspect of the contradictions between the old and new theory of functions, even though they are both present in the same man, Euler, the protagonist of this transformation. In fact, in the classical conception, a function is thought of as associated with the *totality* of the domain in which it "exists," yet the distinction between "continuous" and "discontinuous" functions already foreshadows the idea of associating distinct intervals with the different "pieces" of a "discontinuous" function independently of the (algebraic) form of the function. This is a decisive step, if we consider that it anticipates, in its essentials, the modern idea of the "domain" of the function. Moreover, Euler's remark shows that he "has no idea that what we now call a nonanalytic function may be represented in a finite range by a trigonometric series" (Truesdell, 1960, p. 261).

Bernoulli's reply is based in particular on the possibility of operating on the infinite number of constants α, β, γ, ... present in (1.3.8) to determine the curve. Bernoulli indeed writes that "the resultant curve will enclose an infinite number of arbitrary quantities which one can use for making the final curve pass through as many given points as one wishes, and thus for identifying this curve with that proposed with as great a degree of precision as one wishes" (1758, p. 165). Yet he does not show how one can concretely determine the coefficients of the series. This was precisely the objection that Euler made on a later occasion when he asserted that determining the coefficients α, β, γ, ... seemed to him to be "without doubt very difficult, not to say impossible" (1767c, p. 430).

D'Alembert was also opposed to Bernoulli's proposed solution and adopted Euler's argument on the impossibility of representing arbitrary functions by means of sine series. He further denied the possibility of representing any periodic function by means of such series.

In regards to this Lebesgue has written,

If Bernoulli's assertion were correct it would be necessary for a trigonometric series be equal to one linear function in one interval [as occurs in the case where a polygonal figure is taken as the initial position of the string] and to another linear function in another interval; or, if you wish, it would be necessary that two analytic expressions be equal in one interval and unequal in another. This seemed impossible (1906, p. 21).

This was still believed for a long time (see §§2.3 and 6.2b). As

Lebesgue aptly remarks, "until Weierstrass, who showed that two analytic expressions of a complex variable can be equal in a domain without being equal everywhere (see §7.3), one generally admitted that this Eulerian continuity belonged to every function of a complex variable defined by an analytic procedure" (1906, p. 21 n. 2).

The debate was joined at this point by a young and at that time almost unknown Italian mathematician, J. L. Lagrange (1736-1813).[21] In 1759 he published a paper on the nature and propagation of sound which shared D'Alembert's opinion on the nature of the solution to the problem of a vibrating string, at least in part. Lagrange wrote,

> It seems unquestionable that the consequences which follow from the rules of the differential and integral calculus will always be illegitimate in all the cases where this law [of continuity] is not taken to occur. It follows from this that, since Euler's construction is derived immediately from the integration of the given differential equation, this construction is by its very nature applicable only to continuous curves which can be expressed as any function whatever of the variables t and x. I therefore conclude that all the proofs which one can provide to decide such a question, supposing at first that the ordinate of the curve y is a function of t and x, as D'Alembert and Euler have done until now, are completely unsatisfactory, and that it is only by a calculus like that we have in view, in which one considers the movements of the points of the string individually, that one can hope to arrive at a conclusion which will be secure from all attacks (Lagrange, 1759, p. 68).

Lagrange nevertheless believed that the solution D'Alembert had found was applicable to Euler's "discontinuous" curves (and he thereby also aligned himself with Euler's point of view). He sought to find the solution of (1.3.2), as indicated in the preceding passage, by considering the movement of every point "individually." As we have already seen Johann Bernoulli do, he then studied the behavior of the discrete "model" of the loaded string by considering the oscillations of n bodies fixed on an inextensible string fixed at the ends, and then obtained the solution for the homogeneous vibrating string by "passing to the limit" in the solution for the loaded string.

This "passage to the limit" is anything but rigorous, and although it allows Lagrange to arrive at (1.3.7), from the point of view of rigor it does not justify the conclusions he claimed he had obtained from it. Lagrange himself wrote, "This construction is evidently the same one that Euler devised for the same hypothesis. Thus was the theory of this great geometer put beyond all doubt and based on clear and direct principles which do not rest in any way on the law of continuity that D'Alembert required" (1759, p. 107).

In fact, it was precisely the criticisms raised from the viewpoint of mathematical rigor, both by Daniel Bernoulli and by D'Alembert, the latter in a letter that was written to Lagrange on September 27, 1759[22] and subsequently published as a response to Euler and Bernoulli, (1761b, p. 65) which pushed Lagrange towards the search for another method. He published this in his treatise on the nature and propagation of sound (1760-61).

The starting point for integrating the equation

(1.3.9) $$\frac{\partial^2 y}{\partial t^2} = \frac{\partial^2 y}{\partial x^2}$$

is to multiply both members by a suitably determined function $M(x)$, and then reduce it by integration to

(1.3.10) $$\int_0^{\ell} \frac{\partial^2 y}{\partial t^2} M(x)dx = \left[M \frac{\partial y}{\partial x} - y \frac{dM}{dx} \right]_0^{\ell} + \int_0^{\ell} y \frac{d^2 M}{dx^2} dx.$$

Here y must vanish at $x = 0$ and $x = \ell$. Lagrange requires the same condition for $M(x)$. Then, from (1.3.10), he obtains

(3.1.11) $$\int_0^{\ell} \frac{\partial^2 y}{\partial t^2} M(x)dx = \int_0^{\ell} y \frac{d^2 M}{dx^2} dx$$

and, by also requiring that

(1.3.12) $$\frac{d^2 M}{dx^2} = -k^2 M$$

(where k is a constant) he obtains from (1.3.11)

(1.3.13) $$\int_0^{\ell} \frac{\partial^2 y}{\partial t^2} M(x)dx = -k^2 \int_0^{\ell} y\, M\, dx.$$

By introducing a new variable,

(1.3.14) $$s = \int_0^{\ell} yM\, dx,$$

this is transformed into

(1.3.15) $$\frac{d^2 s}{dt^2} = -k^2 s.$$

Thus the integration of the partial differential equation (1.3.9) is reduced to the integration of the ordinary differential equation (1.3.12) and of a second equation of the same type (1.3.15), which are linked to each other by (1.3.14).

Lagrange justifies using equation (1.3.13) in place of (1.3.9) in this way:

I first imagine that in place of the simple general equation $(\partial^2 y/\partial t^2) = (\partial^2 y/\partial x^2)$, which belongs to all the moving points [of the string], there are an infinite number, each of which represents the movement of each point in particular; a movement that depends, however, on all the other [points], since the differential d^2y that one takes by allowing only x to vary, expresses the second difference of the values of y for three consecutive points. I therefore multiply each of these equations by an indeterminate coefficient M, or rather by the quantity $M \, dx$, taking M as a variable which one can apply to all equations in general, and I take the sum by an integration indicated in the usual manner.

Now, since it is necessary to join together the coefficients of each value of y which correspond to each moving point, I transform my integral equation in such a way that the differentials of y which depend on x vanish (1760-61, p. 177).

Lagrange's approach is similar to the way in which one introduces "weak" solutions by means of an integral identity, even if Lagrange never talks of a new notion of solution (see note 18). On the contrary, he instead alternates between accepting D'Alembert's point of view (taking a function that is at least twice differentiable for the initial shape of the string, as he suggested in a letter to D'Alembert in November, 1764) and that of Euler, which he defended here and in the second edition of the *Mécanique analytique* (1811) many years later, after the debate had completely ceased.[23]

There is a final observation to make on the first solution to the problem of vibrating strings which Lagrange proposed. After setting up the equation of finite differences for n bodies,

$$\frac{d^2 y_k}{dt^2} = c^2 (y_{k+1} - 2y_k + y_{k-1}) \qquad k = 1, 2, \dots, n$$

he obtains, after a long and difficult analysis that occupies 20 sections of the paper,[24]

(1.3.16)
$$y(x,t) = \frac{2}{\ell} \sum_{r=1}^{\infty} \sin \frac{r\pi x}{\ell} \sum_{q=1}^{\infty} \sin \frac{r\pi X_q}{\ell} \, dX_q \cdot$$

$$\cdot \left[Y_q \cos \frac{c\pi rt}{\ell} + \frac{\ell}{r\pi c} V_q \sin \frac{c\pi rt}{\ell} \right],$$

where Y_q and V_q give the initial position and velocity of the qth mass, $X_q = q/(n+1)$ and $dX_q = \ell/(n+1)$. He then replaces Y_q and V_q with $Y(X)$ and $V(X)$ respectively in order to find the behavior of the homogeneous string.

Lagrange also considers

$$\sum_{q=1}^{\infty} \sin \frac{r\pi X_q}{\ell} Y_q dX_q \quad \text{and} \quad \sum_{q=1}^{\infty} \sin \frac{r\pi X_q}{\ell} V_q dX_q$$

to be integrals; that is, in place of (1.3.16) he takes (after interchanging the $\int$ and Σ symbols)

$$y(X,t) = \frac{2}{\ell} \int_0^\ell Y(X)dX \sum_{r=1}^\infty \sin\frac{r\pi X}{\ell} \sin\frac{r\pi x}{\ell} \cos\frac{r\pi ct}{\ell}$$

(1.3.17)

$$+ \frac{2}{\pi\ell} \int_0^\ell V(X)dX \sum_{r=1}^\infty \sin\frac{r\pi X}{\ell} \sin\frac{r\pi x}{\ell} \sin\frac{r\pi ct}{\ell}.$$

Many historians of mathematics have observed that, from this latter equation, Lagrange could have derived the representability of an arbitrary function $y(x)$ by means of a "Fourier" series (see §2.3b). To do this it was only necessary to interchange the integral and series symbols and let $t = 0$. Formally, Lagrange was only one step from Fourier's result, but the fact that he did not take this step "is a very instructive example of how easily an author fails to draw a seemingly nearby conclusion when he has before his eyes an objective lying in a completely different direction" (Burkhardt, 1901-8, pp. 32-3). Lagrange was in fact seeking to prove Euler's conclusion about a *physical* problem and did not have in mind the (theoretical) analytical problem of representing an arbitrary function by means of series. Moreover, the equations (1.3.17) thus obtained introduced divergent series. For Lagrange it required long, exhausting (and less than rigorous) calculations to convert (1.3.17) into Euler's solution (1.3.7).

1.4. Changes in the Concept of a Function

One of the most interesting results of the debate over vibrating strings was to center the attention of mathematicians on Euler's definition of a function, and particularly on his notions of "continuous" and "discontinuous" functions, as they were given in his *Introductio* (1748).

At an early stage of the debate, in the *Institutiones calculi differentialis* (1755b) Euler had given a fairly general definition of a function. In this work he wrote,

> If some quantities so depend on other quantities that if the latter are changed the former undergo change, then the former quantities are called functions of the latter. This denomination is of the broadest nature and comprises every method by means of which one quantity could be determined by others. If, therefore, x denotes a variable quantity, then all quantities which depend upon x in any way or are determined by it are called functions of it (1755b, p. 4; trans. Yushkevich, 1976, p. 70).

It seems quite reasonable to argue that the great generality of this definition was suggested to Euler by his research on vibrating strings. If the mathematical "objects" that satisfy equation (1.3.7) are functions, then it is certainly not the case, for Euler, to think of them as given by any "analytic expression." If, in (1755b) he limits himself to the latter (or at most to functions that are nonanalytic in isolated points), it is because the most common functions reduce to this. Nevertheless, this does not imply an *a priori* limitation to them.

Euler resumed his discussion of discontinuous functions which arise in the integration of partial differential equations in a paper that was presented to the St. Petersburg Academy in 1763 (1767b). Here he defined a continuous function in the above manner, adding however that the curve which represents it is such that "all the parts of the curve are firmly connected with each other in such a way as to make impossible any change in them without disturbing the connection of continuity" (Euler, 1767b, pp. 75-6; trans. Yushkevich, 1976, p. 67).

For Euler, discontinuous curves were those "traced by a free stroke of the hand" (*ibid.*, p. 76), even if these curves extend without interruption [*etiamsi continuo procedant*]. As Yushkevich has noted, "If we disregard the empirical fact that ideal geometrical figures cannot be traced, *discontinuous* functions thus correspond to our arbitrary piecewise continuous functions with piecewise continuous derivatives of both the first and the second order" (1976, p. 68).

In the remainder of this paper Euler emphasizes the fact that such "discontinuous" functions arise in the integration of partial differential equations. This is one of the most notable results of this new field of research, to which Euler dedicated a large part of the third volume of his *Institutiones calculi integralis* (1770).[25]

Despite Euler's numerous and fundamental contributions, the discussion over the character of the functions obtained by integrating partial differential equations remained open even after his death. Thus, in 1787 the Academy of St. Petersburg offered a prize for the best solution to the problem of determining "Whether the arbitrary functions which are obtained from the integration of equations with three or more variables represent any curves or surfaces whatsoever, either algebraic or transcendental, either mechanical, discontinuous, or produced by an arbitrary movement of the hand; or whether these functions include only continuous curves represented by an algebraic or transcendental equation" (Burkhardt, 1901-8, p. 44).

The prize was won by a paper written by L. F. Arbogast (1759-1803) (Arbogast, 1791). Among its most interesting aspects was the distinction made between *discontinuous* and *discontiguous* functions. According to Arbogast,

The law of continuity consists in that a quantity cannot pass from one state to another without passing through all the intermediate states that are subject to the same law. Algebraic functions are regarded as continuous because the different values of these functions depend in the same manner on those of the variable; and, supposing that the variable increases continually, the function will receive corresponding variations; but it will not pass from one value to another without passing through all the intermediate values. Thus the ordinate y of an algebraic curve, when the abscissa x varies, cannot pass abruptly from one value to another; there cannot be a jump (*saut*) from one ordinate to another which differs from it by any assignable quantity; but all the successive values of y must be linked together by one and the same law which makes the extremities of these ordinates make up a regular and continuous curve.

This continuity may be destroyed in two ways:

(1) The function may change its form, that is to say, the law by which the function depends on the variable may change all at once. A curve formed by the assemblage of many portions of different curves is of this kind. ... It is not even necessary that the function y should be expressed by an equation for a certain interval of the variable; it may continually change its form, and the line representing it, instead of being an assemblage of regular curves, may be such that at each of its points it becomes a different curve; that is to say, it may be entirely irregular and not follow any law for any interval however small.

Such would be a curve traced at hazard by the free movement of the hand. These kinds of curves can neither be represented by one nor by many algebraic or transcendental equations. ...[26]

(2) The law of continuity is again broken when the different parts of a curve do not join with each other (Arbogast, 1791, pp. 9-11; trans. Jourdain, 1913, pp. 675-676).

The latter is what Arbogast called "discontiguous" curves, what in today's terminology are called "discontinuous" curves. The arbitrary functions that appear in the integration of differential equations are, for Arbogast, of both types. For example, Arbogast says, in the equation of vibrating strings we can think of a "jump" in the values of $\partial^2 y/\partial x^2$, provided that the same jump occurs in the values of $\partial^2 y/\partial t^2$.

Arbogast's distinction between "discontinuous" and "discontiguous" curves tends to mirror the modern distinction between discontinuous and continuous curves, the latter being potentially defined in different intervals by different functional dependencies.

The emergence of a concept of discontinuity closer to that of the modern was accompanied by a diffuse criticism of the traditional concept of a function as an analytic expression. Among those who first recognized the generality of Euler's (1755b) definition of a function was M. A. Condorcet (1743-1794), the secretary of the *Académie des sciences*, but a man better known as a politician and educational reformer than as a scientist and mathematician.

Condorcet was the author of a *Traité du calcul intégral* that was published in Paris in 1765, and of a second work of the same name which was presented to the Academy in 1778-82 but remained unpublished. It was in this second work, numerous printed pages of which circulated in Paris and became known to mathematicians, that Condorcet gave the following definition of a function.

I suppose that I have a certain number of quantities $x, y, z, \ldots, F$, and that for every determinate value of $x, y, z, \ldots$, etc., F has one or more determinate values which correspond to them; I say that F is a function of x, y, z. ... Finally, if I know that when x, y, z are determinate, F will also be, even if I know neither the manner of expressing F in x, y, z, nor the form of the equation between F, x, y, z; I will know that F is a function of x, y, z" (In: Yushkevich, 1974, p. 134).

This is what Condorcet means by an "analytic function" -- a function that is completely arbitrary by nature, where the term "analytic" serves to indicate that the study of such quantities occurs within the context of analysis. In addition, Condorcet distinguishes three different types of functions: (1) functions for which the form is known, in other words, explicit functions; (2) functions introduced by equations between F and $x, y, z, \ldots$, that is to say, implicit functions; and (3) functions that are only given by means of certain conditions, such as differential equations. Moreover, Condorcet brought the concrete study of functions back to the Taylor series, which, as we will see, (§2.2) became the foundation for the differential calculus with Lagrange. The determination of the coefficients of series further led Condorcet to the concept of "differential functions," the analog of Lagrange's "derivative functions."[27]

Among those who saw the unpublished pages of Condorcet's work was S. F. Lacroix (1765-1843), the author of numerous treatises, among which is his *Traité* (1797-1800), a work that enjoyed considerable popularity for the entire first half of the nineteenth century (see §2.2).

In the preface to the second edition of this work in 1810, Lacroix gave a brief resume of Condorcet's work and undertook to clarify what was meant by a function. After recalling such classical definitions of a function of a quantity as any power of this quantity, or any algebraic expression containing this quantity, Lacroix wrote,

And finally new ideas introduced by the progress of analysis have given rise to the following definition of functions.

Every quantity whose value depends on one or more other quantities is called a function of these latter, whether one knows or is ignorant of what operations it is necessary to use to arrive from the latter to the first (Lacroix, 1810, p. 1).

This definition clearly recalls those of Euler (1755b) and Condorcet. Nevertheless, it is necessary to emphasize the *nominal* aspect of this definition, in the sense that those functions which were then being used in practice were basically the algebraic and elementary transcendental functions.

These reflections on the concept of a function in the second half of the eighteenth century were necessarily accompanied by an attempt to give a rigorous form to the foundations of the calculus. Even though mathematicians like D'Alembert and Euler did not feel this was an urgent issue, just as the need for the rigorous definitions to which we are accustomed today was not felt to be generally necessary, they did express their conceptions of limits and differentials on many occasions.[28]

Thus, for example, D'Alembert, in an article on "limit" in the *Encyclopédie*, wrote that: "One magnitude is said to be the limit of another magnitude when the second may approach the first within any given magnitude, however small, though the first magnitude may never exceed the magnitude it approaches." He then further claimed that in the concept of limit there lies the "true metaphysics" of the calculus.

As is well known, Euler believed the contrary. He thought that in the calculus we are dealing with quantities that are actually zero. For example, he wrote, "There is no doubt that every quantity can be diminished to such an extent that it vanishes completely and disappears. But an infinitely small quantity is nothing other than a vanishing quantity and therefore the thing itself equals 0. It is in harmony also with that definition of infinitely small things, by which the things are said to be less than any assignable quantity; it certainly would have to be nothing; for unless it is equal to 0, an equal quantity can be assigned to it, which is contrary to the hypothesis" (Euler, 1755b, p. 69; trans. Kline, 1972, p. 429).

If on the continent there was substantial insensitivity to the foundations of the calculus, the lively debate in England that had been generated by Berkeley's criticism of Newton's fluxions had not led to more satisfactory results.

In a 1772 paper presented to the Berlin academy Lagrange sketched the first attempt at founding the calculus in a "rigorously algebraical" manner. In Lagrange's work of the period this paper appears isolated, without immediate successors, but the theme must have stood particularly near to the heart of the Turin mathematician if he convinced the scientific section of the Berlin academy, where he was secretary at the time, to offer a prize in 1784 "for a clear and precise theory of what is called the infinite in mathematics."

"We know," continues the text of the prize announcement, "that higher geometry makes continual use of the infinitely large and the infinitely small. Nevertheless, the geometers and even the ancient analysts carefully avoided any use of the infinite, and the great modern analysts assert that the terms *infinite magnitude* are contradictory" (*Nouv. Mem.*, 1784, pp. 12-3).

The Academy then requested an explanation of "how so many true theorems have been derived from a *contradictory* supposition" and asked that the paper "provide a certain, clear, in a word, a truly mathematical principle capable of being substituted for the infinite" (*ibid.*).

Notwithstanding the requirement that the argument be treated "with all generality, and with all the rigor, clarity and simplicity possible," the papers that were received were far from satisfying these conditions. "They are all more or less lacking the clarity, the simplicity, and above all the rigor that we required," the Academy (i.e. with all probability, Lagrange) observed (*Nouv. Mem.* 1786, p. 8).

The outcome of the competition was therefore disappointing. Among the twenty or more papers that were presented the prize was awarded with little enthusiasm to that of S. L'Huillier (1750-1840), a young Swiss mathematician who took up D'Alembert's theory of limits and presented the old argument on the compensation of the errors as an explanation of the correctness of the results obtained with the calculus.

Some time after the conclusion of this competition Lagrange moved to Paris, accepting the repeated invitations that had been made to him by the *Académie des sciences*.

The problem of the foundations of the calculus thus remained open. As we will see in the next chapter, a more coherent attempt to establish a theory of functions, and at the same time to found the infinitesimal calculus in an incontestable manner, was achieved by Lagrange with a treatise that in a certain sense symbolizes the state of analysis and its foundations at the end of the eighteenth century.

Notes to Chapter 1

[1] The difficult problem of the influence of medieval thinkers on the birth of modern science in the seventeenth century has been the subject of numerous studies. See for example the volumes of Crombie (1959-61) and Clagett (1959). For a more detailed study of the history of the concept of function, see the discussion in Yushkevich (1976).

[2] C. B. Boyer (1959) is still the primary work on the history of the calculus. Following the appearance of Boyer's work there was an imposing number of additional works on this subject, among which must be mentioned in the first place the edition of the *Mathematical Papers of Isaac Newton* by D. T. Whiteside. See also Whiteside (1961) and Hofmann (1949).

[3] Leibniz used the term "function" as early as 1673 in a manuscript titled *Methodus tangentium inversa seu de functionibus*. Some years later, in a manuscript of 1679 he introduced the distinction between algebraic curves (called "analytic curves") and transcendent ones, while in 1684 and 1686 he called "algebraic curves" those represented by an equation of a certain finite degree and "transcendent" ones the curves represented by equations of an infinite degree. See also Yushkevich (1976, p. 59).

[4]See *Leibnizs mathematische Schriften*, C. I. Gerhardt (ed.). (Berlin/Halle, 7 vols., 1849-63), vol. 5, pp. 268 and 306. Also vol. 3, pp. 506-7.

[5]*Op cit.* (note 4), vol. 3, p. 150; also pp. 324, 526, 527. For a discussion of this topic see Yushkevich (1976, pp. 57-60).

[6]As is well known, the development of the calculus followed different paths in England and on the continent. Chained to the Newtonian tradition and its fluxional symbolism, English mathematicians in the eighteenth century were unable to produce the extraordinary quantity of new results and techniques that the greater flexibility and fecundity of the Leibnizian approach permitted continental mathematicians to make. The English remained more or less faithful to the geometrical conception of the calculus, supported by the authority of Newton and his *Principia*, and obtained respectable results in the field of synthetic geometry. Chasles (1837) was particularly appreciative of the latter. This situation continued until the end of the century and the beginning of the nineteenth century.

The turning point in English mathematics is traditionally dated from the founding of the Analytical Society at Cambridge in 1813. Formed by Herschel, Peacock, Babbage, and others, this society had among is aims making the continental developments in infinitesimal analysis known to the English mathematical community. An important issue in this struggle was the polemic against the advocates of "dot-age," i.e. Newton's point notation, and in favor of Leibniz' differential notation.

This traditional interpretation of the Analytical Society has, however, been recently questioned by Enros (1983). According to him, the Society "actually played no real part in the movement to reform Cambridge mathematics" (1983, p. 26). At that time, "much more was involved in the transformation of mathematics at Cambridge than simply a switch in notation and methods" (*ibid.*). Enros also stresses "the three related features" of the Society: "debate over the methods of analytical versus synthetical mathematics, ideas about the purposes of higher education, and expectations concerning mathematics and science indicative of the growing professionalism" (*ibid.*).

However this may be, the English mathematicians of the time in large part adhered to the algebraic methods of Lagrange even while these methods were being subjected to radical criticisms in France (see §§2.2 and 3.3). In fact, in the first half of the nineteenth century, English mathematics took on a strongly algebraic character with the development of important branches of modern algebra, such as the theory of forms, the theory of invariants, and symbolic logic. This was apart from the principal currents in the development of analysis. This is the primary reason why we will hardly encounter any English mathematicians in the course of our story, with the exceptions of Green, Stokes, and Thompson.

[7]Equally vast is the bibliography relative to Euler's scientific work. For an overview of Euler's life and his various fields of research, as well as a virtually complete bibliography, see the volume, *Leonhard Euler, 1707-1783: Beiträge zu Leben und Werk. Gedankband des Kantons Basel-Stadt* (Basel: Birkhäuser, 1983).

[8]"Since all numbers which it is possible to conceive are either greater or less than 0, or are 0 itself, it is evident that we cannot rank the square root of a negative number amongst possible numbers, and we must therefore say that it is an impossible quantity. In this manner, we are led to the idea of numbers, which from their nature are impossible; and therefore they are usually called *imaginary quantities*, because they exist merely in the imagination." This is the way Euler expressed it in Part I, Chapter XIII, §143 of his *Einleitung* (Euler, *Opera* (1) 1; trans. J. Hewlett, *Elements of algebra* (London, 1840), p. 43; repr. Springer, 1984).

[9]The mathematician, philosopher, and "encyclopedist" Jean le Rond D'Alembert was one of the most influential figures in the science and culture of the eighteenth century. For a biography see Hankins (1970).

[10]The essential content of these two papers is summarized by Yushkevich and Taton in their preface to (Euler, *Opera* (4a) 5, pp. 17-18). This offers a useful guide to the reading of Euler's correspondence with Clairaut, D'Alembert, and Lagrange. It is, at the same time, an excellent summary of the principal themes of eighteenth-century mathematics.

[11]This is what D'Alembert says at the beginning of a long article in which he explained in detail his (unfounded) considerations, which were based on the following definition of logarithms: "One calls logarithms a sequence of numbers in *any* arithmetic progression which corresponds to a sequence of numbers in *any* geometric progression," where, for convenience, one sets log 1 = 0 (D'Alembert, 1761a, p. 181). Yushkevich and Taton are certainly correct in saying that, "He completely lacks a good definition of logarithms" (In: Euler, *Opera* (4a) 5, p. 19).

[12]D'Alembert's demonstration is often cited but rarely presented in the history books. D'Alembert's reasoning is usually translated in modern terms in the following way:
Given the algebraic equation

(1) $y - P_n(z) = 0$ where $P_n(z) = \sum_{i=1}^{n} a_i z^i$ ($a_i \in \mathbb{R}$),

let z_0 and y_0 satisfy (1).

Then in the neighborhood of a point (z_0, y_0) Puiseux's expansion is valid:

$$(2) \qquad z - z_0 = \sum_k a_k (y - y_0)^{r_k}, \text{ where } r_k \in \mathbb{Q} \text{ and } r_k > r_{k+1}.$$

Consequently, every y near enough to y_0 is the image of some z for (2). It is then necessary to further show that $y = 0$ is an image of some z. From the expansion of (2) it follows that for $|y_1| < |y_0|$ there exists a z_1 such that $y_1 - P_n(z_1) = 0$. The demonstration of the theorem follows immediately by reduction to the absurd, utilizing a compactness argument that was certainly unknown to D'Alembert.

The latter is certainly to our eyes a weak point in D'Alembert's original demonstration, as is the assumption without demonstration of the expansion of (2). Nevertheless, D'Alembert had hit on a fundamental idea which would be clarified in the successive development of the theory of algebraic functions. In fact, considering equation (1) and the expansion in series (2), by means of analytic continuation one obtains a function $z(y)$ that satisfies (1) for real and complex values. Such a function is not in general single-valued and will have a finite number of singular points . The expansion of (2) then represents locally one of the branches of the algebraic function defined by (1) (see §4.6 below). For a discussion of D'Alembert's demonstration see Petrova (1974) and Gigli (1924-27, pp. 189-192).

[13]Contrary to the paper of 1747b, on this occasion D'Alembert's work was not awarded the prize. With all probability Euler's opposition was responsible for this. The awards committee, of which Euler was a member, declared that the papers which had been presented did not satisfactorily meet the requirements set by the Academy and the question was again posed in 1752. D'Alembert, offended by this check, immediately had his paper published in Paris and refused to compete for the new prize.

This affair caused a deep break in the relations between Euler and D'Alembert, which was aggravated by the dispute over vibrating strings that also broke out at the same time. The correspondence between the two mathematicians did not resume until a decade later, in 1763. For the details of this intricate story, see Yushkevich and Taton (In: Euler, *Opera* (4a) 5, pp. 312-314).

[14]For an historical account of fluid mechanics at that time, see Truesdell (1954). On D'Alembert's paper in particular, see pp. l-lviii.

[15]Many historical works have been devoted to this affair. In addition to sections in works of a general character, such as Kline (1972), see for example Burkhardt (1901-8), Jourdain (1913), Truesdell (1960), Grattan-Guinness (1970b), and Yushkevich (1976).

[16]Euler must have been particularly interested in this subject, for,

contary to his usual habit, he published both a Latin version of this article (1749) and a French translation of it (1750). In reality, the latter had been written in 1748, but was published only two years later. This had also happened with the papers of D'Alembert (1749a and 1749b). In the references in this book, as far as the works of D'Alembert, Euler, and Daniel Bernoulli are concerned, I have not used the official dates given in the memoirs of the academy but the actual year of publication.

[17]This is, for example, how Euler stated it in a letter to Lagrange on February 16, 1765 (*Opera* (4a) 5, pp. 452-454).

[18]The classical solution is obtained from the initial conditions

$$y(0,t) = F(x)$$

$$\frac{\partial y}{\partial t}\bigg|_{t=0} = G(x)$$

from which it follows

$$y(t,x) = \frac{F(x - at) + F(x + at)}{2} + \frac{1}{2a} \int_{x-at}^{x+at} G(z)dz.$$

The weak solutions can be thought of as introduced in two ways:

(a) as the limit of a series of functions $\{f_{(k)}(x)\}$ (the classical solutions of (1.3.2)) uniformly convergent in the domain D: $\{t > 0, x - at > 0, x + at < \ell\}$; where every $f_{(k)}(x)$ is a function that is twice differentiable in $[0,\ell]$ and is obtained from D'Alembert's formula;

(b) by means of an integral identity obtained from the integral extended to D:

$$\iint_D \left[\frac{\partial^2 y}{\partial t^2} - a^2 \frac{\partial^2 y}{\partial t^2} \right] z(x,t)dxdt = 0,$$

where $z(x,t)$ is a function twice differentiable in D, equal to zero on the boundary. Integrating by parts we obtain

$$\iint_D y\left[\frac{\partial^2 z}{\partial t^2} - \frac{\partial^2 z}{\partial x^2} \right]dxdt = 0$$

which satisfies, other than every classical solution, also a large class of functions called "weak solutions." See also Euler's solution for the problem of a nonuniform vibrating string (*Opera* (3) 11).

[19]Sometime later Euler himself gave a convincing confirmation in an article on the subject (Euler, 1767a). In particular, in the study of the solutions of functional equations that describe the propagation of sound in a finite tube open at each end, he was led to consider functions that have the value zero everywhere except at

one point. He then showed that by means of these "pulse functions" -- which are obviously discontinuous in the modern sense -- it is possible to give an elegant description in geometrical terms of the propagation and reflection of plane waves. For a detailed discussion of this matter, see Truesdell (1956, pp. lxi-lxii).

[20]In this connection, however, Riemann, in his lectures on partial differential equations given at Göttingen in 1854-55, observed that, "Daniel Bernoulli's argument, that he had infinitely many coefficients and only needed to find them, was in general *a priori* incorrect. Instead he should have indicated that every coefficient continually approaches a given value" (Riemann, 1869, p. 50).

[21]Beginning in 1754, at the age of 18, Lagrange was in contact with Euler by letter. The following year he sent him his research on the method *de maximis et minimis*, which marked the beginning of the modern calculus of variations. For a detailed analysis of Lagrange's contributions to this field see Fraser (1985). A biography of Lagrange (which is, however, weak in the mathematical aspects) can be found in Burzio (1942).

[22]D'Alembert wrote, "In regard to the method by which you can pass from an indefinite number of vibrating bodies to an infinite number, it does not seem obvious to me as to you, but it would take too long to tell you all my difficulties on this subject ... Ten years ago I also had a dispute by letters with Euler on this subject" (In: Lagrange, *Oeuvres* 13, p. 4).

[23]For the development of this discussion, see the letters between Lagrange and D'Alembert (In: Lagrange, *Oeuvres* 13, pp. 13-17).

[24]For the mathematical details, see Burkhardt (1901-8, pp. 27-34). See also Truesdell (1960, pp. 269-70).

[25]At the very end of his life, even D'Alembert (1780) changed his mind and admitted the possibility of considering "discontinuous" functions in the integration of partial differential equations of any order. See Yushkevich (1976, p. 71).

[26]This is nothing but a detailed exegesis of Euler's repeated statements on "discontinuous" functions.

[27]A study of Condorcet's manuscript *Traité* is found in Yushkevich (1974).

[28]The argument has been fully studied in numerous works. A rapid allusion is sufficient here. Further references can be found in Boyer (1959). See also Bos (1974) and Grabiner (1981, Chap. 2).

Chapter 2
THE LAGRANGIAN CALCULUS AND FOURIER SERIES

2.1. Mathematics in France after the Revolution

The passage from the eighteenth to the nineteenth century was marked by a profound rupture in the political, social, and economic fabric of Europe. This was a result of the French revolution, an event that also had a decisive impact on the history of mathematics. In fact, mathematics emerged from the radical transformations caused by the revolution profoundly changed. The role of mathematicians in society and the orientations of their research had been completely altered.

In the eighteenth century mathematicians had pursued their activities primarily within the confines of the academies, freed from all obligations to teach and assured of their livings by the patronage of princes and sovereigns. Examples of these were the Academy of Berlin, where Euler worked for many years, as did Lagrange before he left Germany for Paris; and the Academy of St. Petersburg, where Euler again and Daniel Bernoulli were active.

It was often questions of prestige that spurred the academies to seek out the best mathematicians. Since research mathematics was essentially nonexistent in the universities, it was to the academies that the true development of mathematics was entrusted, and it was to their *actes* and *mémoires* that the dissemination of new results was committed. At least this was formally true. In fact, papers were often published with considerable delay and direct correspondence between mathematicians was common enough. The letters of Euler, D'Alembert, Lagrange, and Bernoulli, to name only a few, are virtual mines of information about the state of mathematics at the time, usually far more interesting and stimulating than the papers that were published in the official journals.

After the revolution this situation changed radically, at first in France and then in the rest of Europe. The first significant upheaval in the status of science in revolutionary France occurred in 1793, when the Convention decreed the suppression of the *Académie des sciences.* In a scourging pamphlet, Marat had called the academicians "modern charlatans" while in the eyes of the Jacobins the Academy was a nest of intrigues and personal self-interests, of corruption and servitude to the *ancien régime.* Thus the members of the Academy were deprived of their privileges, and the Academy was closed.

In place of the "corrupt" academician who was subservient to the monarchy, the revolutionaries sought to substitute a new kind of scientist, one who would be sensitive to the needs of the new society and the political choices of the new state. The scientists were also called to defend their nation in its time of danger, besieged, as it then was, by the other European powers.

In order to support the scientific and cultural renewal of the nation, new institutions of research and teaching were established, such as the *Musée d'histoire naturelle,* the *Collège de France,* the *Lycée des arts,* and the *Bureau des longitudes.* Many of the most eminent scientists of the day continued to participate in the Commission on Weights and Measures, which had first been organized in 1790, while many former members of the Academy were appointed to the Commission on the Arts, which was set up by the Convention in the same year the Academy was suppressed.

In the complex political intrigues surrounding the place of science during this period, it was important that the powerful Committee for Public Safety included scientists like Prieur de la Cote d'Or and Lazare Carnot (1753-1823), the "organizer of the victory" won in 1794. A general and mathematician, Carnot had published a treatise on machines and participated in the prize problem offered by the Academy of Berlin on the infinite in mathematics (see §1.5), though without success. His paper, much expanded and extended, was published as a pamphlet a few yers later, in 1797, when he was forced into political exile in Switzerland after his expulsion from the Directory (Yushkevich, 1971). His *Réflexions sur la métaphysique du calcul infinitésimal,* while not particularly original, had a certain success with many reeditions and foreign translations.

Alongside the official institutions, numerous independent scientific societies also sprang up during this period, such as the *Réunion des sciences* under the initiation of Lalande, the *Société philomatique,* a club of young scientists, and, a few years later, the famous *Société d'Arcueil,* whose principal organizer was Laplace (1749-1827).

In 1795 the Convention created the *Institut national des arts et sciences,* which was designed to take up a role similar to that of the recently disbanded Academy. Organized according to a model proposed by the chemist Fourcroy, the Institute signaled the predominance of the scientific world over the literary in a way that clearly distinguished it from the traditions of the old Academy. This began a tendency that endured for a long time in the organization of French culture. Moreover, the influence of the mathematicians within the community of French scientists was particularly great. In 1799, at the beginning of the Napoleonic period, the number of mathematicians (including, as was then usual, astronomers and mathematical physicists) who were members of the Institute was comparable only to that of the engineers and technicians, the men who had been trained in the new schools, the "*grandes écoles*" created by the revolution (Dhombres, 1981).

The first of these *grandes écoles* was established in 1794 under the political inspiration of Carnot and the scientific direction of Gaspard Monge (1746-1818), the "genial geometer" and inventor of descriptive geometry. Significantly, it was given the name of *Ecole centrale des travaux publiques*, but this was changed the following year to *Ecole polytechnique*. Along with it was also set up the *Ecole normale*, while the *Ecole des ponts et chausees* and the *Ecole des mines* were reorganized.

The French *écoles* were founded with the specific objective of creating a large class of engineers and technicians which could serve the military and economic needs of revolutionary France (and later Bonapartist France as well, since Napoleon favored and protected the development of science in France, and it was certainly not only for his love of knowledge). The most prestigious mathematicians of the day were called to become teachers in these schools: Monge himself, Legendre (1752-1833), Lacroix, Lagrange, and Laplace, who easily converted from the politics of the Directory to those of Napoleon. Lagrange was also assigned to direct the commission for the reform of weights and measures, whose work was completed in 1799 with the adoption of the metric system and the relative unification of measurement.

The type of instruction adopted in these schools, all of which were infused with a severe military spirit, was based on mathematics, both "pure" and "applied." Thus courses on analysis were found alongside those on mechanics and descriptive geometry, the new discipline created by Monge which proved to be of great strategic and military interest. The plan of studies demanded a severe effort on the part of both teachers and students. The former were required not only to give lectures (and the number of hours of instruction was quite high), but to participate in seminars, give exams, and more importantly, to write textbooks for their respective courses. For their part, the students were required to follow the predetermined path of a rigid educational program.

With the foundation of the *grandes écoles* the former academicians became the new professors, a change that caused an irreversible shift in the way of viewing and exercising the role of the mathematician. From this time on the activity of teaching became a fundamental component of the mathematicians' work (with the exception of Gauss, who was in many ways an eighteenth-century mathematician). Many new methods and results first appeared in the pages of textbooks that collected the lessons given at the *écoles*, or later at the universities (which was more common in Italy and Germany), rather then in specialized treatises, as had formerly been the case.

Since the most important French mathematicians were called to teach in these schools, it is not surprising that the level of their instruction rose almost immediately to a high level. This is reflected in the large number of treatises on higher mathematics written in the first decades of the century, which were basically motivated by the teaching requirements of the French *écoles*.

In connection with the teaching programs of the French schools, there was an enormous production of manuals and treatises on mathematics in Paris at the beginning of the nineteenth century, which perhaps has no equal in the history of mathematics. Indeed, the French capital became home to a thriving literary industry of bookshops, printers, and publishers who specialized in scientific texts, all of whom could count on selling large numbers of copies. The manuals of Lacroix and Legendre were perhaps the most famous. They went through an incredibly large number of reprintings and reeditions in the first half of the century.

In addition to the manuals and the *Mémoires* of the Institute, mathematical papers began to appear in specialized journals associated with schools, such as the *Journal de l'Ecole polytechnique*, or with a society, like the *Bulletin de la Société philomatique*. These provided means for the more rapid diffusion of mathematical results.

Thus, in the course of a few decades, France attained a position of leadership in the field of physico-mathematical research.[1] It could boast an impressive number of scientists of the highest level, to the degree that Paris became the fundamental point of reference for every mathematician. As examples of this we can recall the trip that Abel made to Paris in 1826, precisely in order to make contact with the most advanced center of research; and the formative period in the life of his contemporary, Lejeune-Dirichlet, which resulted from his contact with Fourier, at that time the *secrétaire perpétuel* of the *Académie des sciences* and one of the most prestigious figures in French science.

The fact that, after the beginning of the nineteenth century, mathematicians also became professors, was to have important results on the character of mathematics, particularly as regards the rigorous organization of theory for didactic purposes. Above all, the "polytechnic" character of the instruction, with its considerable attention to the "applied" aspects of the subject, was to remain a distinctive feature of French mathematics for a long time. It also helped to assure for analysis a privileged role among the various branches of mathematics.

2.2. The "Algebraicization" of Analysis

Teaching in the *Ecole polytechnique* led Lagrange to reflect again on the foundations of the calculus, and the need to provide his students with a textbook spurred him to publish his lectures as the *Théorie des fonctions analytiques* in 1797.

In this work Lagrange took up the considerations he had briefly treated in an earlier work (1772) concerning the ability to present the principles of the calculus in a systematic manner without making any reference to infinitesimals, evanescent quantities, differentials or limits. He instead stressed the need to reduce the calculus to simple algebraic manipulations of finite quantities.

The object of the *Théorie*, as he declared in the subtitle to the work, was thus to present the theory of functions and the principles of the differential calculus "freed from every consideration of infinitesimals, of vanishing quantities, of limits and fluxions, and reduced to the algebraic analysis of finite quantities."

The volume begins with the following definition of a function: "We call a *function* of one or more quantities every expression of the calculus in which these quantities enter in any way whatsoever, mixed or not with other quantities that one takes as having given and invariable values, while the quantities of the function can have all possible values" (Lagrange, 1797, p. 1; 1813, p. 15).

This definition recalls the one Euler had given in the *Introductio in analysin infinitorum* about forty years earlier, which Lagrange asserts was directly inspired by Leibniz and Johann Bernoulli. They "were the first to use it with this general meaning," he says, "and it has today been generally adopted" (1797, p. 2; 1813, p. 15). In the *Introductio* Euler had based the expansion of elementary functions into series on it. For Lagrange, as for Euler, this definition became the indispensable starting point for his entire theory. But Lagrange used it with even greater generality, since his next step is to show that *any* given function can be expanded as a series.

He in fact writes,

> We therefore consider a function $f(x)$ of any variable x. If in place of x we put $x + i$, i being any indeterminate quantity whatever, it becomes $f(x + i)$ and, *by the theory of series* [my emphasis] we can expand it as a series of this form
>
> $$f(x) + pi + qi^2 + ri^3 + \cdots ,$$
>
> in which the quantities $p, q, r, \dots$, the coefficients of the powers of i, will be new functions of x, derived from the primitive function x and independent of the [indeterminate] quantity i ... (1797, p. 2; 1813, pp. 21-2).

The formation and calculation of these different functions are, to tell the truth, the true object of the new calculus, that is to say, the calculus called *differential* or *fluxional* (1797, p. 2).

In seeking to base the infinitesimal calculus on an algebraic foundation, Lagrange first discusses the inadequacy of the earlier conceptions of the principles of the calculus. He concludes that "the true metaphysics" of the calculus lies in the fact that the errors resulting from neglecting infinitesimals of higher degrees were "corrected or compensated" by the procedures of the calculus themselves, when they were limited to infinitesimals of the same degree.

This is also what we find in Euler and D'Alembert, Lagrange says, notwithstanding their attempt to confront the fundamental fault "by showing, through particular applications, that the differences one takes to be infinitely small must be absolutely zero and that their ratios, the only quantities that really enter into the calculus, are nothing other than the limits of the ratios of finite or indefinite differences" (1797, p. 3; 1813, p. 16).

As for Newton, his idea of considering mathematical quantities to be generated by motion "to avoid the supposition of infinitesimals" seems much clearer to many, Lagrange says,

> because everybody has or believes he has an idea of velocity. But, on the one hand, to introduce movement into a calculus that only has algebraic quantities as its object is to introduce a foreign idea which obligates one to think of these quantities as the lines traversed by a moving body. On the other hand, it must be granted that we do not even have a really clear idea of what the velocity of a point is at every instant when this velocity is variable (1797, pp. 3-4; 1813, p. 17).

Maclaurin's *Treatise on Fluxions* (1742), Lagrange adds, shows very well how difficult it is to render this method rigorous. This is why in the *Principia* Newton preferred to substitute for fluxions the method of ultimate ratios between evanescent quantities, a method that, according to Lagrange, has the same defects of obscurity and imprecision as that of limits.

From Lagrange's point of view, attempts in the right direction were made by Landen (1719-1790) in 1764 and by Arbogast in a 1789 paper that remained unpublished. Now, the aim of his work, Lagrange says, is to "consider the functions that arise from the expansion of any function whatsoever" (1797, p. 6) and to apply these "derived" functions to the resolution of the problems of analysis, of geometry, and of mechanics in which the differential calculus is involved. By freeing it from "every illicit supposition" and "all metaphysics" and basing it on the method of primitive functions and derivatives, Lagrange declares he is able to give to the solution of these problems "the rigor of the ancient demonstrations" (*ibid.*). Nevertheless, the crucial point in his entire construction is his assertion that it is possible to expand *any* function in a series of ascending powers of an indeterminate increment i.

But, so as not to put anything forward without proof, we will begin by considering the form of the series that must represent the expansion of every function $f(x)$ when one substitutes $x + i$ in place of x, and which we have supposed must contain only integral and positive powers of i.

This supposition is in effect verified for the expansion of the various known functions, but no one that I know has sought to demonstrate it *a priori*, which to me seems nevertheless to be even more necessary, in that there are particular cases where it cannot occur (1797, p. 7; 1813, p. 22).

In order to assure himself of the generality of his assumption about the expansion of a function $f(x)$ in series, Lagrange shows that, when x and i remain indeterminate, the series cannot contain fractional or negative powers of i. Nevertheless, and this is the decisive point, there is not in fact any proof of the *existence* of such an expansion for any given function; there is only the fact that the series contains only positive powers of i, after which, "being thus certain of the general form of the expansion of the function $f(x + i)$" (1797, p. 8; 1813, p. 23), we see what each of its terms mean. In the first place we have

$$f(x + i) = fx + iP,$$

where P is a new function of x and i,

$$P = \frac{f(x + i) - fx}{i}.$$

Separating P in so far as it is independent of i (which does not vanish when $i = 0$), and calling it p, he shows that

$$P = p + iQ$$

and hence

$$f(x + i) \ = fx + iP$$

$$P = p + iQ$$

$$Q = q + iR$$

etc.

By repeating with Q the reasoning made on P and substituting, we have

$$f(x + i) \quad = fx + iP$$

$$= fx + ip + i^2 Q$$

(2.2.1)

$$= fx + ip + i^2 q + i^3 R$$

$$= \ldots$$

for which, Lagrange says, we can take i so small that any given term of the series will be larger than the sum of all the following terms. For Lagrange this proposition must be included among the main principles of the theory.

There are questions about the procedure followed, Lagrange is quick to recognize, but these will fall to the wayside when the general form of the remainder of the series is given. He then makes a significant addition.

Moreover, it is easy to see that the method which we have just given for successively finding the terms of the series that represents a function of $x + i$, expanded according to the powers of i, can only be applied, in general, to the expansion of a function of x and of i as long as this function can be reduced to a series that proceeds according to the positive and integral powers of i; for the reasoning ... by which we have proved that every function of $x + i$ is, generally speaking, capable of this form, cannot be applied to any function whatever of x and i. But, *in the cases where this reduction is possible*, [my emphasis] one can always apply to the series resulting from the expansion of the increasing powers of i the consequence that we have drawn ... (1797, pp. 12-3; 1813, pp. 29-30).

This conclusion is certainly true, but clearly reveals the vicious circle in which Lagrange was moving.

At this point in the *Théorie* Lagrange takes the crucial step, that of identifying the functions p, q, r, ... of the expansion by $f'x$, $(f''x/2)$, $(f'''x/3!)$, ..., respectively, where $f'x$, $f''x$, $f'''x$, etc., are the successive derivatives of the function fx.

The technique used is *purely algebraic*. In the series (2.2.1) he supposes that x becomes $x + o$, where o is a quantity independent of i and indeterminate otherwise. When we proceed to calculate $f(x + i + o)$, we obtain the same result when we substitute $i + o$ for i as when we put $x + o$ in place of x.

In the first place we obtain,

$$fx + p(i + o) + q(i + o)^2 + r(i + o)^3 + \cdots,$$

or, by expanding according to the powers of $(i + o)$,

$$fx + pi + qi^2 + ri^3 + \cdots$$

$$+ \, po + 2qio + 3ri^2o + \cdots$$

In the second case, if we indicate by $fx + f'xo + \cdots$, $p + p'o + \cdots$, $q + q'o + \cdots$ the results of substituting $x + o$ for x in the functions fx, p, q, ..., we have,

$$fx + pi + qi^2 + ri^3 + \cdots$$

$$+ \, f'xo + p'io + q'i^2o + \cdots \, ,$$

Then by identifying the corresponding terms, we obtain

$$p = f'x, \qquad 2q = p', \qquad 3r = q', \; ... \; .$$

From this Lagrange can deduce without difficulty,

$$p = f'x, \qquad q = \frac{f''x}{2}, \qquad f = \frac{f'''x}{3!}, \text{ etc.,}$$

where $f'x$, $f''x$, $f'''x$, etc. indicate the first, second, third, etc. derivatives of fx. This is the means by which Lagrange *formally* obtains the Taylor series of a function. In fact, by substituting into the expansion of the function $f(x + i)$, we obtain

$$(2.2.2) \qquad f(x + i) = fx + if'x + \frac{f''x}{2}\,i^2 + \frac{f'''x}{3!}\,i^3 + \cdots \, .$$

This new expression [Lagrange writes] has the advantage of showing how the terms of the series depend upon each other, and above all how, when one knows how to form the first derivative from any primitive function whatever, one can form all the derivative functions that the series contains (1797, p. 14; 1813, p. 33).

There is no doubt that this is true if the function has the right properties,[2] but the delicate point consists precisely of supposing that *every function* is expandable as a series of the type (2.2.1), since the criteria for this expansion presuppose the existence of the derivatives of fx, which is what Lagrange instead seeks to demonstrate. Moreover, as had already been known since the origins of the infinitesimal calculus, the crucial point is precisely to obtain $f'x$.

The weakness of Lagrange's argument did not escape the author himself. In the second edition of the *Théorie* (1813) he commented on this point, "Furthermore, for as far as one knows the differential calculus, one must see that the derivative functions y', y'', y''', ..., relative to x, coincide with the expressions df/dx, d^2f/dx^2, d^3f/dx^3, ..." (1813, p. 33). These words recall the tone that Leibniz had used to present the calculus of differentials in 1684. Whoever knows a little geometry will know how to operate with the new calculus, Leibniz had then said. Whoever knows the first concepts of the calculus will immediately know what the derivative functions $f'x$, $f''x$, ... are, Lagrange now affirmed, with an air of certainty that was basically unfounded.

Finally, in conformance with the spirit of the epoch, we do not find in Lagrange any attempt to assure himself of the convergence of the series (2.2.1). Moreover, Lagrange himself had commented on the use of series in analysis several years before, at the time of his solution to the problem of vibrating strings. In discussing the solution in trigonometric series that had been proposed by Daniel Bernoulli, he wrote,

> Now I ask if, every time that an algebraic formula is found for example in an infinite geometric series, such that $1 + x + x^2 + x^3 + \cdots$, one will not have the right to substitute $1/(1-x)$, *although* [my emphasis] this quantity is only equal to the sum of the proposed series by supposing that the last term x^∞ is zero. It seems to me that one cannot challenge the exactitude of such a substitution without overturning the most common principles of analysis (1760-61, p. 323).

But, as would become clear to mathematicians at the beginning of the nineteenth century, the substitution is valid *only* when $|x| < 1$, which is what Lagrange meant when he said in the language of his time that x^∞ is zero.

The *Théorie* is presented, in Lagrange's style, as a work of grand scope, an organic and rational systematization of both the knowledge of analysis obtained until then and original results. One of its most significant new results is surely the remainder theorem, a "theorem, new and remarkable for its simplicity and generality" (1797, p. 49; 1813, p. 83), according to which, in Lagrange's symbolism, $fx = f + xf' + (x^2/2)f'' + (x^3/2 \cdot 3)f'''u$ and, in particular, $fx = f + xf'u$, where f indicates the value assumed by the function at 0 and u is any value included between 0 and x.

At the time of its appearance the *Théorie* was accepted with general favor. It clearly inspired Arbogast, in his *Calcul des derivations*, and the young "polytechnicians" like Francoeur, Francais, and Servois to elaborate Lagrange's ideas in a purely formal direction by developing an original calculus of operators. In England Babbage, Herschel, and Peacock, the organizers of the Cambridge Analytical Society, found Lagrange's work to be an ideal point of reference for their critique of the Newtonian tradition in the calculus and for the development of purely analytical techniques for calculating with formal series and operators.

Of course, there was no absence of criticism, as the numerous attempts to "give a rigorous demonstration of Taylor's theorem" in France in the early years of the century implicitly testify (see also

§3.6). This theorem, as we have seen, constituted the keystone of Lagrange's entire theoretical edifice.

Notwithstanding the authority and the prestige of "the first geometer of the Empire," as Lagrange was called during the Napoleonic period, it was nevertheless clear that his attempt to give a rigorous and definitive systematization to the foundations of the calculus required further work.

In addition, at the beginning of the century Lagrange's theory of derived functions coexisted with the Leibnizian/Eulerian practice of using infinitesimals and differentials, which had never been abandoned. We can obtain an idea of the variety of views adopted at this time with respect to the principles of the calculus by studying the manuals that were then being used for teaching in the French schools. One of the most important of these, which was widely used for decades, was the summary edition of Lacroix's monumental, three volume *Traité du calcul différentiel et du calcul intégral*, the first edition of which appeared in 1797-1800.

In this work, Lacroix presented both the Leibnizian/Eulerian method of differentials, the method of limits that had been given by D'Alembert in the *Encyclopédie*, which proclaimed that the "true metaphysics" of the calculus lay in the concept of limit, and Lagrange's method of derived functions. Certainly the latter was much more rigorous in Lacroix's eyes, but there is no doubt that in the practice of teaching and the calculus itself infinitesimals and differentials were more flexible and fruitful instruments. This is in substance what Lacroix again affirmed concerning the principles of the calculus in the 1810 edition of the *Traité*.[3]

But the very next year the Lagrangian theory was the object of a determined attack by Hoëne Wronski (1778-1853), a bizarre figure in mathematics and the founder of a philosophical movement called "messianism," which took on the character of a religious sect. It was based on a revelation of the universal and rational "Absolute," whose manifestation was supposed to be Wronski himself. Although Wronski's "philosophy" may seem peculiar to us, it seems to reflect the ideas that had been put forth by the Cult of the Absolute and the Cult of Reason during the Jacobin period.

Wronski's first attempt to apply his ideas to mathematics was set out in his *Introduction à la philosophie de la mathématique et technie de l'algorithmie*, which was presented to the Institute in 1811. The commissioners gave it a lukewarm approval, combined with an invitation to better clarify its arguments.

In this work, after more than two hundred pages of philosophical and mathematical reflections, Wronski arrives at the statement that every function can be given in the form

$$F(x) = A_0\Omega_0(x) + A_1\Omega_1(x) + A_2\Omega_2(x) + \cdots,$$

where the $\Omega_i(x)$ are any functions of x and the determination of the coefficients A_i of the series depend on the determinants that are today called "Wronskian." The Introduction to this work contained an unveiled although attenuated criticism of Lagrange's theory, which Wronski took up with considerably more vigor the following year in a work bearing the unequivocal title, *Réfutation de la théorie des fonctions analytiques de Lagrange.*

Even though the "philosophical" arguments that Wronski adopted were nebulous and scarcely convincing, the criticisms that he brought against Lagrange's ideas were penetrating. Lagrange's theory, Wronski observed, is based on the following two assumptions:

(2.2.3) $f(x + i) = A + iB + i^2C + \cdots$

(2.2.4) $f(x + i) = f(x) + iP.$

But, Wronski asks, from where do we obtain our knowledge of (2.2.3)? What allows us to suppose that such an expression is possible? And again: Lagrange asserts that (2.2.4) is true because it is verified for $i = 0$. But what can we say when $i \neq 0$? In the *Théorie* Lagrange writes that $f'(x)$ is defined as the coefficient of the second term of the expansion (2.2.3), Wronski observes, "but this clearly does not mean anything; the nature of functions is not immediately given by their position in the series."

In the face of these criticisms, it is certainly not surprising that the *Réfutation* was rejected by the two commissioners, Lacroix and Arago, who were assigned to review it. In response to Wronski's impertinent work, Poisson (1781-1840), the brilliant and successful student of Lagrange at the *Ecole polytechnique*, hastened to publish a new "rigorous demonstration of the Taylor series and Taylor's theorem" in number 3 of the *Correspondence de l'Ecole polytechnique.* This supposed that

(2.2.5) $f(x + h) = f(x) + ph^a + qh^b + rh^c + \cdots$

where a, b, c, ... are positive numbers and p, q, r, ... are functions of x. He then sought to prove that the series (2.2.5) is actually that of Taylor.

But Poisson's supposition "is only a *petitio principii*, "Wronski promptly objected, and his "demonstration" does not escape the same difficulties that encumbered Lagrange's.

Although Wronski's critique was prompt and convincing, the solution that he proposed was vague and nebulous. Enmeshed in the obscure language of his "algorithmie," it was destined to be quickly forgotten with the rapid fall in the "philosophical" fortunes of its inventor.

But rejecting Wronski's arguments certainly did not mean adopting Lagrange's point of view. As Wronski himself maliciously pointed out, a paper on generating functions and their applications to probability which Laplace read to the Institute in person in April,

1811 reviewed the recent advances in the calculus without devoting even one word to the theory of his illustrious colleague!

At the same time, as we will see below, the adequacy of Lagrange's entire theoretical framework was being questioned by Fourier (1768-1830), the prefect of the *departement* of Isère at Grenoble. In his works on the propagation of heat, Fourier had in fact asserted that a trigonometric series, and not a series of integral powers, could represent *any* function in a given interval. This, however, raised doubts on Lagrange's part, as we read in the report on Fourier's paper he prepared for the Institute (see below, §2.3b).

The favor shown towards Lagrange's theory of derived functions declined rapidly in the face of ever more specific and convincing criticisms of the use of infinite series in the absence of sufficient criteria to assure their convergence. The search for such criteria was the task to which the mathematicians of the nineteenth century, beginning with Abel and Cauchy, set themselves.

The fortune of the *Théorie* is further shown by Serret's preface to the third edition of 1847, where he lamented the fact that Lagrange's work was "unfortunately little read in our days." But the demand for rigor in analysis, which was led above all by Cauchy, had in fact shown that the "algebraic analysis" of Lagrange was insufficiently rigorous in the eyes of mathematicians and an inadequate foundation for the infinitesimal calculus.

Nevertheless, there is an element that Lagrange emphasized on various occasions which strongly characterizes the development of analysis in the nineteenth century -- its increasing independence from geometrical references and the consequent need to frame autonomous and coherent logical criteria.

In the *Traité de mécanique analytique*, which remained a fundamental text for more than a generation of mathematicians, Lagrange emphasized, not without satisfaction, that

> You will not find any figures in this work. The methods that I present do not require either constructions, nor geometrical or mechanical reasonings, but only *algebraic operations* [my emphasis], ... Those who love analysis will view with pleasure mechanics becoming a new branch of it, and will wish me well for having thus extended its domain (1788, pp. xi-xii).

Lagrange retained the same point of view when he treated the foundations of the calculus and the theory of functions. His discussion remained on a purely algebraic plane, making no recourse to geometrical intuitions, as, for example, in order to "clarify" the meaning of the derivative.

In this Lagrange can therefore be considered a coherent and knowing pioneer of a tendency that would become prevalent from the time of Cauchy. Even though the foundation he set out, the algebraic analysis of finite quantities, was found to be inadequate, he was nevertheless moving on the same road, far from any reference to geometrical evidence, and directed towards finding in the framework of analysis, and exclusively in it, the deepest foundations, the "metaphysics" of the infinitesimal calculus.

2.3. Fourier Series

Lagrange's *Théorie des fonctions analytiques* was, in a certain sense, a synthesis of eighteenth-century analysis, in the same way that his *Mécanique analytique* and Laplace's *Mécanique céleste* had served as syntheses of their fields.   Towards the end of the century there was a widespread conviction that mathematics was moving towards a kind of dead end, with no exit in sight. The many theories that had arisen in the course of the century had been extensively developed and the difficulties posed by the extreme complexity of the calculations seemed to be insurmountable. The dominant opinion was that everything there was to discover had already been discovered, that one could not go beyond the point reached by the Bernoullis, Euler, Clairaut, D'Alembert, Laplace, and Lagrange.

Lagrange himself, for example, had written to D'Alembert on September 21, 1781, saying that,

> It seems to me that the mine [of mathematics] has already gone too deep, and that unless someone discovers new veins it will be necessary sooner or later to abandon it. Physics and chemistry now offer more brilliant riches and are more easily exploited; even the taste of the century seems to have entirely turned in this direction, and it is not impossible that the place of geometry in the academies will one day become like that which is currently occupied by the chairs of Arabic in the universities (Lagrange, *Oeuvres* 13, p. 368).

This opinion was taken up with even more authority by J. B. Delambre (1749-1822), the *secrétaire perpétuel* of the section on mathematics and physics of the *Institut de France*. In his *Rapport historique sur le progrès des sciences mathématiques depuis 1789 et leur état actuel*, which he read to the Institute in February, 1808, he asserted that

> It would be difficult and rash to analyze the chances which the future offers to the advancement of mathematics; in almost all its branches one is blocked by insurmountable difficulties; perfection of detail seems to be the only thing which remains to be done. All of these difficulties appear to announce that the power of our analysis is practically exhausted (In: Kline, 1972, p. 623).

Among the "details" that remained to be perfected were two that only a man with Delambre's optimism could consider as such. The first was the foundations of the calculus, a question that remained opened and current, even after the publication of Lagrange's *Théorie*. The second was the related problem of the foundation of mechanics, and in particular the role of the "principle of virtual velocities," which Lagrange had used as "a kind of axiom" in the *Mécanique analytique*.

In this work Lagrange had expressed his principle of "virtual velocities" as follows. If any system of material points or bodies subject to "any forces whatever" (*puissances quelconques*) is in equilibrium, and if we then subject the system to "any infinitesimal movement whatever" so that every point traverses an infinitesimal space, then the sum of the "forces" multiplied by the space the point traverses in the direction of the force is equal to zero, taking this direction as positive and that opposed to the "forces" as negative.

Apart from the ambiguity in the use of the words "power" and "force" and the implicit use of the concept of "work," the problem here is the nature of the "principle" expressed in Lagrange's statement. In addition, he uses the intuitive idea of an infinitesimal quantity, something that he had sought to banish from mathematics in the *Theorie*.

In 1798, in an article that appeared in the *Journal de l'Ecole polytechnique*, Lagrange sought to explain his "principle" by using the principle of pulleys. In the same issue of the *Journal* R. de Prony (1755-1839), a professor of mechanics at the École, gave a demonstration of it by applying the decomposition of circular movement (De Prony, 1798). For his part, Fourier suggested using some other "principle" whose nature was more immediately intuitive and clearer as the foundation of mechanics, such as the principle of the composition of forces.

Carnot also joined in the discussion with his work on the *Principes fondamentaux de l'équilibre et du mouvement* of 1803. A few years later Poinsot (1777-1859) and Ampère (1775-1836) also called attention to the principle, the first in the context of a long paper on the general theory of equilibrium and the movement of rigid systems (Poinsot, 1806), the second with a "demonstration" of the principle that did not make any use of infinitesimals (Ampère, 1806).[4]

This lively discussion over the principles of mechanics in the early years of the century thus ended by entwining itself with the question over the foundations of the calculus, which was no less challenging and in many respects more important. This is a fundamental point if we want to fully understand the nature of the debate over the foundations of analysis that took place at this time.

It is customary in presenting the history of the infinitesimal calculus during the early years of the nineteenth century to omit any reference to the problems involved in the debate over the "principle of virtual velocities." In my view, this is a serious historical error. It is wrong to introduce interpretative categories that are common today, such as the distinction between "pure" and "applied" mathematics, into periods of mathematics where they do not apply. This distinction did not exist at this time, and would not do so for several decades to come. In addition, the "polytechnical" training of the French mathematicians pushed them towards a unified conception of science.

This conception also seems to animate the exhortation with which Delambre concludes his report. Notwithstanding the "insurmountable difficulties" which the future seems to hold for mathematics, "the spectacle of analysis and mechanics in our time" convinced Delambre to express the hope that "the generations to come will not see anything impossible in what remains to be done." This wish was to be splendidly realized in the extraordinary development of mathematics during the immediately following years.

2.3.a. Fourier's Equation for Heat Diffusion

A little less than two months before Delambre read his report to the Institute, Fourier had presented it with a long paper that sought to confront one of the things "that remained to be done" -- to study in a mathematical way the propagation of heat in bodies. This was a problem of great theoretical and practical interest, which had attracted the interest of many physicists, mathematicians, and practical "technicians" toward the end of the eighteenth century.

There is no doubt that at this time heat was coming to be seen ever more clearly as a form of energy which could be used as an aid in production. For confirmation we need only look to the increasing use of steam engines in industrial processes, especially in England and France. But if it is the *practical interests* that are best expressed in the English textile mills, it is the *theoretical aspect* that particularly engaged the French scientists.

In 1784 Laplace and Lavoisier (1743-1794) published a joint paper containing the results of the experiments they had conducted to determine the specific heat of various substances. Some time later, in 1804, J. B. Biot (1774-1862), a student of Laplace, published his *Mémoire sur la propagation de la chaleur* in the *Journal des Mines*. At the same time Fourier began to occupy himself with the question, but in a completely original manner. His researches were to become a source of discussions and results in analysis for several decades through the works of men like Dirichlet, Riemann, Heine, Cantor, and Lebesgue.

The studies undertaken by Fourier thus seemed, in a certain sense, to confirm the prophetic words which Condorcet had written in 1781. Objecting to Lagrange's view that it would soon be necessary to abandon mathematics as a mine whose riches had been exhausted and search out new paths for the advancement of knowledge, he wrote,

we are far from having exhausted all the applications of analysis to geometry, and instead of believing that we have approached the end where these sciences must stop because they have reached the limit of the forces of the human spirit, we ought to avow rather we are only at the first steps of an immense career. These new [practical] applications, independently of the utility which they may have in themselves, are necessary to the progress of analysis in general; they give birth to questions which one would not think to propose; they demand that one create new methods. Technical processes are the children of need; one can say the same for the methods of the most abstract sciences. But we owe the latter to the needs of a more noble kind, the need to discover the new truths or to know better the laws of nature (In: Kline, 1972, pp. 623-4).

A similar view of the development of mathematics animated Fourier. His point of view is clearly expressed in the *Discours préliminaire* that opens his *Théorie analytique de la chaleur*. "Primary causes are unknown to us," he writes, "but are subject to simple and constant laws, which may be discovered by observation, the study of them being the object of natural philosophy" (1822, p. xv).

Through the works of Newton, he continues, we have found the laws that regulate the entire system of the universe. But these laws, in spite of their vast applications, do not explain phenomena "of a special order" that are observed in "the effects of heat."

Fourier wrote that, following a prolonged and attentive study,

I have concluded that to determine numerically the most varied movements of heat, it is sufficient to submit each substance to three fundamental observations. Different bodies in fact do not possess in the same degree the power to *contain* heat, to *receive or transmit it across their surfaces*, nor to *conduct* it through the interior of their masses. These are the three specific qualities which our theory clearly distinguishes and shows how to measure (1822, p. xvi f.).

Fourier is well aware of the practical applications of his results and writes, "It is easy to judge how much these researches concern the physical sciences and civil economy, and what may be their influence on the progress of the arts which require the employment and distribution of fire" (1822, p. xvii), where the word "arts" here has the meaning generally associated with the phrase "arts et métiers," which originated in the sixteenth century.

These researches, moreover, have a "necessary connection with the system of the world," but the object of a true and proper theory is analytic by nature. This point is fundamental to Fourier's entire theoretical approach: once a certain number of observations have been obtained by means of experimental observations, it is necessary to translate the relations between them *into a mathematical form*, and more precisely, into differential equations. This is the path that must be followed in order to advance our knowledge of nature.

The differential equations of the propagation of heat [Fourier writes] express the most general conditions, and reduce the physical questions to problems of pure analysis, which is the proper object of theory. ...

After having established these differential equations their integrals must be obtained; this process consists in passing from a common expression to a particular solution subject to all the given conditions. This difficult investigation requires a special analysis founded on new theorems. ... The method which is derived from them leaves nothing vague and indeterminate in the solutions, it leads them up to the final numerical applications, a necessary condition of every investigation, without which we should only arrive at useless transformations....

The profound study of nature is the most fertile source of mathematical discoveries. Not only has this study, in offering a determinate object to investigation, the advantage of excluding vague questions and calculations without issue; it is in addition a sure method of forming analysis itself, and of discovering the elements which it concerns us to know, and which natural science ought always to preserve: these are the fundamental elements which are reproduced in all natural effects (1822, p. xxi f.).

From this follows the conclusion that retains a certain interest even today, when mathematics seems to venture into regions ever farther removed from reality: "Considered from this point of view, mathematical analysis is as extensive as nature itself" (1822, p. xxiii).

In the first chapter Fourier takes up these same themes, writing in the very first lines,

The effects of heat are subject to constant laws which cannot be discovered without the aid of mathematical analysis. The theory that we are about to explain has as its object the demonstration of these laws; it reduces all physical researches on the propagation of heat to problems of the integral calculus whose elements are given by experiment. No subject has more extensive relations with the *progress of industry and the natural sciences* [my emphasis] (1822, p. 1).

The *Théorie analytique* appeared in 1822 when Fourier was the *secrétaire perpétuel* of the reestablished *Académie des sciences*. In reality it constitutes the extension and completion of an earlier memoir that Fourier had presented to the *Institut de France* in 1807, which remained in manuscript and has only recently been published (Grattan-Guinness, 1972).

The object of this investigation, Fourier writes, is the propagation of heat, concerning whose nature we can only advance "uncertain hypotheses" consistent with the philosophical approach outlined in the *Discours préliminaire*. But "the knowledge of the mathematical laws to which its effect are subject is independent of every hypothesis" (1822, p. 15) and requires only the accurate examination of experimental facts.   Fourier's researches, nevertheless, are rich in implications and results that go well beyond the theme proposed. They constitute a substantial amplification of the physico-mathematical domain that ever since Newton's time had been coextensive with rational and celestial mechanics.

In the introduction to his edition of Fourier's manuscript, Grattan-Guinness writes,

> By Fourier's time, the mathematical analysis of physical problems was largely based on the construction of a partial differential equation representing the phenomena under investigation together with initial and boundary conditions. Fourier enriched and modified this procedure in a variety of ways that have been important ever since.   Firstly, he distinguished between two kinds of physical behavior -- action at an "interior" point and action on a "surface" boundary of the material involved -- and formulated separate equations for them. Secondly, he expressed these equations in a coordinate system appropriate to the geometrical and physical symmetry of the problem at hand... .   Thirdly, he added to the interior and surface equations explicit statements of initial conditions which would allow the calculation of the unknown constants of the solution and hence give a complete description of the phenomenon under analysis. (Grattan-Guinness, 1972, p. vii)

In integrating the differential equations he obtains, Fourier makes considerable use of trigonometric series, opportunely determining the coefficients (the "Fourier series"). Series of this type had already been known in mathematics, as we have already seen, but "without the background of the physical phenomenon, these questions would have remained mathematical curiosities; ... in Fourier's work, advances in mathematical physics and advances in mathematical analysis were intimately connected, providing mutual stimulus and reinforcement" (Grattan-Guinness, 1972, p. viii).

But the mathematical part of Fourier's 1807 paper, and in particular his use of trigonometric series, provoked objections from the members of the Institute who had been appointed to judge it (Lagrange, Laplace, Monge, and Lacroix).[5]

Fourier replied to these objections in an article on the convergence of trigonometric series that he sent to Lagrange himself,[6] as well as in a note on the same argument that he sent to the Institute in 1808.

The question of the propagation of heat still remained open. In accordance with the customs of the period, in 1811 the *Institut de France* announced a prize for the following problem: "Provide a mathematical theory of the laws of the propagation of heat and compare the results of this theory with exact experiments" (In: Fourier, *Oeuvres* 1, p. vii).

Fourier participated in the competition and was awarded the prize, but he again encountered objections from the viewpoint of rigor, as we read in the report the judges presented to the Institute:

> This work contains the true differential equations of the transmission of heat, both in the interior of bodies and at their surface; and the novelty of the subject, combined with its importance, has caused the Class [of mathematical and physical sciences of the Institute] to award it the prize, observing, however, that the manner in which the author arrives at his equations is not without difficulties, and that his analysis, to integrate them, still leaves something to be desired both as regards generality and rigor (In: Fourier, *Oeuvres* 1, p. vii).

This is probably the reason why the prize-winning paper, contrary to custom, was not published and appeared in print only in 1824 and 1826 when, as had been said, Fourier was himself the *secrétaire perpétuel* of the *Académie des sciences*.[7] A brief note on his initial researches in the memoir of 1807 was published by Poisson in the *Bulletin de la Société philomatique*, but it was not particularly enthusiastic.

Fourier's first approach to the problem of the propagation of heat between discrete bodies dates from the early years of the century and is represented in a series of papers that culminated in the published volume of the *Théorie*. His first model was very simple, consisting of two bodies of equal masses and different initial temperatures that exchange heat at infinitely small distances. This model was then extended to n bodies. Fourier resolved the problem without any particular difficulties, both for when the n bodies are arranged in a straight line and when they are on the circumference of a circle.

Using the theoretical approach common at the time (which we have already seen Lagrange use for the vibrating string) the next step would have been that of taking n to infinity to find the solution for the corresponding continuous bodies, in this case a rectilinear bar or a ring.

But in his initial research, apparently about 1802-3, Fourier stopped at this point, and did not consider the extension to continuous bodies. Instead, he seemed to completely abandon the question until he was stimulated to take it up again by a work J. B. Biot sent him in 1805.

Biot had obtained certain experimental results on the propagation of heat and had sought a mathematical theory for organizing them. Adopting a Newtonian approach,[8] in which the loss of heat from a body to its surroundings is proportional to their difference in temperature, Biot introduced two different coefficients of conductivity, one for the conduction of heat within the body and one for that with the external environment.

Biot considered the case of a bar heated at one of its extremities.

Every point in the bar then receives heat from every "preceding" point and transmits it to those "following," while at the same time losing heat to the exterior. In a situation of thermal equilibrium the phenomenon could be represented by an ordinary differential equation of the second degree, but when the temperature is variable there is a net difference between the interior and the exterior that can be expressed by a partial differential equation.

This is Biot's reasoning, but he did not actually present an equation as such. In seeking to explain his motives, Grattan-Guinness suggests that "the reasons are traceable to his philosophical views on physics, where he followed Laplace in regarding all physical phenomena as the products of Newtonian actions between nearby, *but not necessarily adjacent*, molecules" (Grattan-Guinness, 1972, p. 84).

This was the view that Laplace had presented in his *Exposition du système du monde* (1794) and again in vol. 4 of his *Mécanique céleste* (1804). Laplacian physics was of course dominant in France at this time.[9] On the other hand, Biot himself clarified his approach ten years later, explaining that in his earlier work he had given only "the structure of the formulae without proof" because of his dissatisfaction with "the difficulty of analysis regarding homogeneity."

Biot again clarified what he meant by this, saying, "When we come to form this equation, we find that the laws of homogeneity which govern differentials cannot be satisfied if we suppose that each material and infinitesimally small point of the bar receives heat only by contact with the point which precedes it and transmits it only to the point which follows it. *This difficulty can be set aside only by assuming, as M. Laplace did, that a particular point is influenced not only by those that touch it but also by those that are only a small distance away from it, ahead and behind* [my emphasis]. Then homogeneity is reestablished, and all the rules of the differential calculus are observed" (Biot, 1816, IV, pp. 667-8; trans. Truesdell, 1980, p. 50).

Biot wrote this long after Fourier had presented his prize-winning paper to the Institute. Biot limited himself to remembering this in a note, writing that Fourier "*reproduced* the partial differential equation" [my emphasis]. On the other hand, when Biot published his *Traite*, the discussion over the nature and propagation of heat was as lively as ever.

In 1815, inspired by Laplace's ideas, Poisson presented a long paper to the Institute on the propagation of heat in solids. This was followed by the publication of articles and notes in the *Journal de physique* and the *Bulletin de la Société philomatique*. For his part, Fourier expounded his ideas of heat as a fluid in various articles that appeared in the *Annales de chemie et de physique*, and claimed his own priority in the most polemical fashion by writing, "Poisson has too much talent to apply it to the work of others, or to use it to discover what is already known to others." The rivalry between Poisson and Fourier was more or less latent, but never lessened. Whereas Fourier collected his own results in a systematic manner in the *Théorie*, Poisson's research culminated in his *Théorie mathématique de la chaleur* of 1835, which was enriched by a large supplement in 1837.[10]

In this climate, it is difficult to competently evaluate Biot's justifications for his initial uncertainties, or the influence of his early work on Fourier. However, it is highly probable that Fourier took the idea of internal and external conductivity and of different coefficients of conductivity for each from Biot's article of 1805. But unlike Biot, Fourier considered a transverse section of the bar, of width dx, whose temperature was influenced only by the sections adjacent to it and by the atmosphere in contact with the exterior surface of the section. Using these hypotheses, and indicating the external and internal coefficients of conductivity with h and k respectively, he obtained

$$k \frac{d^2t}{dx} - ht = 0,$$

an equation that is still dishomogeneous, but which becomes homogeneous with the simple substitution of k/dx in place of k. The motivation for this lies in the fact that the ability to conduct heat increases when the size of the section is decreased. Thus

$$k \frac{d^2t}{dx^2} - ht = 0,$$

whose integral is

$$t = Me^{-x\sqrt{h/k}} + Ne^{x\sqrt{h/k}},$$

where M and N are arbitrary constants.

From the physical nature of the problem (the propagation of heat along an infinite bar), t must become infinitely small as x increases (and therefore $N = 0$). If we set as the initial conditions, $x = 0$, $t = A$, then the solution is given by

$$t = Ae^{-x\sqrt{h/k}}.$$

In presenting this result, Fourier only mentions Biot's research.

Taking up the question of the change in temperature over time (which Biot had simply stated and not actually studied), and indicating the temperature with v, Fourier obtained in a like manner

$$k \frac{\partial^2 v}{\partial x^2} - hv = \frac{\partial v}{\partial t}.$$

This is the equation that Fourier had found for the case of the infinite bar in his manuscript of 1805, while for the plate the equation of the propagation of heat is

$$k \left[\frac{\partial^2 v}{\partial x^2} + \frac{\partial^2 v}{\partial y^2} \right] - hv = \frac{\partial v}{\partial t},$$

and for a solid body,

$$(2.3.1) \quad k\left[\frac{\partial^2 v}{\partial x^2} + \frac{\partial^2 v}{\partial y^2} + \frac{\partial^2 v}{\partial z^2}\right] - hv = \frac{\partial v}{\partial t}.$$

But in both the memoir of 1807 and in the edition of the *Théorie* the term $-hv$ is (correctly) omitted. What is the significance of this omission?

Fourier took into account the fact that the dispersion of the heat of the body (on which the factor $-hv$ depends) involves only the molecules at the surface of the body and not those in the interior. In fact, as a comment to (2.3.1) Fourier adds the following note:

> When we better know from the results of experiments what this property is that all bodies have to spontaneously dissipate a part of their heat, we will know whether the last term hv should enter into the equations relative to the interior of the body or only into the equations that relate to their surfaces (Grattan-Guinness, 1972, p. 111 n. 6).

This distinction was to become of great importance in physics. Through it, Fourier was in a position to obtain the general equation for the propagation of heat within a body, reserving for a later time the study of the conditions relating to the surface. In article 27 of the 1807 paper, relative to the general equation for the movement of heat, we read:

> The equation just obtained represents the successive states of any solid whose different points are constantly changing temperature. The coefficient h, which is the measure of the exterior conductivity, does not enter into this equation at all. But it remains to express the situation relative to the surface, which will be explained in the rest of the paper. It is in this part of the calculation that the coefficient h is introduced. The general equation for the propagation of heat in solids is thus the following:

$$(2.3.2) \quad \frac{\partial v}{\partial t} = \frac{K}{CD}\left[\frac{\partial^2 v}{\partial x^2} + \frac{\partial^2 v}{\partial y^2} + \frac{\partial^2 v}{\partial z^2}\right],$$

> where v is a function of the three coordinates x, y, z and the time t, K is the measure of the conductivity of the substance of which the solid is formed, and D is its density. The true meaning of this equation lies in the fact that the function v must satisfy the general condition that is expressed by it. But independently of this general condition, in all cases there are many other special conditions that depend on the shape of the body, the nature and state of the surface, the action of one or more souces of heat (*foyers*), and of various other circumstances

that can present themselves in individual questions (Grattan-Guinness, 1972, p. 126-7).

2.3.b. The Introduction of Trigonometric Series

In seeking to integrate equation (2.3.2) Fourier begins by treating the case of a rectangular lamina of infinite length (which is why the term $\partial^2 v/\partial z^2$ is missing in (2.3.2)). He initially considers the case of "stationary temperature," that is, the situation of thermal equilibrium (which implies $\partial v/\partial t = 0$). As a boundary condition he simply requires that the temperature be 1 at the extremities of the lamina and 0 at its two sides. Thus we have to first integrate the equation

$$(2.3.3) \quad \frac{\partial^2 v}{\partial x^2} + \frac{\partial^2 v}{\partial y^2} = 0,$$

which Fourier does by using the method of the separation of variables. That is, he supposes that

$$v = \phi(x)\psi(y),$$

from which, by differentiating,

$$\phi''(x)\psi(y) + \phi(x)\psi''(y) = 0,$$

or,

$$\frac{\phi(x)}{\phi''(x)} = -\frac{\psi(y)}{\psi''(y)} = A,$$

where A is a constant.

We see from this [Fourier writes] that we can take for $\phi(x)$ a quantity of the form e^{mx} and for $\psi(y)$ the quantity $\cos(ny)$. Then by supposing

$$y = ae^{mx}\cos(ny),$$

and substituting into the proposed, we will have the equation of condition

$$m^2 = n^2$$

... [which] will always be satisfied if we take for the value of v the quantity $ae^{nx}\cos(ny)$ or $ae^{-nx}\cos(ny)$ (Grattan-Guinness, 1972, p. 137-8).

Discarding the first for physical motives relative to the nature of the problem (v would become infinitely large as x increases), we then reduce the preceding solution to $v = ae^{-nx}\cos(ny)$, n being any positive number whatever and a an indeterminate constant. We will then form the general solution by writing

$$v = a_1 e^{-n_1 x} \cos n_1 y + a_2 e^{-n_2} \cos n_2 y + \cdots \quad (ibid., \text{ p. } 138).$$

In order to completely determine the solution it is consequently necessary for Fourier to determine the pair of infinite arbitrary constants

$$a_1, a_2, ..., a_k, ...$$

$$n_1, n_2, ..., n_k, ... \,.$$

To find the n_k, Fourier uses the condition that on both sides of the lamina the temperature be zero. He further supposes that it is possible to "cut" the lamina longitudinally along the x axis into two equal parts, and in addition requires that half the width of the lamina be 1. This, for every value of x, must make $v = 0$ when $y = 1$ and $y = -1$, which implies that the constants n_1, n_2, ..., n_k, ... are equal to the odd multiples of a fourth of the circumference, that is, $n_1 = \pi/2$, $n_2 = 3\pi/2$, ..., etc.

In order to find the constants a_k, Fourier utilizes the remaining condition, that is, that $v = 1$ for $x = 0$ for any whatever y. This leads to the equation

$$(2.3.4) \qquad 1 = a_1 \cos \frac{\pi}{2} y + a_2 \cos \frac{3\pi}{2} y + a_3 \cos \frac{5\pi}{2} y + \cdots,$$

which, by setting $\frac{1}{2}\pi y = u$, can be written

$$(2.3.5) \qquad 1 = a_1 \cos u + a_2 \cos 3u + a_3 \cos 5u + \cdots$$

with u included between $-\pi/2$ and $\pi/2$.

In order to determine the coefficients, Fourier considers the series (2.3.5) and its derivatives, obtained by differentiating the series term by term, and sets $u = 0$. This leads to the equations

$$1 = \sum_{k=1}^{\infty} a_k$$

$$0 = \sum_{k=1}^{\infty} (2k - 1)^2 a_k$$

$$0 = \sum_{k=1}^{\infty} (2k - 1)^4 a_k$$

$$\cdots,$$

that is, to a system of infinite linear equations in infinite unknowns (the coefficients a_k). This is how Fourier intends to resolve the system:

The indeterminate a_1, for example, will have one value for the case of two unknowns, another for the case of three unknowns, and so on for four, etc. It will be the same for the indeterminate a_2, which will have as many different values as many times as one makes the elimination. All of the other indeterminates are likewise susceptible to an infinity of different values. Now, the value of one of these unknowns, for the case where their number is infinite, is the limit towards which the different values that arise from the successive eliminations continually tend (Grattan-Guinness, 1972, p. 149).

In other words, Fourier proposes to consider only the first k equations in k unknowns, for finite k. For example, to determine the value of the coefficient a_1, he resolves the system by successively eliminating the unknowns $a_2, \dots, a_k$ and consequently obtaining the values of the remaining $k - 1$ unknowns, $a_2, \dots, a_k$, relative to the first k equations. Finally, passing to the limit as k tends to infinity, he obtains the "true" values of the coefficients.[11]

In this way he finds the following expression for the coefficient a_1,

$$a_1 = \frac{3 \cdot 3 \cdot 5 \cdot 5 \cdot 7 \cdot 7 \cdot 9 \cdot 9 \cdot 11 \cdot 11 \cdot 13 \cdot 13 \dots}{2 \cdot 4 \cdot 4 \cdot 6 \cdot 6 \cdot 8 \cdot 8 \cdot 10 \cdot 10 \cdot 12 \cdot 12 \cdot 14 \dots}$$

which, from Wallis, is known to be equal to $4/\pi$.

Substituting the values obtained for a_1 into the system and obtaining the values of a_2, a_3, etc. by a similar procedure, Fourier finds the following expressions for the coefficients:

$$a_1 = 2 \cdot \frac{2}{\pi}; \quad a_2 = -2 \cdot \frac{2/\pi}{3}; \quad a_3 = 2 \cdot \frac{2/\pi}{5};$$

$$a_4 = -2 \cdot \frac{2/\pi}{7}; \text{ etc.}$$

from which, finally, we find,

$$(2.3.6) \quad \frac{\pi}{4} = \cos u - \frac{1}{3} \cos 3u + \frac{1}{5} \cos 5u - \cdots .$$

Fourier continues,

Since these results seem to stray from the ordinary consequences of the calculus, it is necessary to examine them with care and to interpret them in their true sense.
We consider the equation

$$y = \cos u - \frac{1}{3} \cos 3u + \frac{1}{5} \cos 5u - \frac{1}{7} \cos 7u + \cdots$$

as that of a line where u is the abscissa and y the ordinate. We have already seen from the preceding remarks that this line must be composed of the separate parts aa, bb, cc, dd, ... where each one is parallel to the axis [of the abscissa] and equal to half the circumference. These parallels are placed alternately above and below the axis at a distance of 1 and *joined by the perpendiculars ab, cb, cd, ed, ... , etc., which are themselves part of the line* [my emphasis].

In order to form an exact idea of the nature of this line, it is necessary to suppose that the number of terms of the function

$$\cos u - \frac{1}{3}\cos 3u + \frac{1}{5}\cos 5u - \frac{1}{7}\cos 7u + \frac{1}{9}\cos 9u \cdots$$

has at first a determinate value. In this case, the equation

$$(2.3.7) \qquad y = \cos u - \frac{1}{3}\cos 3u + \frac{1}{5}\cos 5u - \frac{1}{7}\cos 7u + \frac{1}{9}\cos 9u \cdots$$

belongs to a curved line that passes alternately above and below the axis, cutting it every time that the abscissa u becomes equal to one of the quantities $\pm \frac{1}{2}\pi$ ... $\pm \frac{3}{2}\pi$ ... $\pm \frac{5}{2}\pi$ ... etc. As the number of terms of the equation increases, the curve which it represents tends more and more to approximate the preceding line composed of parallel straight lines and perpendicular straight lines (*ibid.*, pp. 158-9).

The figure which Fourier has in mind, but which he does not draw, is the following (Figure 2):

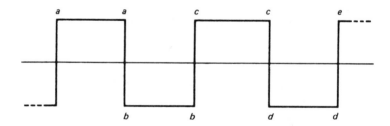

Figure 2

From the modern point of view, this drawing certainly does not represent a function, unless one thinks of it as a *discontinuous* curve, that is, *without* the perpendicular segments *ab, cd, ed, ...* , which Fourier himself expressly considers to be integral parts of the curve.

This leads to the delicate and difficult question of the continuity of a function, a concept about which the ideas of mathematicians at the beginning of the nineteenth century became very confused.

Even Fourier's own approach oscillates between that of the classical, Leibnizian acceptance of continuity and a position closer to the "modern" view. The so-called "genetic" description which he gives of the function (2.3.7) clarifies what his initial point of view was: that y is considered to be a function of u and m (the number of terms taken in the expansion of the series) and therefore that the curve of Figure 2 is thought of as the "limit", as m increases indefinitely, of a succession of continuous curves, each one corresponding to

$$y = \cos x - \frac{1}{3} \cos 3x + \cdots - \frac{1}{2m - 1} \cos(2m - 1)x$$

for every integral value of m.

The same observations, Fourier says, apply, for example, to the graph of the function given by the series

(2.3.8) $\sin x - \frac{1}{2}\sin 2x + \frac{1}{3}\sin 3x - \cdots$

which, from Euler, is known to converge to $x/2$.

In general [Fourier writes], these series present themselves, and it is easy to form them by various means; but *the essential point* [my emphasis] is to identify the limits within which one can take the value of the variable. For example, the equation given by Euler [2.3.8] occurs only so long as the value of x is within 0 and π or within 0 and $-\pi$. For all the other values of x the second term has a determinate value very different from $x/2$.

One must use considerable care in the calculations that produce these series if the limits beyond which the equation ceases to exist are not known.   In fact, since these [limits] are not the same for all equations, one can obtain very erroneous results by combining different series.   It is this observation that explains the contradictory consequences that arise in combining different series of sines and cosines (Grattan-Guinness, 1972, p. 169).

This was one of the points about which Lagrange raised objections. In reply, Fourier wrote him a private communication showing that, by substituting $\pi - x$ for x, the convergence of the series

(2.3.9) $\frac{\pi - x}{2} = \sin x + \frac{1}{2}\sin 2x + \frac{1}{3}\sin 3x + \cdots$

follows from that of (2.3.8). Lagrange must have objected that, by differentiating (2.3.9) you obtain the series

$$-\frac{1}{2} = \cos x + \cos 2x + \cos 3x + \cdots,$$

and then by integrating it

$$C - \frac{1}{2}x = \sin x + \frac{1}{2}\sin 2x + \frac{1}{3}\sin 3x + \cdots.$$

Then setting $x = 0$ (which implies $C = 0$), you have the series

$$-\frac{1}{2}x = \sin x + \frac{1}{2}\sin 2x + \frac{1}{3}\sin 3x + \cdots,$$

which disproves Fourier's result.

Fourier, however, observed that the value $x = 0$ does not render (2.3.9) valid, as is easily verified by a substitution, and this answers Lagrange's observation.

This is the way in which Fourier treats infinite series. He does not have at his disposition any sufficient *criterion* to guarantee the convergence of a series and therefore *calculates* the first m terms directly, using the tricks of elementary trigonometry or the techniques of the infinitesimal calculus, and then takes it to the limit as m grows to infinity.

In its substance, this approach is radically different from that of the eighteenth century. Fourier utilizes the enormous body of results on numerical series obtained by sixteenth- and seventeenth-century mathematicians to find confirmation of general results in particular cases, but he pioneers in the care with which he poses the question of the convergence of infinite series, and consequently also the way he defines its sum.

What distinguishes Fourier's approach from that later used by Cauchy, which is now standard, is the absence of *existence criteria* for the limit that gives the sum of the series. In order to know if a series converges, Fourier has no other way than to *show concretely* what the sum is by using a suitable process of calculation. The physical nature of the problems he treats assures him of "excluding vague questions and calculations without aim," as he writes in the *Discours préliminaire* to the *Théorie* (1822).

Thus equation (2.3.7) presents itself to him in the search for integrating the differential equation that gives the propagation of heat in a lamina. But the study of the nature of this equation and of others like it appears to him to be "the true core of the question."

"It follows from my research on this subject," Fourier writes, "that even discontinuous arbitrary functions can always be represented by expansions into sines or cosines of multiple arcs [i.e. a sine or cosine series] ... A conclusion that the celebrated Euler always rejected" (Grattan- Guinness, 1972, p. 183).

"The expansions in question thus have this in common with partial differential equations," Fourier adds, "that they can express the properties of completely arbitrary and discontinuous functions. It is for this reason that they arise naturally in the integration of these latter equations" (*ibid.*, p. 185).

What does Fourier mean here by the "discontinuity" of a function? Surprisingly, the first example of a "discontinuous line" is that of the line aa, bb, cc, ... of Figure 2 without reference to the perpendicular segments ab, bc, cd, ... This corresponds to the "modern" definition of continuity (and of discontinuity) that Cauchy gave in 1821.

We could certainly understand the contrasting conceptions of continuity that are found within the space of a few pages if, as Fourier states in a note at the end of this section of the manuscript, his work had been "frequently interrupted and the calculations made at various times." But the great ambiguity and theoretical oscillation between the old and new conceptions remain when he immediately afterwards declares that a "discontinuous" line is "composed of arcs of parabolas and segments of a line." This is entirely consistent with Euler's old definition. The approach presented in the definitive version of the *Théorie* is no different.

Thus, for example, in studying the propagation of heat in a bar, when he obtains the expression

$$\frac{2}{\pi} \int_0^\infty \frac{\cos qx}{1 + q^2}\, dq$$

for the initial state $t = 0$, he comments:

It follows that the line whose equation is

$$y = \frac{2}{\pi} \int_0^\infty \frac{\cos qx}{1 + q^2}\, dq$$

is composed of two symmetrical branches which are formed by repeating to the left of the axis of y the part of the logarithmic curve that is on the right of the axis [of y], and whose equation is $y = e^{-x}$. We see here a second example of a discontinuous function expressed by a definite integral. This function, $(2/\pi)\int_0^\infty(\cos qx)/(1 + q^2)dq$, is equivalent to e^{-x} when x is positive, but it is e^x when x is negative (1822, pp. 395-6).

At this point Darboux, the editor of Fourier's *Oeuvres*, justly comments: "This is not a truly discontinuous function, but rather a function expressed by two different laws according to whether the variable is positive or negative" (Fourier, *Oeuvres* 1, p. 396 n. 1).

Within the margins of this ambiguity we can place Fourier's claim to be able to expand "entirely arbitrary and discontinuous functions" into series of sines and cosines of multiple arcs. In order to accomplish this, he first expands a function that contains only the odd powers of the variable and that can consequently be thought of as expandable into a series of odd powers of x. If we indicate this function with $\phi(x)$, the problem consists of finding the coefficients $a, b, c, d, \dots$ of the expansion

$$\phi(x) = a \sin x + b \sin 2x + c \sin 3x + d \sin 4x + \dots$$

After a long and complicated analysis, he finally arrives at the "remarkable result" given by the equation

$$\frac{1}{2} \pi\phi(x) = \sin x \int \sin x\phi(x)dx$$

(2.3.10)
$$+ \sin 2x \int \sin 2x\phi(x)dx + \cdots$$

$$+ \sin ix \int \sin ix\phi(x)dx + \cdots,$$

where the integrals of the second member are taken between $x = 0$

and $x = \pi$. In this case, Fourier concludes, the series always gives the required expansion for the function $\phi(x)$.

Having based his construction of the representation by a sine series on a process of elimination obtained by confronting two series, that of the expansion of $\phi(x)$ into a Taylor series and the series $a \sin x + b \sin 2x + c \sin 3x + \cdots$, it can appear, Fourier says, that the result is limited to odd functions.

To show that this is not the case he immediately shows that (2.3.10) is valid for any function $\phi(x)$.

He therefore sets

$$(2.3.11) \quad \phi(x) = \sum_{i=1}^{\infty} a_i \sin ix$$

and finds the coefficients a_i.

To do this he multiplies both members of (2.3.11) by $\sin nx$ and tacitly assumes, as was then usual, that the series is integrable term by term. Integrating from 0 to π he obtains:

$$\int_0^{\pi} \phi(x)\sin nx \, dx = \sum_{i=1}^{\infty} a_i \int_0^{\pi} \sin ix \sin nx \, dx.$$

It can be easily shown, Fourier says, that[12]

(1) the integral $\int_0^{\pi} \sin ix \sin nx \, dx = 0$ for $i \neq n$,

(2) the value of the integral is $\pi/2$ for $i = n$.

It follows that the coefficient a_i is given by the integral

$$\frac{\pi}{2} a_i = \int_0^{\pi} \phi(x) \sin ix \, dx$$

in accordance with (2.3.10).

By using a similar analysis Fourier proves that one can also expand any function whatever into a series of cosines of multiple arcs. In fact, taking

$$(2.3.12) \quad \phi(x) = \sum_{i=0}^{\infty} a_i \cos ix$$

to determine the coefficients a_i, he proceeds in a similar manner to that just seen, multiplying both members by $\cos nx$ and integrating from 0 to π. Since

$$\int_0^{\pi} \cos ix \cos nx \, dx = \begin{cases} 0, & \text{for } i \neq n \\ \pi/2, & \text{for } i = n \neq 0 \\ \pi, & \text{for } i = n = 0 \end{cases}$$

he obtains for a_i $(i \neq 0)$

$$\frac{\pi}{2} a_i = \int_0^{\pi} \phi(x) \cos ix \, dx,$$

and for a_0 the value

$$\pi a_0 = \int_0^\pi \phi(x)dx,$$

and finally for the function $\phi(x)$ the representation in series,

(2.3.13) $\quad \dfrac{\pi}{2}\,\phi(x) = \dfrac{1}{2}\displaystyle\int_0^\pi \phi(x)dx + \sum_{i=1}^\infty \cos ix \int_0^\pi \phi(x) \cos ix\, dx.$

At this point in the 1807 manuscript Fourier limits himself to considering a number of special cases that illustrate some of the expansions obtained in series. In the *Théorie* of 1822 (Art. 231), however, he moves towards more general considerations by trying to demonstrate that by means of similar expansions in series we can represent "arbitrary" functions of a variable.

"The two general equations that express the expansion of any function whatever, in cosines or sines of multiple arcs, give rise to several remarks that explain the true meaning of these theorems, and the direct application of them," Fourier writes (1822, p. 224). In fact, if in the series

$$\Sigma\, a_i \cos ix,$$

x is included between 0 and $-\pi$, the series remains the same, as it also does if we augment the variable by any integral multiple of 2π. Then in (2.3.13) $\phi(x)$ is periodic and represents a curve composed of equal arcs, each of which corresponds to an interval of 2π on the x axis.

For example, suppose we draw an arbitrary line that corresponds to an interval equal to π. Then, if we substitute any value X for x between 0 and π, the value of $\phi(X)$ is completely determined (Figure 3). The integrals that appear in the coefficients in the expansion (2.3.13) are always determinate, Fourier says, and have a value like "that of the total area of $\int \phi(x)dx$ between the curve $\phi(x)$ and the x axis." The arc $\phi\phi a$ is completely arbitrary, but the same thing cannot be said for the other parts of the line, which, on the contrary, are determinate. "The arc $\phi\alpha$, which corresponds to the interval from 0 to $-\pi$, is the same as the arc $\phi\phi a$, and the whole arc $\alpha\phi\phi a$ is repeated on consecutive parts of the axis at intervals of 2π."[13]

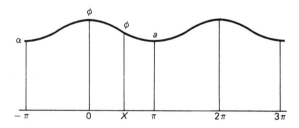

Figure 3

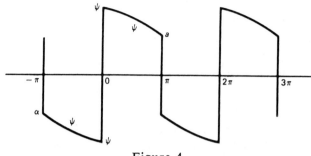

Figure 4

In a similar manner, if we have a function whose graph is given in the above manner (Figure 4), we can utilize the expression (2.3.10) and take as the limits of integration for the integral in the second member $-\pi$, π, in place of 0, π, provided that we write $\pi\psi(x)$ instead of $(\pi/2)\psi(x)$.

Then in the first case the equality of the function entails

$$\phi(x) = \phi(-x),$$

and in the second

$$\psi(x) = -\psi(-x).$$

Now, Fourier says, "Any function whatever $F(x)$, represented by a line traced arbitrarily between $-\pi$ and $+\pi$, may always be divided into two functions such as $\phi(x)$ and $\psi(x)$." The argument on which he bases this assertion is by nature geometrical (Figure 5).

On the basis of evident geometrical symmetry, Fourier concludes that

$$F(x) = \phi(x) + \psi(x) \quad \text{and} \quad f(x) = \phi(x) - \psi(x) = F(-x);$$

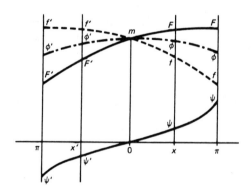

Figure 5

hence

$$\phi(x) = \frac{1}{2}F(x) + \frac{1}{2}F(-x) \quad \text{and} \quad \psi(x) = \frac{1}{2}F(x) - \frac{1}{2}F(-x),$$

from which it follows that

$$\phi(x) = \phi(-x) \quad \text{and} \quad \psi(x) = -\psi(-x).$$

Thus the two functions $\phi(x)$ and $\psi(x)$, whose sum is equal to $F(x)$, can be expanded, the first in cosines of multiple arcs, the second in sines.

By then applying (2.3.10) to the first and (2.3.13) to the second and taking the integral from $-\pi$ to π, we have

$$\pi[\phi(x) + \psi(x)] = \pi F(x) = \frac{1}{2}\int \phi(x)dx$$

$$+ \sum_{i=1} \cos ix \int \phi(x) \cos ix \, dx$$

$$+ \sum_{i=1} \sin ix \int \psi(x) \sin ix \, dx.$$

It is not difficult, given the particular properties of $\phi(x)$ and $\psi(x)$, to see that in the preceding formula we can substitute $\int F(x)\sin ix \, dx$ for $\int \phi(x)\cos ix \, dx$ (and similarly $\int F(x)\sin ix \, dx$ for $\int \psi(x)\sin ix \, dx$) in order to obtain

(2.3.14)
$$\pi F(x) = \frac{1}{2}\int F(x)dx + \sum_{i=1} \cos ix \int F(x) \cos ix \, dx$$

$$+ \sum_{i=1} \sin ix \int F(x)\sin ix \, dx,$$

which is the complete "Fourier series" for the function $F(x)$. Readily assuming an exchange of the Σ and $\int$ symbols, it is only a question of trigonometrical conversions to then show, as Fourier does in Art. 235, that it can be written in the more compact form

(2.3.15) $$F(x) = \frac{1}{\pi}\int F(\alpha)d\alpha \left\{ \frac{1}{2} + \sum_{i=1} \cos i(x - \alpha) \right\}.$$

At this point Fourier interposes the old question of the vibrating string, one that nevertheless had not ceased to interest mathematicians. Secure in the results obtained, Fourier writes:

If we apply the principles that we have just established to the question of the movement of vibrating strings, we can resolve all the difficulties that the analysis employed by Daniel Bernoulli presented. In fact, the solution proposed by this great geometer does not seem at all applicable to the case where the initial figure of the string is that of a triangle or a trapezoid, or is such that only one part of the string is set in motion while the other parts blend with the axis. The inventors of the analysis of

partial differential equations even took this application to be
impossible (Grattan-Guinness, 1972, pp. 250-1).

The objections D'Alembert and Euler raised to Bernoulli's proposed
solution, Fourier adds,

> show how necessary it is to demonstrate that any function can
> always be expanded as a series of sines or cosines of multiple
> arcs. Of all the proofs of this proposition, the most complete is
> that which consists of actually resolving an arbitrary function
> into such a series and assigning the values of the coefficients
> (*ibid*., p. 252).

Bernoulli's mistake, Fourier says, lay in not concretely demonstrating
how one can determine the coefficients, but in simply affirming that it
was possible to do so. "But the geometers," Fourier observes, "only
admit that which they can't dispute."
Finally Fourier concludes this brief note on vibrating strings with
an observation that is extremely significant from a methodological
point of view.

> In order to usefully apply these equations, it is necessary to
> give their integrals a form appropriate to the exact nature of the
> question one is treating, and restrain or extend the generality of
> the integrals so that they correspond perfectly to that of the
> question. In the theory with which we are occupied, the form of
> the integrals is determined by the nature of the physical
> conditions, as we will see in the rest of this paper. All research
> into other integrals would be entirely unfruitful here, but it was
> necessary to make the results coincide with a given initial state.
> The resolution of the following question is based on the
> expansion of an arbitrary function into sines and cosines of
> multiple arcs. We will carry the results to the most general level
> when the nature of the question requires it[14] (*ibid*., p. 253).

2.4. Fourier's 'Program'

In the *Théorie analytique*, Fourier went on to consider the problem of
the propagation of heat in solid homogenous bodies, such as rings,
spheres, cylinders, cubes, and so on. By studying the propagation
along an infinite line, he showed how a function can also be
represented in the following manner:[15]

$$(2.4.1) \quad \pi\phi(x) = \int_{-\infty}^{+\infty} d\alpha\phi(\alpha) \int_{0}^{+\infty} dq\, \cos(qx - q\alpha).$$

This representation of $\phi(x)$ by means of integrals, as Fourier
himself admitted, had been unknown to him in 1807 when he
presented his first paper on the propagation of heat to the Institute.

He first found it two years later, in a paper that Laplace published in the *Journal de l'Ecole polytechnique* (1809). With respect to the memoir of 1807, the element of greatest novelty and interest in the *Théorie* is found in Chapter IX. Here Fourier uses (2.4.1) with considerable lack of restraint, presenting it in different forms and showing how it can be obtained from (2.3.15).

In regards to this Fourier comments, "When, in the convergent series which this analysis furnishes, we give to the quantities which denote the dimensions infinite values; each of the terms becomes infinitely small, and the sum of the series is nothing but an integral. We might pass directly in the same manner and without any physical considerations from the different trigonometrical series which we have employed ... to definite integrals" (1822, p. 345).

Fourier's study of the representation of "arbitrary" functions by means of integrals intertwines with Poisson's simultaneous work on the "Poisson integral" (see §5.1). It inaugurates one of the most interesting chapters in the mathematical analysis of the nineteenth century.[16]

But for Fourier, as he himself repeatedly stressed, this research was not spurred by a drive towards the ever greater generalization of results. Instead, generality was an issue raised by the problems he faced, and he was pushed in this direction only in so far as it was needed to completely resolve these problems. For Fourier, analysis remained a very concrete science.

This approach to mathematics remained more or less dominant in French mathematics during the last century. The impulse that Fourier gave with his *Discours préliminaire* and the whole of the *Théorie analytique* can be seen as the enunciation and first achievement of a "research program" towards whose fulfillment many generations of French mathematicians worked, more or less consciously. The distinctive elements of this program can be seen in the strong physical motivations that drove mathematical research, in the rejection of Lagrangian formalism, and, more generally, in the conception of science as an element of civil progress.

The first argument was certainly not new. The greater part of eighteenth-century mathematics was initially motivated by questions of a physical nature, from the origins of the calculus in Newton to the problem of vibrating strings, the theory of gravitational potential, and the propagation of waves. But what was new and conscious was the role played by physics in the development of mathematics.

The context in which Fourier worked lay outside the framework of Newtonian theory. This explains the initial uncertainty in the statement of the problem both by Biot and by Fourier himself. In addition to stimulating research, physics also acted as a "control" on its results.

Thus the claims for representation in series that seem to unfold on an entirely analytic level, are for Fourier motivated precisely by the *necessity* of resolving physical problems. Without these, he says, we could not take a single step forward in our research. From this it also follows that these techniques are "legitimized" by the physical nature of the problems being resolved.

This final argument also clarifies the character of the mathematics used by Fourier, and of its complete novelty with respect to the "algebraic" form of Lagrange's *Théorie analytique*, for example. It seems to me that there is a fundamental contrast between the two that emerges in a specific problem, that of the representation of functions by trigonometric series.

Even if we do not find in Fourier a *definition* of a function (and to be fair, the *Théorie* was not meant to be a treatise on analysis), it is clear that his conception of a function was not forced into the straightjacket imposed by Lagrange. Nor was his idea of "analysis" that of Lagrange's *algebraic* study of finite quantities. In the place of Lagrange's "rigor," for which Fourier was accused of exercising insufficient care, he substituted a different, *non-formal* conception.

Particularly interesting in this regard is his insistence on the fact that a function can be defined "by parts" or represented in a certain interval by a series that is divergent elsewhere or converges to another function. Let us take for example the case of series, which was a crucial issue in mathematics at the beginning of the nineteenth century. For Fourier, it is not so rigorous to treat functions by means of the Taylor series (even if this is formally unexceptionable), as to make an *actual calculation* of the first n terms, which shows what the limit for infinite n must be, and then to verify the results on the basis of the mathematical or physical data from which it began.

He uses the Taylor series in the first steps of his research on the representation of arbitrary functions by trigonometric series, but he soon frees himself from this and instead is not satisfied until he succeeds in *directly* exhibiting the expansion he seeks. In the process he glimpses, among other things, the property that trigonometric functions are, in modern terminology, an orthonormal base of a suitable Hilbert space of functions.

Mathematics must find its stimulus and verification, and therefore its proper criteria of rigor, in external physical reality. This is the message that clearly emerges from Fourier's work.

Mathematics is an integral part of "natural philosophy" and this in turn is the instrument we have at our disposal to understand nature. This is the "philosophy" that rests at the foundation of Fourier's research and which shapes his concept of science. He does not see it as an abstract and formal theory, but as a branch of knowledge that has strong practical interests. The concreteness of Fourier's conception of science is the element that distinguishes mathematics and analysis in France in the first decades of the nineteenth century from the ideas found elsewhere. In Germany, for example, apart from Gauss we can compare the ideas put forward by men like Jacobi (1804-1851) and F. E. Neumann (1798-1895).

The mathematical faculties in the German universities grew up in the shadow of the faculties of philosophy and philology (*mathematicus non est collega!*). As a result, for a long time they preserved a strongly "abstract" idea of "pure mathematics." Far from the English industrial revolution and the French political revolution, German mathematicians elaborated a conception of mathematics separate from practice, dedicated in the Kantian manner to "the honor of the human spirit."[17]

In the 1820's Fourier was the *secretaire perpetuel* of the *Académie des sciences*, and there formed around him a circle of young intellectuals and mathematicians which included men like Liouville (1809-1882), Sturm (1803-1855), and Dirichlet, as well as Auguste Comte, the philosopher of Positivism.

Indeed it was the latter who, in his inaugural lecture of 1829, dedicated his *Cours de philosophie* to Fourier, the author of the *Théorie*. In Comte's opinion, this work was a paradigm for scientific research.

Notes to Chapter 2

[1]The history of mathematical physics in France at this time is extremely interesting and covers a wide range of topics, many aspects of which have not yet been fully studied. For an introductory survey which gives an initial, schematic picture of the complexity of the subject, accompanied by an ample bibliography, see Grattan-Guinness (1981).

[2]A function $f(x)$ is expandable as a Taylor series (2.2.2) if it is holomorphic in a finite domain that includes x. In particular, the expansion is valid in every circle with center x in which f is holomorphic.

[3]On the other hand, Lagrange himself, in the preface to the second edition of the *Mécanique analytique* (1811) wrote, "When one fully understands the spirit of this system and one is convinced of the exactitude of his result by the geometrical method of first and last reasons, or by the analytical method of derived functions, one can use infinitesimals as a sure and handy instrument to shorten and simplify the demonstrations" (1811, p. xiv). For a stimulating discussion of Euler's and Lagrange's influence on analysis and mechanics in France at the beginning of the nineteenth century, see Grattan-Guinness (1983).

[4]For a history of the principle of virtual velocities, see Lindt (1904).

[5]With all probability, Monge judged Fourier's work favorably. He had been his protector since the time of the *École polytechnique* and the Egyptian expedition. (Fourier owed his part in the expedition to Monge and became one of the most influential persons in the affair.) We can say the same thing for Lacroix and Laplace, but not for Lagrange, whose opinion was markedly different and apparently decisive.

[6]According to Grattan-Guinness (1972, p. 172), the manuscript is in the library of the *Institut de France* in Paris, among Lagrange's papers.

[7]For a complete bibliography of Fourier's works, including his manuscipts and unedited papers, see Grattan-Guinness (1972).

[8]A detailed study of the origins and development of Newton's researches on this subject can be found in Ruffner (1966).

[9]An exhaustive picture of the fortunes of Laplacian physics is given by Fox (1974).

[10]As far as integrating the equation of heat is concerned, Poisson preferred the solutions in terms of integrals in Laplace's manner, criticizing the use of trigonometric series that we will see Fourier propose later (§5.1). For a discussion of the works of Biot, Poisson, and Fourier in the context of the history of thermodynamics, see Truesdell (1980). For the person and works of Poisson, see the volume of contributions edited by Métivier, Costabel, and Dugac (1981). For a biography of Fourier see Herivel (1975).

[11]"This point is by no means evident and will have need of demonstration," Darboux justly adds in a note on Fourier's procedure (Fourier, *Oeuvres* 1, p. 150, n. 1). In fact, it is not said that such a limit exists in general. This method of resolving a linear system of infinite equations in infinite unknowns, forgotten for almost a century, was taken up and used by Riesz (1880-1956) in his book of 1913. For a discussion of the systems of infinite linear equations in the context of functional analysis, see Monna (1973a, pp. 7-21).

[12]The following identities had already been noted by Euler in 1777.

[13]It is in these circumstances that Fourier introduces and explains the meaning of the symbol $\int_a^b$ that has since become standard.

[14]It is perhaps useful to remember that the origins of the research in question lie in the integration of equation (2.3.3) with given initial conditions.

[15]As Darboux observes, this "celebrated equation," which has ever since remained linked with Fourier's name, has a rigorous meaning only under special conditions.
"The double integral

$$J(A,B,h) = \frac{1}{\pi} \int_B^A \phi(\alpha)d\alpha \int_0^h \cos q(x - \alpha)dq$$

$$= \frac{1}{\pi} \int_0^h dq \int_B^A \phi(\alpha) \cos q(x - \alpha)d\alpha,$$

where h is a positive number and $B < A$, tends towards a determinate limit $J(A,B,\infty)$ when, for A and B fixed, h increases without limit" and the function $\phi(x)$ remains limited and has a finite number of discontinuities and of maxima and minima when x varies from $-\infty$ to $+\infty$ along the x axis. "It follows from this," Darboux continues, "that $J(A,B,\infty)$ has a determinate limit $J_0(x)$ when B and A tend respectively towards $-\infty$ and $+\infty$, and that this determinate limit is equal to $\frac{1}{2}[\phi(x + 0) + \phi(x - 0)]$ or to $\phi(x)$ when $\phi(x)$ is continuous for the value of x being considered.
"The preceding propositions stand even when the function $\phi(x)$ becomes infinite for certain values of x, limited in number, provided that the integral $\phi(x)dx$ remains finite for the values of x that render $\phi(x)$ infinite" (In: Fourier, *Oeuvres* 1, p. 409 n. 1).

[16]To obtain an idea of these developments, consult the work of Dini (1880), which offers a rigorous and exhaustive panorama of the results obtained at the time.

[17]In a letter to Legendre, Jacobi wrote: "It is true that M. Fourier was of the opinion that the principal aim of mathematics was its public utility and the explanation of natural phenomena; but a philosopher such as himself should have known that the unique aim of science is the honor of the human spirit, and that in this regard, a question concerning numbers is worth as much as a question about the system of the world" (*Werke* 1, pp. 454-5).
The foundation of the "mathematical and physical seminar" at Königsberg in 1835 by Jacobi and F. E. Neumann, together with the formation of the so-called "school of Königsberg," can be seen as the definitive moment of takeoff for German mathematics to a European level.
The turning point can be seen in the foundation of Crelle's *Journal für die reine und angewandte Mathematik* in 1826.
The state of mathematical research at this time was described by Abel at the beginning of his study trip through Europe thus: "[Crelle] also speaks a great deal of the feeble level of mathematics in Germany, and says that the knowledge of the majority of mathematicians amounts to a little geometry and to something that they call analysis, but which is nothing other than the theory of

combinations.　Nevertheless, it seems to him that a very happy period is now going to begin for mathematics in Germany. ..."

"It is extraordinary in this regard how the young mathematicians here in Berlin and, from what I hear, everywhere in Germany, lift Gauss to the skies, so to speak.　For them he is the substance of all mathematical perfection ..." (Abel, 1902, Correspondence, pp. 10-11).

And yet Gauss' direct influence was not very great, as appears from what Abel writes a little later in the same letter.　Anticipating his trip to Göttingen, where Gauss worked at the observatory, he writes, "Göttingen has a good library, it is true, but that is all; because Gauss is the only one who knows anything, [a somewhat rash and unfair judgement on Abel's part!] and he is absolutely unapproachable" (*ibid.*, p. 12).

Chapter 3
NEW TRENDS IN RIGOR

3.1. Problems with the Foundations of Analysis

In October of 1826 Abel (1802-29) was in Paris, the next stop after Berlin on his study trip through Europe. He wrote to his old teacher Holmboe (1795-1850) in Christiania [Oslo].

Although I am in the noisiest city on the continent, I feel as if I were in a desert.

I know hardly anybody. This is due to the fact that during the summer everybody lives in the country, and is consequently invisible.

Up to now I have only made the acquaintance of Legendre, Cauchy, and Hachette, plus a few secondary but very able mathematicians, M. Saigey, the director of the *Bulletin des sciences [mathématiques]*,[1] and Herr Lejeune Dirichlet, a Prussian, who came to see me the other day taking me for a compatriot. He is a very acute mathematician. He demonstrated at the same time as Legendre the impossibility of resolving in whole numbers the equation $x^5 + y^5 = z^5$ and other nice things.

Legendre is an extremely amiable man, but unfortunately "as old as the stones." Cauchy is crazy and there is nothing to be done with him, even though at the moment he is the mathematician who knows how mathematics must be done. His works are excellent, but he writes in a very confusing way. At first I understood virtually nothing of what he wrote, but now it goes better. He is currently publishing a series of papers with the title, "Exercises des mathématiques." I buy them and read them assiduously. Nine fascicules have appeared since the beginning of the year. Cauchy is intensely Catholic and bigoted. A strange thing for a mathematician. He is moreover the only one today working in pure mathematics. Poisson, Fourier, Ampère, etc. etc. occupy themselves with nothing other than magnetism and other physical matters. Laplace hardly writes any more. The last thing he did was a supplement to the *Théorie des probabilités*. ... Everyone works by himself without interesting himself in others. Everyone wants to teach and no one wants to learn. The most absolute egoism reigns everywhere (Abel, 1902, Correspondence, pp. 45-6).

The picture that Abel described to his friend and the first-hand information he provided him about the Parisian scene offer an excellent insight into the men and ideas that animated the city that was, at the beginning of the last century, the most active center of scientific research in Europe.

Perhaps Abel was a bit cavalier in describing the activities of the Parisian mathematicians, but his approach to mathematics was substantially different from the "polytechnic" spirit of the French scientists and their "physical matters;" his research themes belonged to the domain of pure analysis. He had just finished "a long paper on a certain class of transcendental functions,"[2] and was then working on the theory of equations, his "favorite subject." In addition, he was occupied with "imaginary quantities, for which there is much to do, the integral calculus, and most of all the theory of infinite series whose base is so little established" (*ibid.*, p. 47).

His natural point of reference was therefore the works of Cauchy (1789-1857), the man who, in his opinion, was the only pure mathematician worthy of the name. His judgement may have been a bit severe, but it well expresses the new orientations in the field of mathematics, and in analysis in particular.

Indeed (and this was the thing that most impressed the young Abel), for some time Cauchy had been working on a critical revision of the principles of analysis, whose manifesto was the publication of the *Cours d'analyse* in 1821, the lectures that Cauchy had given at the École Polytechnique. From it Abel had accepted the need for a new rigor in analysis. It was a problem that worried the Norwegian mathematician and appears frequently in his letters. "I want to dedicate all my efforts to bring a little more clarity into the prodigious obscurity that one incontestably finds today in analysis," he had earlier written to his professor Hansteen in Christiania [Oslo] (*ibid.*, p. 23).

It lacks at this point such plan and unity that it is really amazing that it can be studied by so many people. The worst is that it has not at all been treated with rigor. There are only a few propositions in higher analysis that have been demonstrated with complete rigor. Everywhere one finds the unfortunate manner of reasoning from the particular to the general, and it is very unusual that with such a method one finds, in spite of everything, only a few of what may be called paradoxes. It is really very interesting to seek the reason.

In my opinion that arises from the fact that the functions with which analysis has until now been occupied can, for the most part, be expressed by means of powers. As soon as others appear, something that, it is true, does not often happen, this no longer works and from false conclusions there flow a mass of incorrect propositions that link together. ...

Until one uses a general method, this will continue to be true; but I must be extremely careful, because propositions once admitted without rigorous demonstration (which means, without demonstration) become so strongly rooted in me, that I am at every instant exposed to using them without looking more closely (*ibid.*, p. 23).

An example of this approach is provided by the study of infinite series, one of the most crucial questions in analysis at the time. On January 16, 1826, Abel wrote to Holmboe,

Another problem that has greatly concerned me is the summation of the series

$$\cos mx + m \cos(m - 2)x + \frac{m(m - 1)}{2} \cos(m - 4)x + \cdots .$$

If m is a positive integer, the sum of this series, as you know, is $(2 \cos x)^m$, but if m is not an integer, this is no longer the case unless x is smaller than $\pi/2$.

There is no other problem that has occupied mathematicians as much as this in the recent past. Poisson, Poinsot, Plana, Crelle, and many others have sought to resolve it, and Poinsot is the first who has found an exact sum, but his reasoning is completely false, and no one as yet has been able to find out why. I have succeeded with complete rigor. ...

I have found that

$$\cos mx + m \cos(m - 2)x + \cdots = (2 + 2 \cos 2x)^{m/2} \cdot \cos mk\pi$$

$$\sin mx + m \sin(m - 2)x + \cdots = (2 + 2 \sin 2x)^{m/2} \cdot \sin mk\pi,$$

where m is a quantity lying between -1 and $+\infty$, k is an integer, and x is a quantity lying between $(k - 1/2)\pi$ and $(k + 1/2)\pi$. If you let $k = 0$ in the second formula, you have the curious formula:

$$\sin mx + m \sin(m - 2)x + \frac{m(m - 1)}{2} \sin(m - 4)x + \cdots = 0$$

for all the values of x lying between $-\pi/2$ and $+\pi/2$. If m lies between -1 and $-\infty$, the two series are divergent, and consequently have no sum. Divergent series are in their entirety an invention of the devil and it is a disgrace to base the slightest demonstration on them. You can take out whatever you

want when you use them, and they are what has produced so
many failures and paradoxes. Can one think of anything more
hideous than to say that

$$0 = 1 - 2^n + 3^n - 4^n + \text{etc.}$$

where n is a positive integer? *Risum teneatis amici.* [Let us hold
our laughter friends.] I have become extremely sensitive to all
this because, except for cases of the most extreme simplicity, for
example geometric series, there is hardly anywhere in the whole
of mathematics a single infinite series whose sum is determined
in a rigorous manner. In other words, that which is the most
important in mathematics is without foundation. Most things are
exact, this is true; and it is extremely surprising. I strain myself
to find the reason. An exceedingly interesting subject.

I do not think you can show me many propositions where
infinite series appear, where I cannot make fundamental
objections against their demonstration. Do it, I reply. Even the
binomial formula is not yet rigorously demonstrated.

I have found that

$$(1 + x)^m = 1 + mx + \frac{m(m - 1)}{2} x^2 + \cdots$$

for all values of m when x is smaller than 1. If x is equal to
$+1$, one obtains the same formula in the case where m is > -1,
and only then, but if $x = -1$, the formula is invalid, unless m is
positive. For all the other values of x and m the series $1 + mx +$
etc. is divergent.[3] The Taylor theorem, the basis of all of higher
mathematics, is also poorly founded. I have found only one
rigorous demonstration, that of Cauchy in his *Résumé des lecons
sur le calcul infinitesimal.* He there demonstrates that

$$\phi(x + \alpha) = \phi x + \alpha\phi'x + \frac{\alpha^2}{2}\phi''x + \cdots$$

whenever the series is convergent (but one would sooner use it in
all cases). In order to show by a general example (*sit venia
verbo*) how poorly one can reason and how it is necessary to be
prudent, I will choose the following example:

Let

$$a_0 + a_1 + a_2 + \cdots$$

be any infinite series. You know that a very common manner of
finding the sum is to take the sum of

$$a_0 + a_1x + a_2x^2 + \cdots$$

and then to let $x = 1$ in the result. This is OK, but it seems to
me that we cannot accept it without demonstration, because if

we prove that

$$\phi(x) = a_0 + a_1 x + a_2 x^2 + \ldots$$

for all values of x smaller than 1, it does not follow that one can say the same for $x = 1$. It is very possible that the series $a_0 + a_1 x + a_2 x^2 + \cdots$ approaches a completely different value of $a_0 + a_1 + a_2 + \cdots$ as x approaches 1. This is clear in the general case where the series $a_0 + a_1 + a_2 + \cdots$ is divergent, because then it does not have any sum. I have demonstrated that this is exact when the series is convergent. The following example shows how one can err. One can rigorously demonstrate that

$$\frac{x}{2} = \sin x - \frac{1}{2} \sin 2x + \frac{1}{3} \sin 3x - \cdots$$

for all values of x smaller than π. It seems that consequently the same formula must be true for $x = \pi$; but this will give

$$\frac{\pi}{2} = \sin \pi - \frac{1}{2} \sin 2\pi + \frac{1}{3} \sin 3\pi - \text{etc.} = 0.$$

One can find innumerable examples of this kind.

In general the theory of infinite series, up to the present, is very poorly established. One performs every kind of operation on infinite series, as if they were finite, but is it permissible? Never at all. Where has it been demonstrated that one can obtain the derivative of an infinite series by taking the derivative of each term? It is easy to cite examples where this is not right, for example

$$\frac{x}{2} = \sin x - \frac{1}{2} \sin 2x + \frac{1}{3} \sin 3x - \cdots .$$

By taking the derivatives, one has:

$$\frac{1}{2} = \cos x - \cos 2x + \cos 3x - \text{etc.}$$

A completely false result, because this series is divergent.

It is the same for the multiplication, division, etc. of infinite series.

I have begun to review the most important rules that are (today) admitted under this relation, and to show in which cases they are correct and which not (Abel, 1902, Correspondence, pp. 15-19).

Abel's long letter provides vivid evidence of the unease mathematicians were beginning to feel about the foundations of their science in the 1820s; its precariousness was becoming ever more clear. Abel spent a great deal of time with problems involving infinite series because these were the essential instrument of analysis as it had been constructed in the eighteenth century, when Lagrange had made them the foundation of the infinitesimal calculus. In the *Théorie de la chaleur* Fourier had begun to treat the issue of convergence with greater precision, but it was far from being completely clear. Nor were the observations of Legendre in his *Exercices* (1811-17) any more illuminating, or those of Laplace, who added a *Remarque générale sur la convergence des séries* to the end of Book I of his *Théorie des probabilités*.

In opposition to the common opinion, Abel maintained that, in order to differentiate a function given by a series, it is in general not permissible to differentiate the series term by term, thinking that in this way one can obtain the derivative of the function represented by the initial series, something that we have seen Fourier do repeatedly and *without difficulties* when he determined the coefficients of his series. But Abel did not follow out his intuitions; he instead limited himself to showing the need for further research by means of a counter-example. Although penetrating, the assertions of the Norwegian mathematician do not seem to justify the claims of those who see here the idea of the uniform convergence of a series of functions (see §3.5 below).

Abel's arguments concerning the need for rigor in mathematics were becoming ever more widely diffused at the time and were accepted by many mathematicians. But in order to fully understand the motives for and the strength of this new attitude towards the foundations of mathematics, it is necessary to ask ourselves how and why *the methods* of doing mathematics, and consequently the accepted criteria of mathematical rigor, were seen to be inadequate in the eyes of mathematicians at the beginning of the century, an opinion that Abel stated most clearly.

It is difficult and probably impossible to find a single answer. There were many factors both internal and external to mathematics that came together to form the new point of view. One of them was forcefully emphasized by Abel himself -- working mathematicians encountered errors and paradoxes when they held to the point of view that had been accepted until then.

On one side there was the old and unresolved problem of the infinitesimal calculus, which, despite its universal application and the immense quantity of results that it had produced, remained problematic in its principles. On the other hand, there was the fact that new researches, like those of Fourier, had shown clearly that not even the fundamental concepts, such as that of a function, seemed to be adequately defined.

For example, one did not know what happened to a function that was the sum of an infinite series of functions. If these are all continuous at a point x_0, will the sum also be so, and under what conditions? Furthermore, one did not know precisely what was meant, beyond the intuitive visual image, by continuity. Problems like these arose almost immediately when one faced, as Fourier did, concrete phenomena and not simply "abstract" questions of rigor.

Consequently, from practice there arose specific indications that there were open questions whose current formulation was unsatisfactory.

Other factors then entered into play. The first, which might be called philosophical in nature, was made clear by Bolzano (1781-1848), even though it was already implicit in the "metaphysics" of analysis put forth by Lagrange. It is essentially this: analysis must not seek its fundamental principles and its proper criteria of rigor in other sciences or in other branches of mathematics, and specifically not in physics and geometry. This is the road that would lead, after about fifty years, to the so-called "arithmetization of analysis," the formation of analysis on the foundations provided by the arithmetic of the natural numbers. This tendency was accompanied by an increasing specialization and division of labor within mathematics itself. In the place of the "geometers" of the eighteenth century, the nineteenth century substituted the cultivators of pure analysis, geometry, or algebra. Men who were capable of wholly dominating the diverse and varied fields of mathematics became ever more rare.

The second factor, which is more strictly tied to the problem of rigor, has motives more clearly external to mathematics proper. This is the fact that, at the beginning of the century, the great majority of "militant" mathematicians, that is, with the exception of Gauss and a few others, almost the totality of French mathematicians, were engaged in teaching in the *grandes écoles*. This is to say that they were involved in reorganizing mathematical theory *for didactic purposes*.

This meant isolating the fundamental principles of theory (in analysis typically the concepts of function, continuity, derivative, etc.) and from these deriving theorems in a deductive fashion, one that shows clearly how the various propositions are connected with each other. This can be seen in the large number of textbooks written for students at that time.

There is certainly a sense in which it can be said that, from Cauchy on, the decisive step in the conceptualization of rigor and the organization of theory that became dominant in the nineteenth century originated with didactic questions, or at least were present in them.

The case of the real numbers is representative, as Dedekind himself testified when he recalled the unease he had felt as he sought to explain continuity in his course on analysis at the Zurich polytechnic.[4] Weierstrass' lectures were also celebrated in this regard. In them Weierstrass presented the definitions of real number, continuity, limit, derivative, etc., that are standard today (see §7.2 and 3 below).

All of these factors, inextricably connected, came together in various ways at this time. They are still frequently to be found today, although with different "philosophies" and purposes, helping to form a new point of view in analysis.

3.2. Gauss' "Geometrical Rigor"

During his stay in Berlin Abel had learned that Gauss (1777-1855),
who was then the director of the Göttingen observatory, was
working on a large treatise of astronomical physics, a topic that lay
far from the interests of the young Norwegian mathematician. But
Gauss' research actually embraced far more diverse fields of
mathematics, from theoretical astronomy to analysis, higher
arithmetic, and geometry. In fact, for many years Gauss had been
developing a conception of rigor that was not unlike what we have
seen Abel express to Holmboe.

Even before the age of twenty, during the years he was studying
at Brunswick and Göttingen, Gauss had begun to record his
discoveries in his scientific diary. The first entry was that of
March 30, 1796, "The principles on which the division of a circle is
based, and its geometric divisibility into 17 parts, etc." (Gauss, 1981,
p. 41), which anticipated the contents of the last section of his
master work, the *Disquisitiones arithmeticae* (1801).

Mingled with theorems of higher arithmetic, the topic Gauss most
preferred at that time, there are numerous remarks on the theory of
series. For example, in August of 1796 Gauss wrote, "The sum of the
infinite series $1 + (x^n/n!) + (x^{2n}/2n!) + \cdots$," (1981, p. 44) while in
December he noted, "Trigonometric formulae expressed by series,"
(*ibid.*, p. 46) and in April, 1800, "That the series $a \cos A + a'\cos(A + \phi) + a''\cos(A + 2\phi) +$ etc. converges to a limit if a, a', a'', etc. form a
progression converging continuously to 0 without changing sign has
been demonstrated" (Gauss, 1981, p. 54). At the same time he also
discovered the expansion in series of the elliptic integral

$$\frac{1}{\pi} \int_0^\pi \frac{d\phi}{(1 - k^2\cos^2\phi)^{1/2}}$$

and the relations between this and the theory of the arithmetic-
geometric mean, a subject that Gauss claimed he had studied since
the age of 14 (see Cox, 1984).

Gauss did not limit himself to committing his ideas on series to the
pages of his diary; he also gave some glimpses of them in the
dissertation which concluded his university studies (Gauss, 1799) (see
§4.1 below). Criticizing D'Alembert's demonstration of the
fundamental theorem of algebra "with geometrical rigor," Gauss in
fact accused the French mathematician of "not correctly handling
infinite series." In a note he added, "I here note in passing that there
is a large number of these series which at first sight seem especially
convergent" (Gauss, 1799, p. 10). Among these, he observed, were
many of those which Euler had used in his *Institutiones calculi
differentialis* "in order to determine the sum of other series as
closely as possible ... which, as far as I know, nobody has heretofore
seen. Wherefore," Gauss continued, probably with the problem of
asymptotic expansions in mind, "it is highly desirable that it be
clearly and rigorously shown why series of this kind, which at first

converge very rapidly and then ever more slowly, and at length diverge more and more, nevertheless give a sum close to the true one if not too many terms are taken, and to what degree such a sum can safely be considered as exact" (*ibid.*).

In the early years of the century, during the period of his most intense and imaginative research activity, Gauss began to search for the answer to this question by dedicating himself to the study of hypergeometric series. This time too Euler provided his starting point. Euler had in fact considered the linear differential equation of the second order,

$$(3.2.1) \quad x(1 - x)y'' + [\gamma - (\alpha + \beta + 1)]y' + \alpha\beta y = 0,$$

one of whose integrals is the hypergeometric series

$$(3.2.2) \quad F(\alpha,\beta,\gamma,x) = 1 + \frac{\alpha\beta}{1!\gamma}x + \frac{\alpha(\alpha + 1)\beta(\beta + 1)}{2!\gamma(\gamma + 1)}x^2 + \cdots .$$

He had found several transformations for it and had shown how one could express it in the form of an integral.[5]

Gauss began to work on the problem of the convergence of the series $F(\alpha,\beta,\gamma,x)$ about 1805 and in 1812 presented a paper on it to the Society of Göttingen, a work that marks the beginning of the study of series with a modern sense of rigor.

"The transcendental functions always have their true sources in the infinite, openly or hidden," he wrote in the notice of this work. "The operations of integrating, of summing infinite series, of expanding infinite products in infinite continued fractions, or even the operation of approaching to a limit which can be continued according to particular laws without end -- this is the true ground on which the transcendental functions will be raised" (In: *Werke* 3, p. 198).

This had also been Euler's opinion (§1.1), but Gauss worked with these objects with a different sense of rigor and a different concern for the study of operations repeated an infinite number of times.

The series in question was one of the greatest generality, Gauss observed, since by varying the parameters α, β, γ it could represent algebraic functions, like the sum of the binomial series, or transcendental ones, like the logarithmic, trigonometric, and higher transcendental functions. But an indispensable and prerequisite condition, Gauss added, was "the restriction to these cases where the series converges" (*ibid.*).

To this end he considered the ratio of the coefficients of x^m and x^{m+1} (where $x \in \mathbb{C}$) and, from D'Alembert's criterion, concluded that the series was convergent for $|x| < 1$ and divergent for $|x| > 1$, "so that its sum cannot be expressed." From the moment that the function is defined by the series, he added, "the inquiry by its nature is limited to those cases where the series actually converges. Consequently it is foolish to ask which value of the series corresponds to a value of x greater than one" (Gauss, 1813, p. 126).

A few months later he wrote to Bessel to say that, "a series that does not always converge, like mine above, can also be taken as a definition only within the limits where it converges" (In: *Werke* 10(1), p. 363).

Gauss left for a later section the discussion of the case $x = 1$ and began with the observation that deriving (3.2.2) term by term "leads to a similar function, which is clearly

$$\frac{dF}{dx} = \frac{\alpha\beta}{\gamma}F(\alpha + 1, \beta + 1, \gamma + 1, x)" \quad (1813, \text{ p. } 126).$$

In order to find which functions can be represented by a hypergeometric series, Gauss began to study the "related series" that can be obtained by varying each one of the three parameters by one. For every series $F(\alpha,\beta,\gamma,x)$, one then obtains the six series: $F(\alpha + 1, \beta, \gamma, x)$, $F(\alpha - 1, \beta, \gamma, x)$, $F(\alpha, \beta + 1, \gamma, x)$, $F(\alpha, \beta - 1, \gamma, x)$ $F(\alpha, \beta, \gamma + 1, x)$, $F(\alpha, \beta, \gamma - 1, x)$. Together with $F(\alpha,\beta,\gamma,x)$, these satisfy a linear equation two by two, something that Riemann would later make the basis of his study of Gauss' series and of the differential equation that it satisfies (see §6.2c).

Gauss then went on to consider the expansions in continued fractions that allow one to obtain the quotients of hypergeometric series, showing for example that for suitable values of the parameters one can obtain the expansions in continued fractions of the power of a binomial, of $\log(1 + x)$, and of other elementary transcendentals.

For the case of $|x| = 1$, Gauss then demonstrates "with geometrical rigor" that the series will converge if and only if $\gamma - \alpha - \beta > 0$. To do this he introduces the so-called "Gauss criterion" by considering a series Σu_i in which the ratio u_{i+1}/u_i can be given in the form

$$\frac{m^\lambda + am^{\lambda-1} + bm^{\lambda-2} + \cdots + n}{m^\lambda + Am^{\lambda-1} + Bm^{\lambda-2} + \cdots + N}$$

and shows that the series is convergent if and only if $1 + A - a > 0$. He finds the desired condition for the hypergeometric series by noting that this can be obtained by setting $\lambda = 2$, $A = \alpha + \beta$, $B = \alpha\beta$, $a = \gamma + 1$, and $b = \gamma$ (Knopp, 1947, pp. 297-8). Gauss then expresses $F(\alpha,\beta,\gamma,x)$, as the quotient

$$\frac{\Pi(\gamma - 1) \cdot \Pi(\gamma - \alpha - \beta - 1)}{\Pi(\gamma - \alpha - 1) \cdot \Pi(\gamma - \beta - 1)}$$

where the transcendental function Πz (expressed as an infinite product) "is basically nothing other than Euler's inexplicable function" $\Gamma(z + 1) = z!$.

But Euler's function $\Gamma(z)$, in Gauss' view, had "a clear meaning" only for integers, while his own way of introducing the function was "generally practicable" and had a meaning that was as clear for real values of z as for imaginary ones.

The rest of Gauss' paper was devoted to studying the properties of Πz. He showed how one could easily convert the integral $\int_0^1 x^{\lambda-1}(1-x^\mu)^\nu dx$ to such a function and how all the transformations of the integral found with such difficulty by Euler could be easily obtained from the properties of the functions Πz. He finally considered the derivative

$$\Psi z = \frac{d \log \Pi z}{dz} = \frac{d \Pi z}{\Pi z \, dz}$$

and established several "remarkable properties," as, for example, the fact that $\Pi z - \Pi 0$ if $z \in \mathbb{Q}$ can be expressed by means of logarithms and trigonometric functions.

While he did not see any particular problem in the possibility of deriving a series term by term, in regards to the convergence of the series Ψz Gauss noted that, "This example shows the care that is necessary when one deals with infinite quantities. In our view, these can be admitted in mathematical researches only in so far as they can be reduced to the theory of limits" (1813, p. 159).

The paper finally concludes by showing that $-\Psi 0$ is the Euler-Mascheroni constant 0.5772156649 ..., for which Gauss here calculated the first 23 decimals.

The 1813 paper represented only the first part of an extensive research program on this subject. Gauss pursued it with unpublished studies on the determination of hypergeometric series by means of the differential equation of the second order (3.2.1), as Euler had suggested. The second part of his program was to include a general treatment of differential equations with coefficients given by rational functions in x, while the third part was to be devoted to the theory of elliptic and modular functions.

In April of 1816 Gauss told his friend Schumacher that he intended to publish the results of his studies, together with parts of his research on the arithmetic-geometric mean (*Werke* 10(1), p. 248). But, as so often happened, he did not pursue his plans. Thus his ideas on the fundamental concepts of the theory of series remained in manuscript, (*Werke* 10(1), pp. 390-395) together with two papers devoted to determining the convergence conditions of series (*ibid.*, pp. 407-419).

In the first paper, after defining a sequence in a completely modern manner as an application $f: N \to R$, Gauss introduces the concept of the upper and lower bounds of a bounded sequence. If one then considers a number $< \lambda$ and "allows λ to take continuously all intermediate values, then one must necessarily arrive at the least upper bound L'" (*Werke* 10(1), p. 391). From this Gauss identifies the essential property -- in the sequence there are always terms "that are larger than any other quantity, that is smaller than L'." He characterizes the greatest lower bound in a similar manner. Gauss finally gives the definitions of lim sup u_n and lim inf u_n of a sequence and calls the "absolute limit" of a sequence the common limit when lim sup u_n = lim inf u_n.

Equally interesting is his second paper on the series that give the expansion of periodic functions, which contains among other things the definition of the definite integral as a limit of sums[6] that we will see Cauchy set out in his *Résumé* of 1823 (see §4.4 below).

Aware of the importance of his results, Gauss wrote to Laplace in 1812, "I have in my papers many things for which I could perhaps lose the priority of publication, but you know, I prefer to let things ripen" (*Werke* 10(1), p. 374).

This was in fact what happened with the theory of elliptic functions and non-Euclidean geometry. It was this reticence towards publishing his discoveries that eventually led to various priority disputes with Legendre.

On the other hand, Gauss had a conception of the role of the mathematician that was radically different from that of the French "polytechnicians." In the quiet of the Göttingen observatory he could follow the course of his thoughts, occasionally confiding them in letters to his friends but making them public only when they had attained the necessary perfection and systematization. He did not feel the burning necessity for publication and the sense of professional competition that we have seen Abel describe as dominating the Parisian mathematicians.

Thus, in the end, many new and important concepts of analysis that had been first discovered by Gauss were eventually accepted by mathematicians only through the works of Cauchy.

3.3. Bernhard Bolzano

The new conceptions of analytical rigor that we have seen Gauss express are also found in the works of a Bohemian monk, Bernhard Bolzano, a man who lived and worked far from the principal currents of scientific thought at the time.[7] In a forgotten work that appeared in Prague in 1816, Bolzano in fact demonstrated the binomial theorem, "one of the most important theorems of all analysis." Although many mathematicians, including Euler and Lagrange, had given demonstrations of this theorem, from Bolzano's point of view these were all unsatisfactory from the point of view of rigor.[8] In this 1816 paper, which passed almost unnoticed by his contemporaries and was only brought to light a half century later by Hankel (1871), Bolzano presented a detailed study of the behavior of $(1 + x)^n$ where $n \in \mathbb{Z}, \mathbb{Q}, \mathbb{R}$. He left aside the case where x and n are imaginary and where n is an irrational power of a negative number, justifying these omissions by saying that "these are those conditions where we cannot attain our objectives until various concepts have been clarified more precisely than they have been up to now" (1816, p. 144).

Bolzano's observations on the differential calculus were also interesting. This "is based on the weakest arguments," he said, "for example, on the self-contradictory concepts of infinitely small quantities, on the assumption that even zeros could have a ratio to each other," or on the theory of series (1816, pp. x-xi). The latter appeared in the discussion of his theorem when $n \notin \mathbb{N}$. In regard to this he observed that "the difference

$$(1 + x)^n - 1 - nx - \frac{n(n - 1)}{2} x^2 - \cdots - \frac{n(n-1)\cdots(n-r+1)}{1\cdot 2\cdot \cdots \cdot r} x^r$$

can be made smaller than any given quantity, when one takes the number of terms in the series large enough" (1816, p. vi), with the condition, however, that $|x| < 1$.

Bolzano published a more complete analysis of the convergence of series the following year in a work that had as its object the demonstration, *in purely analytic terms*, of a theorem that was already well known to mathematicians. Indeed, its initial formulation dates to the sixteenth century.

If a real function of a real variable, continuous in an interval $[\alpha, \beta]$, *has values of opposite signs at the two points* α *and* β, *then there exists at least one real root of the equation* $f(x) = 0$ *between* α *and* β.

The natural reference for this theorem is clearly geometrical: a continuous curve cannot pass from one side of the x axis to the other without cutting it in at least one point. It is a proposition that had always been held to be true and which Lagrange himself had stated by saying that this curve, "will, little by little, approach the axis before cutting it, and approach it, consequently within a quantity less than any given quantity" (1813, p. 28). Bolzano did not deny this. Nevertheless, he asserted that in order to accept it it is *necessary* to have a rigorous demonstration, by which he meant one expressed in analytic terms, which does not rely on the geometrical evidence of the problem.

The arguments that Bolzano brought to his demonstration and the motives that he brought to his method of reasoning were completely unusual in the context of mathematics at the time. Conscious of the novelty of his point of view, he hoped that it would be adopted by the most influential mathematicians and thence become current in mathematics, an end that was in fact realized, although apparently without any direct influence from Bolzano's work. Abel did know of Bolzano, as we see from his notes,[9] and Lobachevski (1793-1856) as well. Although the latter is known primarily for his work on non-Euclidean geometry, he also worked on various problems of analysis. Unfortunately his papers, many of which were written in Russian, had no influence whatsoever on his contemporary mathematicians.

There is no evidence that Bolzano's paper, which was published in the Acts of the Royal Bohemian Society of Science, was read by Cauchy. Nor do we find any reference to the Prague mathematician in Cauchy's works or surviving papers. The remarkable fact that similar ideas on continuity and the convergence of series are found almost simultaneously in Bolzano and Cauchy, as we will see, does not provide sufficient grounds for speaking of a direct influence of the first on the second and even less for a charge of plagiarism on Cauchy's part. Continuity and the convergence of series were problems of great interest at the time, and this would not be the first or the only time in which two mathematicians who were working on the same problem arrived at similar conclusions, unknown to each other.[10]

Bolzano opens his paper with the words, "There are two theorems in the theory of equations for which one could until recently say that a completely correct proof was unknown" (Bolzano, 1817, p. 3). The first is the intermediate value theorem, which is the subject of this work. The second is the fundamental theorem of algebra, for which Gauss had first given a proof in 1799 (see §4.1). According to Bolzano, however, this proof contained a fault, namely, "that this purely analytic truth is founded on a geometric consideration" (*ibid.*, pp. 3-4), a defect which he says is absent from Gauss' two more recent demonstrations of this theorem (Gauss, 1816a and b).

Bolzano's aim is thus to prove the first of these theorems. The demonstration is preceded by a critical analysis of the earlier attempts, which had been based primarily on the geometric evidence of the proposition.

Now, Bolzano says, "We cannot raise any objections at all against either the correctness or the evidence of this geometrical proposition." His point is otherwise.

> But it is just as clear that it is an insufferable offense against *right method* to want to derive the truths of *pure* (or general) mathematics (that is, arithmetic, algebra, or analysis) from considerations that belong to a purely *applied* (or special) part of it, namely to geometry (1817, pp. 4-5) [my emphasis].

To say this much means that he must therefore undertake to clarify what we mean by a mathematical *demonstration*.

Bolzano maintains that in science demonstrations cannot be simple procedures for fabricating evidence, but rather *foundations*, in other words, "presentations of every objective reason which the truth to be proved has" (*ibid.*, p. 5).

In this view a geometrical demonstration of the proposition in question creates a vicious circle. This proposition therefore does not require a demonstration that fabricates evidence but instead has need of a *foundation*.

Bolzano's second argument is against demonstrations based on the concept of the continuity of a function in which the concepts of time and motion are allowed to enter. If there are two functions such that $f(\alpha) < \phi(\alpha)$ for $x = \alpha$ and $f(\beta) > \phi(\beta)$ for $x = \beta$, and the functions obey the law of continuity, then the first inequality is valid at the beginning of x's variation and the second at the end. Consequently there is an intermediary instant, etc.

"No one will deny that the concept of time, and even more that of motion, is just as foreign to general mathematics as that of space,"

Bolzano asserts, (1817, p. 6) quoting Lagrange's words almost literally (see §2.2). Their use is at most legitimate as a means of providing examples. But it must be completely clear that we cannot accept examples in place of *demonstrations*, and that the substance of a demonstration can never be based on improperly used linguistic expressions and on the secondary images that they evoke.

Consistent with these premises, Bolzano gave the following definition of continuity: "A function $f(x)$ changes according to the law of continuity for all values of x which lie within or outside certain limits[11] if, when x is any such value, the difference $f(x + \omega) - f(x)$ can be made smaller than any given value when ω can be taken as small as one wishes"[12] (1817, pp. 7-8). Now, it is certainly clear that a continuous function cannot assume a certain value without having first taken all the values that "precede" it, Bolzano says. For example, $f(x + n\Delta x)$ can assume all values between $f(x)$ and $f(x + \Delta x)$ when n is taken arbitrarily between 0 and 1, inclusive, but this "cannot be seen as an explanation of the concept of continuity, rather, it is much more a theorem about it" (1817, p. 8).

In order to arrive at a demonstration of the proposed proposition, Bolzano first makes some considerations on series, introducing a criterion of convergence.

After having set

$$A + Bx + Cx^2 + \cdots + Rx^r = F_r(x)$$

and

$$A + Bx + Cx^2 + \cdots + Rx^r + \cdots + Sx^{r+s} = F_{r+s}(x),$$

he states the following:

Theorem. *When a sequence of quantities*

$$F_1(x), F_2(x), ..., F_n(x), ..., F_{n+r}(x)$$

[i.e. a sequence of partial sums] *has the character that the difference between its nth member $F_n(x)$ and every later $F_{n+r}(x)$, no matter how far distant it may be, remains smaller than every given quantity when n is taken large enough, then there is always a certain constant quantity, and only one, which the terms of this sequence always approach and can come as near to it as one wishes when the sequence is extended far enough* (1817, p. 21).

This is the famous "Cauchy criterion" (see §3.4b below) which, however, Bolzano formulates in terms of a sequence of functions instead of numbers, which he will nevertheless use in his demonstration of the intermediate value theorem. He in fact states what is today known as the property of uniform convergence (see §5.4).

As will also be the case for Cauchy, the absence of a rigorous theory of the real numbers traps Bolzano's demonstration in a vicious circle, something that he himself seemed to recognize afterwards (Rychlik, 1961). The basis of Bolzano's argument was in fact that "there is certainly nothing impossible in assuming the existence of the quantity X which the terms of the series approach arbitrarily closely" and then to demonstrate the unicity of X.

The next step is to introduce the idea of a *least upper bound* of a set of real numbers. Bolzano writes,

> In researches in applied mathematics it often happens that one finds that there is a particular property M which belongs to all the values of a variable quantity x that are smaller than a certain u; without at the same time discovering that this property no longer belongs to values that are larger than u. In such cases there can perhaps still be some u_1 that $> u$ for which it is true, in the same way as for u, that all the values of x standing below it possess the property M; indeed this property can perhaps belong to all x without exception. However, when one discovers only this one, that M is definitely not a property of all x, then from the combination of these two results one can correctly conclude that there is a certain quantity U which is the larger of those for which it can be true that all smaller x possess the property M (1817, pp. 24-5).

Bolzano proves this conclusion with the following theorem.

Theorem. *If a property M does not belong to all values of a variable quantity x, but to all that are smaller than a certain u, then there always is a quantity U which is the largest of those for which it can be said that all smaller x have the property M* (1817, p. 25).

This is the way in which the theorem on the existence of the least upper bound of an infinite upperly bounded set of real numbers is stated.[13]

In his demonstration, Bolzano uses the property of series he had demonstated earlier. Since M is valid for all x smaller than u, but not for all x, then, Bolzano says, there exists a quantity $V = u + D$ ($D > 0$) for which it can be said that M does not belong to all $x < V$. He now considers the quantity $u + D/2^m$ with $m \in \mathbb{N}$.

If M belongs to all x smaller than $u + D/2^m$ for all m, then this u is the largest value for which it is true that all $x < u$ possess the property M.

If, on the contrary, this is not the case, Bolzano, by means of reasoning based on reiterations of the argument, constructs the convergent series

$$u + \frac{D}{2^m} + \frac{D}{2^{m+n}} + \cdots .$$

If U is its sum, then M is true for all $x < U$. He shows without difficulty that M is not true for $x < U + \varepsilon$.

Both of these conclusions are found with purely analytic reasoning and the theorem is thus fully demonstrated.

"The preceding theorem is of the greatest importance and will be used in all branches of mathematics," both pure and applied, Bolzano comments (1817, p. 29). He further shows the difference between the least upper bound and the maximum of a set of variables by asserting that the existence of U does not in fact imply the existence of any largest x for which the property M is true.

At this point he can demonstrate the proposition given in the title of his paper. He states it in this way:

Theorem. *If two functions of x, $f(x)$ and $\phi(x)$ vary according to the law of continuity for all values of x or at least for all that lie between α and β, and further if $f(\alpha) < \phi(\alpha)$ and $f(\beta) > \phi(\beta)$, then there is always a certain value of x lying between α and β for which $f(x) = \phi(x)$* (1817, p. 31).

Bolzano initially supposes that α and β are both positive (although the demonstration can be extended to other possible cases, as Bolzano indeed does without difficulty) and that $\alpha < \beta$, or $\beta = \alpha + i$, for positive i. Since $f(\alpha) < \phi(\alpha)$, then also $f(\alpha + \omega) < \phi(\alpha + \omega)$, with ω arbitrarily small (for the continuity of the function).

"We can then assert for all ω that are smaller than a certain one, that the two functions $f(\alpha + \omega)$ and $\phi(\alpha + \omega)$ stand in the ratio of a smaller quantity to a larger" (i.e. one is smaller than the other) (1817, p. 31).

Bolzano indicates this property with M, a property that obviously does not belong to all ω (for example it is not true for $\omega = i$). Then, on the basis of the theorem of the least upper bound which he has just demonstrated, there is "a certain quantity U which is the largest of those for which it is possible to assert, that all ω which are $< U$ have the property M" (1817, p. 32). U is included between 0 and i, and is such that $f(\alpha + U) = \phi(\alpha + U)$, it not being possible, from the property of U, that either $f(\alpha + U) < \phi(\alpha + U)$ or $f(\alpha + U) > \phi(\alpha + U)$.

The theorem is completely demonstrated, and to have it in the usual form it is enough to set $\phi(x) \equiv 0$.

3.4. Cauchy's *Cours d'analyse*

In response to requests from men like Laplace and Poisson and "for the greater use of the students," in 1821 Cauchy decided to write down and publish the series of lectures on mathematical analysis that he had given at the École Polytechnique. Cauchy had himself formerly been a student at the École, where he had attended the lectures of Poisson, Lacroix, and Ampère. Afterwards he had worked for a time as an engineer at Cherbourg before returning to Paris to devote himself to a scientific career. In the Parisian mathematical environment Cauchy gained early recognition. He was first called to substitute for Biot, and subsequently asked to teach at the École Polytechnique, the Sorbonne, and the Collège de France.

In 1816, following the restoration of the Bourbon kings, the Academy was reestablished. However, many members of the Institute were excluded because of their Bonapartist leanings, including Carnot and Monge. Cauchy was nominated to take the place of the latter, but this was accompanied by violent arguments and protests that did not question Cauchy's scientific qualifications but rather his good sense in occupying in this way the post of a man whose exclusion, in Biot's view, constituted an act of "political inhumanity and a deathtoll for the Academy" (Valson, 1868, I, p. 58).[14]

Although Cauchy's volume was animated by a conception of analysis and rigor not unlike that found in Bolzano, it found a much different reception in the mathematical world than Bolzano's pamphlet. The *Cours d'analyse* became the manifesto of the "new" analysis, a book that, as Abel wrote, "must be read by every analyst who likes rigor in mathematical researches" (1826, p. 221).

The introduction to the *Cours d'analyse* expresses Cauchy's conception of analytical rigor with considerable vigor.

> As for methods, I have sought to give them all the rigor that one demands in geometry, in such a way as never to revert to reasoning drawn from the generality of algebra. Reasoning of this kind, although commonly admitted, particularly in the passage from convergent to divergent series and from real quantities to imaginary expressions, can, it seems to me, only occasionally be considered as inductions suitable for presenting the truth, since they accord so little with the precision so esteemed in the mathematical sciences. We must at the same time observe that they tend to attribute an indefinite extension to algebraic formulas, whereas in reality the larger part of these formulas exist only under certain conditions and for certain values of the quantities that they contain. In determining these conditions and these values, I have abolished all uncertainty. ...
> It is true that, in order to remain continually faithful to these principles, I was forced to admit many propositions that perhaps seem a bit severe at first sight" (1821, p. ii-iv).

The first of these propositions that were "a bit severe" to admit is the fact that "a divergent series does not have a sum" (1821, p. iv). This must have been a very troublesome result in the eyes of his contemporaries if Cauchy had to stress it so firmly. It was in fact clearly opposed to a tradition that had always been dominant in analysis and which had been reasserted in Lagrange's *Théorie des fonctions analytiques*, one of the books that Cauchy had studied the most as a student. But if Cauchy agreed with Lagrange on the necessity of founding analysis in a rigorous manner, without limiting oneself to justifying the methods by their successful application to geometry, physics, etc., he nevertheless clearly distanced himself from Lagrange when it came to setting out the foundations of analysis. Arguments drawn from algebra, Cauchy asserted in opposition to Lagrange, cannot serve as the basis of the "greatly esteemed precision" of analysis.

Infinite series played a decisive role in these questions. Such series must be treated with extreme rigor, Cauchy said, even at the cost of a drastic reduction in the applicability of the formulas used. "Thus, before taking the sum of any series," he wrote, "I must examine in which cases the series can be summed or, in other words, what the conditions of their convergence are. In this regard, I have established the general rules that seem to me to merit some attention" (1821, p. v).

As we will see, the instrument that Cauchy developed in order to achieve his objective of a critical revision of analysis is the theory of limits. It is in fact the limit concept that allows him to define the continuity of a function, the derivative and the integral, the convergence of a series and its sum. We know how important it was for mathematics to have isolated this concept, which had been more or less clearly present in the spirit of every mathematician since the origins of the infinitesimal calculus.

a) Limits and Continuity

The *Cours d'analyse*, as is natural for a general, didactically effective treatise, opens with a series of preliminaries in which Cauchy reviews the various kinds of numbers (natural, rational, etc.), introduces the concept of absolute value (which he calls "numerical value"), calculations with literal quantities, and finally the concept of limit. He defines it thus:

"When the values successively attributed to the same variable indefinitely approach a fixed value in such a way as to end by differing from it as little as one wishes, this latter is called the *limit* of all the others" (1821, p. 19).

It is interesting to note the example that Cauchy gives in order to illustrate this concept. "Thus, for example, an irrational number is the limit of the various fractions that furnish ever closer values of it" (*ibid.*).

The introduction of the limit allows Cauchy to specify in an unequivocal manner the meaning of *infinitesimal* and of positive and negative *infinity*. These definitions have become classic. Thus,

When the successive numerical values of the same variable decrease indefinitely in such a way as to fall below any given number, this variable becomes what one calls an *infinitesimal* or an *infinitely small* quantity. A variable of this kind has zero as a limit.

When the successive numerical values of the same variable
increase more and more in such a way as to rise above every
given number, we say that this variable has *positive infinity* for a
limit, indicated by the sign ∞ if it is a positive variable, and
negative infinity, indicated by the symbol —∞, if it is a negative
variable (1821, p. 19).

Finally Cauchy presents the usual operations of the calculus, sum,
product, etc., and the exponential, logarithmic, and trigonometric
functions.

In the first chapter Cauchy immediately gives the definition of a
function of one or more real variables. "When the variable
quantities are linked together in such a way that, when the value of
one of them is given, we can infer the values of all the others, we
ordinarily conceive that these various quantities are expressed by
means of one of them which then takes the name of *independent
variable*; and the remaining quantities, expressed by means of the
independent variable, are those which one calls the *functions* of this
variable" (1821, p. 31).

In a similar manner he defines functions of many independent
variables and distinguishes between explicit and implicit functions.
These are present "when one only gives the relations between the
functions and the variables, that is to say, the equations which these
quantities must satisfy, as long as these equations are not resolved
algebraically" (1821, p. 32).

After having defined infinitesimals of the first and following
orders by means of limits, Cauchy gives the following definition of
the continuity of a function:

Let $f(x)$ be a function of the variable x, and let us suppose
that, for every value of x between two given limits, this function
always has a unique and finite value. If, beginning from one
value of x lying between these limits, we assign to the variable x
an infinitely small increment α, the function itself increases by
the difference

$$f(x + \alpha) - f(x),$$

which depends simultaneously on the new variable α and on the
value of x. Given this, the function $f(x)$ will be a *continuous*
function of this variable within the two limits assigned to the
variable x if, for every value of x between these limits, the
numerical value of the difference

$$f(x + \alpha) - f(x)$$

decreases indefinitely with that of α (1821, p. 43).

At this point he reformulates the same concept in terms of

infinitesimals.

> In other words, *the function* $f(x)$ *will remain continuous with respect to* x *within the given limits if, within these limits, an infinitely small increase of the variable always produces an infinitely small increase of the function itself* (*ibid.*).

Cauchy had had the opportunity to publicly clarify his ideas on continuity at the end of the preceding year, when he had reported to the Academy on a long paper on projective geometry that had been presented by Poncelet (1788-1867). Cauchy's report, which was published in Gergonne's *Annales*, was reprinted at the beginning of Poncelet's *Traité des propriétés projectives des figures* (1822).

In the *Traité* Poncelet adopted a principle similar to the one that had been used by Carnot in his *Géométrie de position* (1803). This was the so-called "principle of continuity" according to which the "descriptive" properties of "any figure in a general position and indeterminate in some way" are conserved by "insensible" and continuous transformations of the figure, unless a determinate point is attained in which they clearly fail to be true.

There was an apparent ambiguity and uncertainty in Poncelet's formulation of his principle. On the one hand, he asked, "Is it not evident that the properties and the relations found for the first system remain applicable to the successive states of this system, provided that one always considers the particular modifications that can follow from it, as when certain magnitudes vanish, change their meaning or sign, etc., modifications that it will always be easy to recognize at first, and by certain rules?" (Poncelet, 1822, p. xxii).

This, Poncelet adds, is precisely what "our greatest geometers" did in laying the foundations of the infinitesimal calculus and mechanics. One can find a similar law in all those writings where "one seeks a certain generality in the conceptions."

This principle provided Poncelet with a useful tool for his synthetic approach to the study of the projective properties of conics. For example, he demonstrated that two conics which are tangent to each other in two real or ideal points are generated from the projection of two concentric circles, since the first figure can be obtained from the second "by a progressive and continuous movement." This theorem is true in the complex projective plane $P^2(\mathbb{C})$ and shows how Poncelet utilized the principle of continuity and "ideal" entities in order to avoid introducing imaginary elements into geometry, such as imaginary centers of projections or imaginary planes.

Moreover, Poncelet observes, the "principle of continuity is generally admitted in all those studies that are based on algebraic analysis" (1822, p. 67).

Cauchy was clearly of a different opinion. He objected that "This principle is, to tell the truth, a strong induction, with whose aid one extends the theorems that had first been established with certain restrictions to cases where these no longer exist" (In: Poncelet, 1822, p. ix). It is true that the applications given by Poncelet lead to exact results, Cauchy admits, but "nevertheless we think that it should not be generally admitted and indiscriminately applied to all kinds of questions in geometry, nor in analysis. In according too much confidence to it, one could occasionally fall into obvious errors" (*ibid.*). This happens in the case of integration, Cauchy adds, clearly having in mind the results of his paper on definite integrals which at the time had not yet been published (see §4.2 below).

The definition of continuity given in the *Cours d'analyse* must now, in Cauchy's view, eliminate all ambiguity and recourse to geometric intuition.

Cauchy's definition of a function appears to be completely free of the requirement that the dependent variable be expressible by means of "an analytic expression," as it had been for Lagrange. Moreover, Cauchy defines the continuity of a function in "local" terms, as Bolzano had done. Like the Bohemian mathematician, he clearly expresses the idea that a function is considered "within two given limits" when, for example, we want to affirm something about its continuity.[15]

The definitions of a continuous function given by Bolzano and Cauchy appear strikingly similar, something that seems even more remarkable when we remember that at the time it was a completely new way of studying continuity. But while Bolzano seemed to move knowingly towards the distinction between continuity and derivability,[16] Cauchy still seemed to be tied to classical analysis and produced standard examples of continuous functions that were everywhere derivable, such as $a + x$, $a - x$, ax, $\sin x$, $\cos x$, $\log x$, A^x, etc. As a discontinuous function he gave the example of a/x for $x = 0$.

Two years after the *Cours d'analyse*, Cauchy wrote in the *Résumé* of his lectures given at the École Polytechnique that, "The two functions $x^{1/2}$, $1/\log x$, ... become discontinuous in passing from the real to the imaginary as the variable x decreases and passes through zero" (1823a, p. 39).

The sense of this statement seems to be that Cauchy thought that continuous functions were always derivable, and ceased to be so only in points of discontinuity.

In fact, if we take for example the first of the functions proposed by Cauchy, $y = x^{1/2}$, its derivative is $y = 1/2\sqrt{x}$, which is discontinuous at the origin.

This becomes immediately clear when we pass from real to complex values, Cauchy says. In this case, in fact, the point $x = 0$ proves to be a "multiple point" or a "branch point" in modern terminology, and the function changes its value when the variable completes a full turn around the point.[17]

Many years later Cauchy had the opportunity to clarify this idea. In a letter to Coriolis, which was published in the *Comptes rendus de l'Académie des sciences* in 1837, he set out his own method for representing the roots of algebraic equations or the integrals of differential equations which was based on an expansion into convergent series (see §4.5). In regards to continuity, he wrote,

> According to the definition given in my *Cours d'analyse*, a function of a variable is continuous between the given limits when, between these limits, every value of the variable produces a unique and finite value of the function, and this varies by insensible degrees with the variable itself. Having said this, a function that does not become infinite in general only ceases to be continuous by becoming multiple (1837, p. 39).

A full awareness of the novelty inherent in this definition of continuity, with respect to the tradition of Euler and Lagrange, does not appear until much later in Cauchy's writings. In his *Mémoire sur les fonctions continues* of 1844 we read,

> In the works of Euler and Lagrange, a function is called *continuous* or *discontinuous* according to whether the different values of this function, corresponding to the different values of the variable, are or are not subject to the same law, are or are not furnished by one and the same equation. It is in these terms that the continuity of functions is defined by these illustrious geometers when they say that "arbitrary functions, introduced by the integration of partial differential equations, can be continuous or discontinuous functions." Nevertheless, the definition that we have just recalled is far from offering mathematical precision; for, if the different values of a function, corresponding to different values of a variable, depend on two or more distinct equations, nothing will hinder us from diminishing the number of these equations or even from replacing them by one single equation, whose decomposition would furnish all the rest. There is more: the analytical laws to which the functions can be subjected are generally expressed by algebraic or transcendental formulas, and it can happen that different formulas represent, for certain values of a variable x, the same function, or, for other values of x, different functions. Consequently, if one considers the definition of Euler and Lagrange to be applicable to all kinds of functions, whether algebraic or transcendental, a simple change of notation is often enough to transform a continuous function into a discontinuous one, and vice versa. Thus, for example, if x designates a real variable, a function that reduces to $+x$ or to $-x$ according to whether the variable x is positive or negative will for this reason be placed in the class of discontinuous functions, and yet the same function could be regarded as continuous when one represents by the definite integral

$$\frac{2}{\pi} \int_0^\infty \frac{x^2 dt}{t^2 + x^2} \quad \cdots .$$

Thus, the character of the continuity of functions, seen from the point of view where the geometers first stopped, is a vague and uncertain character. But the uncertainty will vanish if, in place of Euler's definition, we substitute that which I have given in Chapter II of the *Analyse algébrique* (Cauchy, 1844a, pp. 145-6).

In Chapter 2 of the *Cours d'analyse*, after having defined the continuity of a composite function, among the "remarkable properties" of continuous functions Cauchy mentioned the natural property of serving to represent the ordinate of a curve. In this connection Cauchy stated the intermediate value theorem: if $f(x)$ is continuous between x_0 and X and if $f(x_0) < b < f(X)$, there exist one or more values of $x \in [x_0,X]$ for which $f(x) = b$. Cauchy's demonstration was based on geometric intuition: given the continuity of the function, the line $y = b$ "could not but meet" the curve between $f(x_0)$ and $f(X)$.

Cauchy himself nevertheless recognized that an argument of this kind did not well suit the rigor "that I had made a law unto myself," since he immediately afterwards refers to a demonstration "by a direct and purely analytic method" that was placed in the appendix to the *Cours* dealing with the numerical resolution of equations.

Here, in order to demonstrate the existence of a zero $a \in [x_0,X]$ of a continuous function in $[x_0,X]$ such that $f(x_0)$ and $f(X)$ are of opposite sign, he makes successive subdivisions of the interval $[x_0,X]$ in order to construct a pair of sequences $\{x_i\}$ and $\{X_i\}$ which satisfy 'Cauchy's criterion,' (see below) such that $x_0 < x_1 < ...$, and $X > X_1 > X_2 > ...$, and furthermore that $f(x_i)$ and $f(X_i)$ are of opposite sign for every $i \in \mathbb{N}$. "One can conclude that the general terms $\{x_i\}$ and $\{X_i\}$ will converge to a common limit" a and, by the continuity of the function, $f(x_i)$ and $f(X_i)$ also will converge to $f(a)$. "It is clear, Cauchy concludes, "that the quantity $f(a)$ is necessarily finite, and cannot differ from zero" (1821, p. 462). This is a demonstration which, like that of Bolzano, becomes completely rigorous once the existence of $\mathbb{R}$ has been assumed.

b) The Convergence of Series

In the *Cours d'analyse*, Cauchy devoted many pages to the study of "singular values of functions in some particular cases" (1821, p. 51).

This involves "one of the most important and most delicate questions of analysis," Cauchy says (*ibid.*), which is to study the limits of functions for $x = \pm\infty$ and $x = 0$. This leads him to single out the so-called "indeterminate forms" of the type $0/0$; ∞/∞, $\infty - \infty$, $0 \cdot \infty$, 0^0, ∞^0, 1^∞.

In the course of this study, Cauchy states and demonstrates theorems that have since become classic.

Theorem I. *If, for increasing values of x, the difference f(x + 1) − f(x) converges towards a certain limit k, the fraction f(x)/x will converge towards the same limit at the same time* (1821, p. 54).

Theorem II. *If, the function f(x) being positive for very large values of x, the ratio f(x + 1)/f(x) converges towards the limit k as x increases indefinitely, the expression $[f(x)]^{1/x}$ will converge towards the same limit at the same time* (1821, p. 58).

It is clear that these two theorems are also valid when $f(x)$ is only defined for integral values of x, and therefore both can be reformulated in terms of sequences. Cauchy later used this to state convergence criteria for series.

These occupy the whole of Chapter 6 of the *Cours d'analyse*. Here Cauchy defines the convergence of a series as, "If, for ever increasing values of n, the sum s_n indefinitely approaches a certain limit s, the series will be called *convergent* and the limit in question will be called the *sum* of the series. If, on the contrary, while n increases indefinitely the sum s_n does not approach any fixed limit, the series will be *divergent* and will no longer have a sum" (1821, p. 114).

After giving the example of the geometric series $1 + x + x^2 + \cdots$, which converges to $1/(1 − x)$ if $|x| < 1$, Cauchy states the famous necessary and sufficient condition of convergence, the so-called 'Cauchy criterion':

In order for the series

$$u_0 + u_1 + u_2 + \cdots + u_n + \cdots$$

to be convergent, it is necessary and sufficient that increasing values of n make the sum

$$s_n = u_0 + u_1 + u_2 + \cdots + u_{n-1}$$

converge indefinitely towards a fixed limit s; in other words, it is necessary and sufficient that, for infinitely large values of the number n, the sums

$$s_n, s_{n+1}, s_{n+2}, \ldots$$

differ from the limit s and consequently from each other, by infinitely small quantities (1821, p. 115).

Cauchy has no difficulty proving that the condition is necessary, but as for its sufficiency, like Bolzano, he limits himself to asserting that, "Reciprocally, when these various conditions are fulfilled, the convergence of the series is assured" (1821, p. 116). (The conditions he refers to are that the general term u_n tends to zero and that the sum of the quantities $u_n + u_{n+1} + \cdots$, taken from the first for an arbitrary number, "end by constantly assuming numerical values inferior to every assignable limit.")

Cauchy holds his assertion to be evident, based on the geometrical intuition of the continuity of the number line, but, as would become clear about fifty years later, the rigorous demonstration of the sufficiency of 'Cauchy's criterion' first requires the construction of the field of real numbers.

On the other hand, thinking that the irrational numbers are defined as "the limit of the various fractions that furnish values ever closer to it," as Cauchy does in the *Cours d'analyse*, (1821, p. 19) leads inevitably to a vicious circle.

But while Bolzano confronted the study of real numbers in an attempt to "arithmetize" analysis,[18] Cauchy never seemed to show any interest in the issue, neither before nor after the composition of the *Cours d'analyse*.

Here, after having applied his condition to show that the harmonic series is divergent and that the exponential series $\Sigma(1/n!)$ converges, Cauchy states the following:

Theorem. *When the different terms of the series [$\Sigma_{n=1}^{\infty} u_n$] are functions of the same variable x, continuous with respect to this variable in the neighborhood of a particular value for which the series is convergent, the sum s of the series is also a continuous function of x in the neighborhood of this particular value* (1821, p. 120).

Cauchy does not mention Fourier in this context, but it is clear that he questions Fourier's entire approach. Since in fact the sine and cosine functions that appear in Fourier series are continuous functions, it follows that, according to this theorem, the Fourier series of discontinuous functions do not converge to the given functions.

Cauchy's "demonstration" is the following: if we use s to indicate the sum of a series like $s = s_n + r_n$ (where r_n is the remainder of the series taken from the nth term), then the three quantities s_n, r_n, and s are functions of x. The first of these is clearly continuous in the neighborhood of the particular value of x at issue, and the given series is convergent at x by hypothesis.

We now consider the increase of the three functions s, s_n, and r_n when we increase x by an infinitesimal quantity α. The increase in s_n will be infinitesimal for every finite n and "that of r_n will become insensible at the same time as r_n if we assign a very large value to n" (1821, p. 120). Consequently, the increase in s will also be infinitesimal, from which Cauchy deduces the continuity of s in the neighborhood of x.

But a more refined analysis of the demonstration shows that Cauchy's conclusion is incorrect and based on the hypothesis that

$r_n(x + \alpha)$ becomes infinitely small for $n > N$ very large and independent of x. In general this will not be true if the series is simply convergent.

Cauchy's mistake results from various kinds of confusion. Above all, the systematic and hasty use of infinitesimals in the demonstration prevents him from grasping the functional dependence of the "very large value of n" being sought.

In the second place, his geometrical conception, which is largely intuitive, leads him to think that, as n increases, the graph of the curve $y = s_n(x)$ comes ever closer to that of $f(x)$, which he takes to exist. This is true in a certain sense, but only in that, by fixing a particular value x in the interval, for every arbitrarily small positive ε, there exists an integer ν for which, for this value of x, $|s_n(x) - f(x)| < \varepsilon$ for $n > \nu$.

This does not mean that the curves end by coinciding geometrically. This can happen not only in the neighborhood of a point of discontinuity of $f(x)$, but also when $f(x)$ is continuous.

This can be clearly seen with a few examples (Carslaw, 1921, pp. 125-6).

Example 1. Consider the series $\Sigma u_n(x)$, where

$$u_n(x) = \frac{1}{(n-1)x + 1} - \frac{1}{nx + 1} \quad (x \geqslant 0),$$

whose partial sums are

$$s_n(x) = 1 - \frac{1}{nx + 1}.$$

Then, when $x > 0$, $\lim s_n(x) = 1$; while when $x = 0$, $\lim s_n(x) = 0$.

When $x > 0$ the curve $f(x)$ consists of the part of the line $y = 1$, excluding the origin, where $f(x) = 0$. Therefore it is discontinuous at the origin while the s_n are not for every finite n.

Example 2. Consider the series $\Sigma u_n(x)$, where

$$u_n(x) = \frac{nx}{1 + n^2x^2} - \frac{(n-1)x}{1 + (n-1)^2x^2}.$$

Hence the $s_n(x)$ are given by

$$s_n(x) = \frac{nx}{1 + n^2x^2}$$

and $\lim s_n(x) = 0$ for all x. The sum of the series is continuous, but the curves $y = s_n(x)$ differ substantially from the curve $y = f(x)$ in the neighborhood of the origin. The first have a maximum at $(1/n, 1/2)$ and a minimum at $(-1/n, -1/2)$. The x coordinate of the maximum tends to zero as n increases (Figure 6).

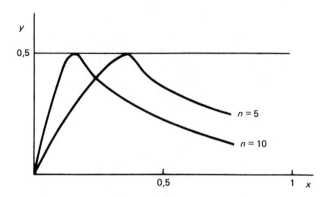

Figure 6

"If we reasoned from the shape of the curves $y = s_n(x)$, we should expect to find the part of the axis of y from $-1/2$ to $1/2$ appearing as a portion of the curve $y = f(x)$," Carslaw concludes (1921, p. 126).

The final source of confusion regards operations with double limits. In fact, if the series is convergent and its sum is $f(x)$, then

$$f(x) = \lim_{n \to \infty} s_n(x).$$

Further, by supposing that there is a limit of $f(x)$ as x tends to x_0, we have,

(α) $\lim_{x \to x_0} f(x) = \lim_{x \to x_0} [\lim_{n \to \infty} s_n(x)].$

On the other hand, if $f(x_0)$ is the sum of the series for $x = x_0$, from the definition we find that this value is given by $\lim s_n(x_0)$.

If the $s_n(x)$ are all continuous in the interval in question,

$$s_n(x_0) = \lim_{x \to x_0} s_n(x),$$

and consequently the sum of the series at the point x_0 can be written as

(β) $\lim_{n \to \infty} [\lim_{x \to x_0} s_n(x)].$

But (α) and (β) give the same value only for the case where $f(x)$ is continuous at x_0.

3.5. Abel's Theorems on Convergence

The first to advance reservations about Cauchy's theorem was Abel. In a note attached to his 1826 article on the convergence of the binomial series he observed: "But it seems to me that this theorem admits exceptions. For example, the series

$$\sin x - \frac{1}{2}\sin 2x + \frac{1}{3}\sin 3x - \cdots$$

is discontinuous for every value $(2m + 1)\pi$ of x, m being a whole number. There are, as we know, many series of this kind" (Abel, 1826, p. 225 n.).

What Abel was referring to was one of the series proposed by Fourier as an example of his expansions into trigonometric series (see §2.3). (In all likelihood Abel had found Fourier's *Théorie* in Berlin, in Crelle's library, together with Cauchy's *Cours d'analyse*.)[19]

It was after reading the *Cours* that Abel saw Fourier series as effective counterexamples to Cauchy's theorem, precisely on the basis of his definition of continuity.

In regards to the delicate question of the behavior of series of continuous functions, Abel set out two theorems in this article. In the first he asserted that, "If the series

$$f(\alpha) = v_0 + v_1\alpha + v_2\alpha^2 + \cdots + v_m\alpha^m + \cdots$$

is convergent for a certain value δ of α, it will also be convergent for every value less than δ and, for ever decreasing values of β, the function $f(\alpha - \beta)$ will indefinitely approach the limit $f(\alpha)$, supposing that α is equal to or less than δ" (1826, p. 223).

With this statement Abel put into question a mathematical procedure that had been common enough at the time, but which had often led mathematicians to completely erroneous statements about numerical series. In fact, in order to calculate their sum, the usual method was to transform the given series into a series of powers of a variable x, to calculate its sum, and then to substitute into the latter a suitable value of x which would give the initial series.

A classic example is the series of Grandi (1671-1742), $1 - 1 + 1 - 1 + \cdots$, whose "sum" caused much discussion among the mathematicians of the eighteenth century. Many argued that the series converged to 1/2 on the basis of the following reasoning: In the geometrical series

$$1 + x + x^2 + \cdots = \frac{1}{1 - x}$$

we set the value -1 for x. We thus obtain the series of Grandi, whose sum is consequently 1/2. But first Bolzano (1817) and then Cauchy (1821) had demonstrated that the convergence of the geometrical series is assured only if $|x| < 1$. To this Abel added that the substitution would be legitimate only if the series converges for

$x = -1$ (that is, $\delta = 1$), which it clearly does not.

Abel's theorem must have raised many difficulties if, many years later, after Dirichlet's death, the French mathematician Liouville had believed it necessary to record his earlier doubts.

It is a matter of proving that, if the series

$$a_0 + a_1 + a_2 + \cdots + a_n + \cdots$$

is convergent and has A as its sum, the sum of the series

$$a_0 + a_1\rho + a_2\rho^2 + \cdots + a_n\rho^n + \cdots ,$$

which will be convergent *a fortiori* when the variable ρ is positive and less than unity, will tend towards the limit A when we let ρ tend indefinitely towards unity.[20] Talking one day with my wonderful and late lamented friend Lejeune Dirichlet, I told him that I had found it exceedingly difficult to set out (and even to understand) the demonstration that Abel had given for this important theorem (In: Dirichlet, *Werke* 2, p. 305).

Liouville's account continued with a presentation of the demonstration that Dirichlet had given "on two feet" and "before his very eyes." Let

$$S = a_0 + a_1\rho + a_2\rho^2 + \cdots + a_n\rho^n + \cdots ,$$

where $0 < \rho < 1$. Setting s_0, $s_1 - s_0$, $s_2 - s_1$, ... in place of a_0, a_1, a_2, ..., respectively, we have

$$S = s_0 + (s_1 - s_0)\rho + (s_2 - s_1)\rho^2 + \cdots ,$$

or

$$S = (1 - \rho)(s_0 + s_1\rho + s_2\rho^2 + \cdots).$$

"We divide S into two parts, one of which includes the first n terms and the other all the rest, and let n increase as ε [$= 1 - \rho$] decreases, but so slowly that the limit of $n\varepsilon$ is zero" (Dirichlet, 1863, p. 306). We easily see that the limit of S, as ρ tends to 1, is given by the sum A of the initial series. "I do not think that anyone in the future can think of requiring further clarifications," Liouville concluded (In: Dirichlet, *Werke* 2, p. 306).

Passing then to consider, in place of v_n, the functions $v_n(x)$, Abel stated the following theorem:

Let

$$v_0 + v_1\delta + v_2\delta^2 + \cdots$$

be a convergent series, in which v_0, v_1, v_2, ... are continuous functions of the same variable quantity x between the limits $x = a$ and $x = b$. The series

$$f(x) = v_0 + v_1\alpha + v_2\alpha^2 + \cdots,$$

where $\alpha < \delta$, will be convergent and a continuous function of x between the same limits" (1826, pp. 223-4).

In the demonstration Abel considers the two functions

$$\phi(x) = v_0 + v_1\alpha + v_2\alpha^2 + \cdots + v_{m-1}\alpha^{m-1}$$

$$\psi(x) = v_m\alpha^m + v_{m+1}\alpha^{m+1} + \cdots,$$

where $f(x) = \phi(x) + \psi(x)$. Now, since we can write

$$\psi(x) = \left[\frac{\alpha}{\delta}\right]^m v_m\delta^m + \left[\frac{\alpha}{\delta}\right]^{m+1} v_{m+1}\delta^{m+1} + \cdots,$$

if

$$\theta(x) = \sup_{n\in N}(|v_m\delta^m + \cdots + v_{m+n}\delta^{m+n}|),$$

consequently we have

$$|\psi(x)| \leqslant \left[\frac{\alpha}{\delta}\right]^m \theta(x).$$

From this[21] Abel concludes that, if we add an infinitesimal increment β to x, $f(x)$ will also change by an infinitesimal quantity and consequently the function $f(x)$ is continuous.

But like Cauchy's 'demonstration,' Abel's too contains an illicit hypothesis. He in fact implicitly supposes that $\sup_{x\in[a,b]}\theta(x)$ is finite, which in general is not true, as Kronecker first pointed out[22] (see §7.1). According to Sylow (1902, p. 53), "there is reason to believe that later [Abel] was not completely satisfied" with his demonstration, since he returned to it again in a paper *Sur les séries*, which was first published in the second edition of his works in 1881. However, his results here were not much different.

Abel hit the mark when he used a counterexample to illustrate the imprecision of Cauchy's theorem on series of continuous functions, but it was otherwise as far as finding the error in the demonstration was concerned. Here Abel limited himself to observing, with the care appropriate for a young man beginning his career facing a distinguished mathematician, that it seemed to him that the theorem admitted "exceptions," which is a roundabout way of saying that a theorem is false and requires additional hypotheses.

On the other hand, as Abel himself acknowledged, Cauchy's *Cours d'analyse* was his theoretical guide. Thus, the techniques that Abel used were those of Cauchy; his definitions of continuity and of convergence are the same, as are his use of infinitesimals in demonstrations.

It is precisely this latter fact that kept both men from finding the weak point in Cauchy's demonstration and from seeing that there is another form of convergence, what is today called the *uniform* convergence of a series of functions. This assumption is necessary in order to render Cauchy's theorem correct. Those mathematicians and historians who have seen the idea of uniform convergence in Abel's theorem are consequently mistaken. Examples are Pringsheim, when he writes that Abel "here directly proves the existence of that property which we today call *uniform* convergence" (1899, p. 35), and Hardy, according to whom "the idea [of uniform convergence] is present implicitly in Abel's proof of his celebrated theorem on the continuity of power series" (1918, p. 148).

In reality, Abel had the same idea of the convergence of series as Cauchy, and the idea of uniform convergence only emerged about twenty years later in a rather different context.

Among the new results that Abel also stated in his 1826 paper, making use of Cauchy's techniques, were the *criteria of convergence* of series.

The enunciation of convergence criteria for series is in fact among the most significant results of Cauchy's *Cours d'analyse*. On the basis of Theorems I and II above, Cauchy stated the *criteria of the ratio* and of the *root* that assure the convergence of the series Σu_n when, respectively,

$$\lim \frac{u_{n+1}}{u_n} = K < 1 \quad \text{and} \quad \lim(u_n)^{1/n} = K < 1.$$

For series that contain as many positive as negative terms, Cauchy establishes the convergence when the series $\Sigma \rho_n$ of their respective absolute values satisfies the above stated criteria, while for a series of terms with alternating signs he shows that it is sufficient that the general term u_n tends to zero in order to guarantee the convergence of the series.

In the case of the series of whole powers of a variable

$$a_0 + a_1 x + a_2 x^2 + \cdots$$

he states the theorem according to which the series is convergent if "the numerical value [i.e. the modulo] of the variable x is less than $1/A$," where

$$A = \lim_{n \to \infty} \sup \sqrt[n]{|a_n|}.$$

"On the contrary, the series will be divergent if the numerical value of x surpasses $1/A$" (1821, p. 136).

This proposition passed unnoticed until it was rediscovered by Hadamard in 1892. Under the name of the 'Cauchy-Hadamard theorem,' it has proved to be of fundamental importance in the theory of analytic functions, allowing one to determine the radius of the circle of convergence of the series that gives any element of the analytic function (see §7.3).

The study and determination of convergence criteria for series reflects a radical change in the way of understanding analysis with respect to the eighteenth-century tradition. Cauchy establishes the convergence of series under suitable hypotheses for the general term u_n *without actually knowing* what the value of the sum is. If for u_n such and such conditions are true, then *there exists* a number S such that $\lim s_n = S$ for $n \to \infty$.

The scheme of Cauchy's criteria heralded a new approach to analysis. Since then ever more sophisticated theorems of an *existential* nature have become common in mathematics and indeed today constitute the fundamental structure of many theories.

3.6. Cauchy's Definitions of Derivative and Differential

Among the most remarkable limits and indeterminate forms that Cauchy considered in the *Cours d'analyse* are those of the type 0/0.

> When the two terms of a fraction are infinitely small quantities whose numerical values decrease indefinitely with those of the variable α, the singular value that received this fraction, for $\alpha = 0$, is sometimes finite, sometimes zero or infinite (1821, p. 64).
>
> Among the fractions whose two terms converge to the limit zero with the variable α, we must place the following
>
> $$\frac{f(x + \alpha) - f(x)}{\alpha}$$
>
> whenever we assign to the variable x a value in whose neighborhood the function $f(x)$ remains continuous (1821, p. 65).

As examples, Cauchy proposes the functions $f(x) = x^2$ and $f(x) = a/x$, for which the two limits in question are $2x$ and $-a/x^2$ respectively. After having demonstrated with the method also used today that

$$\lim_{\alpha \to 0} \frac{\sin \alpha}{\alpha} = 1.$$

Cauchy adds, "The study of the limits towards which the ratios

$$\frac{f(x + \alpha) - f(x)}{\alpha}, \qquad \frac{f(\alpha) - f(0)}{\alpha},$$

converge being one of the principal objectives of the infinitesimal calculus, we will not dwell on it any longer" (1821, p. 67).

This is the only allusion to the concept of derivative that we find in the *Cours*. The principal concepts of the differential calculus instead constitute the subject of the second part of the course Cauchy gave to the students in their first year at the École

Polytechnique, which was published in 1823 as the *Résumé des leçons donnees a l'Ecole royale polytechnique sur le calcul infinitesimal.* Here we find the definitions of derivative, differential, and integral in terms of limits that have since become classic. Here Cauchy gives the first rigorous systematization of the theory of expanding functions in Taylor series, the Lagrangian basis of the theory of functions.

Cauchy is well aware of the profound theoretical novelty of his approach. In the *Avertissement* to the *Résumé* he writes,

> The methods that I have followed differ in many respects from those found in works of the same kind. My principal aim has been to reconcile rigor, which I had made a law unto myself in my *Cours d'analyse*, with the simplicity that results from the direct consideration of infinitely small quantities. For this reason, I have thought it necessary to reject the expansions of functions in infinite series whenever the series thus obtained are not convergent; and I have been forced to return Taylor's formula to the integral calculus. This formula can no longer be admitted as general unless the series that it includes is reduced to a finite number of terms and completed by a definite integral. I am not unaware that the illustrious author of the *Mécanique analytique* [Lagrange] took the formula in question as the basis of his theory of *derived functions.* But, despite all the respect that I have for such a great authority, the majority of geometers now agree in recognizing the uncertainty of the results to which one can be led by the use of divergent series, and I add that in many cases the Taylor theorem seems to furnish the expansion of a function in convergent series, even though the sum of the series essentially differs from the proposed function. For the rest, those who read my book will I hope be convinced that the principles of the differential calculus and its most important applications can easily be set out without the use of series (Cauchy, 1823a, pp. 9-10).

Leaving the diplomacy aside, this is a decisive and radical attack on Lagrange's conceptions of the foundations of the infinitesimal calculus. Only a few years earlier, Lagrange had published a supplement to the lessons he had given on the theory of functions at the Ecole Polytechnique in which he had written,

> Every function of one variable only can always be regarded as an exact derivative; for, if it does not naturally have a primitive function, one can always find one by series ... by resolving the given function in a series of powers of the variable and then taking the primitive function of each term (Lagrange, 1806, p. 364).

In the same volume of the *Journal de l'Ecole polytechnique* that contained Lagrange's supplement, A. M. Ampere (1775-1836) published a long note whose object was "a new demonstration of the Taylor series." In criticizing Lagrange's assumption, Ampère sought to demonstrate that it is possible for every function $f(x)$ to be expanded in a series of increasing powers.[23] He began from a definition of the derivative that was based on a property of derived functions found by Lagrange. This property was stated by Ampère as,

> Let A and K be the values of $f(x)$ corresponding to $x = a$ and $x = k$. One can always suppose that a and k are chosen in such a way that $f(x)$ never becomes infinite in the interval and that neither $A = k$ nor $a = k$, so that
>
> $$\frac{K - A}{k - a}$$
>
> is neither zero nor infinite, and that one can take this quantity for that above or below which one can always take $[f(x+i) - f(x)]/i$, giving to i a value as small as necessary (1806a, p. 151).

From this property Ampère drew the definition of derivative that to him "seemed the most general and the most rigorous possible" (*ibid.*, p. 156).

"The derived function of $f(x)$ is a function of x such that $[f(x+i) - f(x)]/i$ is always included between two of the values that this derived function takes between x and $x + i$, whatever x and i may be (*ibid.*).

Ampère's work and his lectures at the École Polytechnique did not go without influence on Cauchy, who cites Ampère among those to whom he is indebted in both the *Cours d'analyse* and in the *Résumé*. But Cauchy's approach reverses the terms of the problem: by defining the derivative of a function (if it exists) as a suitable limit, what Ampère had given as a definition became a property of the derivative, expressible as a theorem.

This is how Cauchy proceeds. After having repeated the definitions of limit, infinitesimal, and continuity given in the *Cours d'analyse*, he writes,

> When the function $y = f(x)$ remains continuous between two given limits of the variable x and one assigns to this variable a value included between the two limits in question, an infinitely small increase in the variable produces an infinitely small increase in the function itself. As a consequence, if one then sets $\Delta x = i$, the two terms of the *ratio of differences* [*rapport aux différences*]
>
> $$\frac{\Delta y}{\Delta x} = \frac{f(x + i) - f(x)}{i}$$

will be infinitely small quantities. But, while these two terms will indefinitely and simultaneously approach the limit zero, the ratio itself can converge towards another limit, either positive or negative. This limit, *when it exists* [my emphasis], has a determinate value for every particular value of *x*, but it varies with *x*. ... The form of the new function that will serve as the limit of the ratio

$$\frac{f(x + i) - f(x)}{i}$$

will depend on the form of the proposed function $y = f(x)$. To indicate this dependence, we give the new function the name of *derived function*, and designate it, with the aid of an accent, by the notation y' or $f'(x)$ (1823a, pp. 22-3).

In the next lecture Cauchy gives the definition of the differential of the function $f(x)$ as "the limit towards which the first member of the equation

$$\frac{f(x + \alpha h) - f(x)}{\alpha} = \frac{f(x + i) - f(x)}{i} h \quad (i = \alpha h)$$

converges when the variable α approaches indefinitely close to zero while the quantity *h* remains constant" (1823a, p. 27). In the particular case of $f(x) = x$, the equation

$$df(x) = hf'(x)$$

reduces to $dx = h$, from which

$$df(x) = f'(x)dx,$$

or, equivalently,

$$dy = y'dx.$$

This, Cauchy says, allows us to write the first derivative as dy/dx, that is, as the ratio between the differential of the function and that of the variable.

The fundamental theorem from which Cauchy takes the remarkable properties of his calculus of derivatives is stated as follows:

If, the function $f(x)$ being continuous between the limits $x = x_0$, $x = X$, we designate by A the smallest and by B the largest of the values that the derived function $f'(x)$ assumes in this interval, the ratio of increments

$$\frac{f(X) - f(x_0)}{X - x_0}$$

will necessarily be included between A and B (1823a, p. 44)

It is in this context that Cauchy refers in a note to the work of Ampère mentioned above.

Thus, while for Lagrange the Taylor series was the premise for the theory of derived functions, for Cauchy the study of the possibility of transforming "any function of x or of $x + h$ into entire functions of x or of $x + h$ to which are added definite integrals" constitutes the conclusion of his lectures on the infinitesimal calculus (1823a, p. 214).

He here gives the following formula:

$$f(x + h) = f(x) + hf'(x) + \frac{h^2}{2!} f''(x) + \cdots$$

$$+ \frac{h^{n-1}}{(n-1)!} f^{(n-1)}(x) + \int_0^h \frac{(h-z)^{n-1}}{(n-1)!} f^{(n)}(x+z)dx,$$

which is the 'Taylor formula with the remainder in the form of Cauchy.' This formula, Cauchy adds, presupposes that the functions $f(x + z)$, $f'(x + z)$, ..., $f^{(n)}(x + z)$ remain continuous between the limits $z = 0$, $z = h$ (1823a, p. 215).

When the integral that gives the remainder tends to zero for increasing values of n, then it serves to *expand* the function $f(x + h)$ "in series ordered according to the ascending integral powers of the quantities x and h. The remainders of these series are precisely the integrals of which we just spoke" (1823a, 221). Cauchy concludes with an observation of the greatest interest. If we consider the Maclaurin series in particular (which can be obtained from the Taylor series),

$$f(0) + xf'(0) + \frac{x^2}{2!} f''(0) + \cdots$$

one could think that the series always has $f(x)$ as a sum when it is convergent, and that, in the case where its different terms vanish one after the other, the function $f(x)$ itself vanishes; but, to be certain of the contrary, it is sufficient to observe that the second condition will be fulfilled if we suppose $f(x) = e^{-(1/x)^2}$, and the first if we suppose

$$f(x) = e^{-x^2} + e^{-(1/x)^2}.$$

However, the function $e^{-(1/x)^2}$ is not identical to zero, and the series derived from the last supposition does not have the binomial $e^{-x^2} + e^{-(1/x)^2}$ as its sum, but its first term e^{-x^2} (1823a, pp. 229-30).

With this counterexample Cauchy not only clarifies the intrinsic theoretical weakness of Lagrange's construction, but initiates one of the most difficult questions in nineteenth-century analysis -- that of knowing whether and how one can represent functions in series converging to the given function. The issue was of special interest for Fourier series, where the unsolved problem of their convergence was still to be faced.

Abel declared his intention of devoting himself to this task in a letter he wrote to Holmboe in December, 1826, but it was instead taken up by Cauchy, by Poisson, and later by Lejeune-Dirichlet (see Chapt. 5).

Notes to Chapter 3

[1]The *Bulletin des sciences mathématiques, astronomiques, physiques et chimiques* was actually edited by Baron de Férussac. Saigey, to whom Abel refers, was the editor of the mathematics series of the *Bulletin*.

[2]"I showed it to Cauchy, but he scarcely wished to glance at it. And I dare to say without bragging that it is good. I am anxious to hear the judgement of the *Institut*", Abel wrote confidently (1902, p. 46). But the Norwegian mathematician would wait in vain for the Institute's judgement. Fourier, the *secrétaire perpétuel*, referred it to Cauchy and Legendre for a report, and all trace of it was lost until Jacobi recalled Legendre's attention to it after Abel's death.

The paper was then returned to the Academy with Cauchy's opinion favorable to its publication. Nevertheless, reasons of a political nature (connected with the July revolution of 1830) and the delays of the Academy kept Holmboe from obtaining it for publication in his 1839 edition of Abel's works. The paper was finally published by the Academy in 1841.

In this work Abel presented the fruits of his research which, beginning from the idea of the inversion of elliptic integrals, had led him to study transcendental functions "whose derivatives can be expressed by means of algebraic equations, all of whose coefficients are rational functions of the same variable" and to establish "a general property" for them (Abel, 1841, p. 145). This is the so-called "Abel's theorem," which is of fundamental importance for algebraic geometry.

It is not possible to outline here the theory of elliptic and Abelian functions which was inaugurated by Abel and Jacobi. It constitutes one of the most important chapters of complex analysis and algebraic geometry in the nineteenth century, and would fill an entire volume on its own. For a historical account see Brill and Noether (1894) and Dieudonné (1974, Vol. II, pp. 1-113).

But the publication of this paper in 1841 still did not end its tormented history. The manuscript, which at the time was in the possession of Libri (1803-1869), disappeared when, following an enormous scandal, he was forced to take refuge in England in 1848, accused of having stolen books and manuscripts from the public libraries. The manuscript was subsequently rediscovered in 1952 by Viggo Brun (1885-1978) in the Biblioteca Moreniana in Florence, where a number of Libri's manuscripts had eventually landed. For the history of this manuscript and a biography of Abel, see Ore (1957).

[3]Abel here anticipates the basic content of a paper that appeared the same year in Crelle's *Journal* (Abel, 1826).

[4]See the preface to Dedekind's paper (1872).

[5]A general discussion of the equation

$$x^2(a + bx^n)d^2y + x(c + ex^n)dy\,dx + (f + gx^n)y\,dx^2 = 0,$$

for which (3.2.1) is a special case, is found in Chapters 8 and 9 of Euler's *Institutiones calculi integralis*, Vol. II (In: *Opera* (1) **12**, pp. 177-270).

[6]A detailed study of Gauss' contributions to analysis is given by Schlesinger (1933). The whole of Volume 10, Part 2, of Gauss' *Werke* is a collection of special articles on various aspects of Gauss' mathematical activity.

[7]Bolzano's philosophical and mathematical work and his influence on the science of our day were the subject of a conference held on the bicentenary of his birth (Bolzano 1981b). On the same occasion his early mathematical works were reprinted in facsimile (Bolzano 1981a).

[8]Significantly, Bolzano does not mention Gauss' article of 1813, which was probably unknown to him at the time.

[9]"One finds elsewhere in Abel's manuscripts (Cahier III), 'Bolzano is an able man,' a phrase that I could not understand because I only knew of Bolzano as the name of a city. It was especially interesting to see a mathematician Bolzano cited in the *Enzyklopädie der mathematischen Wissenschaften* as having, even before Cauchy, given the well known fundamental criterion of convergence. In the course of his trip, Abel had also read, without any doubt, Bolzano's book, *Rein analytischer Beweis des Lehrsatzes* etc." (Sylow, 1902, p. 13).
Incredibly, this is what Sylow (1832-1918), a mathematician known for his theorems on groups to every student who has taken a course of elementary algebra, wrote in a scientific biography of Abel based on his manuscripts in 1902! This illustrates the meager knowledge of Bolzano's mathematical works that was prevalent even at the beginning of our century.

[10]This problem has recently been the subject of a fierce debate that seems extreme in view of the absence of material evidence.

Grattan-Guinness' suggestion (1970a) that Cauchy *could* have read Bolzano's paper evoked a lively reaction from Freudenthal (1971). According to Grabiner (1981), Lagrange directly influenced both Bolzano and Cauchy and the fact that both mathematicians arrived at similar conclusions should therefore not be surprising. This subject was again taken up by Dauben (1981b). "Whether Cauchy actually read Bolzano or not, however, matters little," he observes, and rightly remembers that Bolzano wrote in regards to continuity in his *Funktionenlehre* (1930), "By the continuity of a function it would be best to understand the terminology introduced by Lagrange, Cauchy, and others," that is, the fact that $|f(x + \Delta x) - f(x)|$ "becomes and remains smaller than any given function $1/N$ if Δx is taken sufficiently small" (Dauben, 1981b, 245). In this way Bolzano avoided any claim to priority.

[11]Bolzano here adds the note: "There are functions that are continuously variable for all values of their roots, for example $\alpha x + \beta x$. But there are other functions that vary according to the law of continuity only within or outside of certain limiting values of their root. In this way $x + \sqrt{(1 - x)(2 - x)}$ varies continuously only for all values of x that $< +1$ or $> +2$, but not for the values that lie between $+1$ and $+2$" (1817, p. 7 n.).

[12]The definition was made more precise by the addition of the absolute value of the difference (for positive ω) in a later work of Bolzano, the *Funktionenlehre* of 1830.

[13]This proposition is equivalent to the so-called "Bolzano-Weierstrass theorem." Bolzano used it but did not demonstrate it in the *Funktionenlehre*. The theorem was enunciated and demonstrated by Weierstrass in his lectures without even knowing of Bolzano's existence.

[14]For a more recent biography of Cauchy see Belhoste (1984).

[15]In order to clarify this point it is perhaps better to turn to modern symbolism. A function $f: [a,b] \to R$ is continuous at every point $x_0 \in [a,b]$ if

$$\forall x_0 \in [a,b] \quad \forall \varepsilon > 0, \ \exists \delta(x_0,\varepsilon) > 0:$$

$$\forall x \in [a,b] | x - x_0| < \delta \Rightarrow |f(x) - f(x_0)| < \varepsilon,$$

while it is uniformly continuous in $[a,b]$ if

$$\forall \varepsilon > 0 \ \exists \delta(\varepsilon) > 0: \forall x, x_0 \in [a,b] \ |x-x_0| < \delta \Rightarrow |f(x) - f(x_0)| < \varepsilon.$$

The ambiguity in Cauchy's language, in my opinion, does not permit us to clarify which of the two definitions he had in mind, even though he in fact utilizes uniform continuity in the definition of the definite integral (see §4.4 below).

[16]In the *Funktionenlehre* Bolzano in fact considered a function continuous in every point of an interval but not derivable in a dense subset of it. Bolzano's ideas, however, remained unedited and the problem of the relationship between the continuity and derivability of a function in fact remained open until the 1860s when Weierstrass published his celebrated counterexample (see §6.3). It is interesting to note that in this circumstance Bolzano criticized a paper of Galois (1830) in which the young student at the Ecole Normale thought he had "demonstrated *a priori* the existence of derived functions" for any function whatever. Ampère (1806a) was also read for many years as an attempt to demonstrate that a continuous function always possesses a derivative (except at isolated points) (see §3.6 below).

[17]For a detailed discussion of this example, see §6.2 below.

[18]These studies, which remained in manuscript, have recently been published by Rychlik (1961), even if the opinion of historians on their adequacy differs.

[19]A. L. Crelle (1780-1855) was a mediocre mathematician but an extraordinary organizer. Abel became acquainted with Crelle in Berlin during his European trip, and his letters frequently refer to the courtesy and hospitality shown him by Crelle. The latter put his library at Abel's disposition and invited him to the weekly meetings he held at his house with other young mathematicians in Berlin, among whom were Martin Ohm (1792-1872), the brother of the more famous physicist Simon Ohm.

[20]It is in this form, where $\delta = 1$, that the theorem is found to be of particular usefulness for the practice of calculating with series.

[21]In reality, Abel does not use the absolute value either here or in the preceding theorem.

[22]For a counterexample see the article by Dugac in Dieudonné (1978, I, pp. 355-92).

[23]An accurate discussion of Ampère's work and of the origins of Cauchy's concept of derivative can be found in Grabiner (1978) and (1981).

Chapter 4
COMPLEX FUNCTIONS AND INTEGRATION

4.1. The Fundamental Theorem of Algebra

Well over half of the *Cours d'analyse* is devoted to complex analysis, which constitutes one of Cauchy's most fundamental contributions to mathematics, and probably his most important. In his hands the calculus of complex quantities became an indispensable instrument of analysis, losing the aura of mystery and inexplicability that had accompanied complex numbers since their appearance in the solution of algebraic equations of the third degree.

But no matter how uncertain the nature of complex numbers and variables was, they had still been widely used by mathematicians in the eighteenth century (see §1.2).

This uncertainty was still dominant at the end of the century, as is witnessed by Gauss' reluctance to take a public position on them in (1799), even though he seems to have been fully aware of the geometric interpretation of complex numbers and his demonstration of the fundamental theorem of algebra would be incomprehensible without them.

"[I] have demonstrated that equations have imaginary roots by a true method," Gauss wrote in his mathematical diary in October of 1797, thus recording the demonstration that was subsequently published in his dissertation of 1799 at the conclusion of his studies at the University of Helmstedt under J. Pfaff (1765-1825).

In this work Gauss wrote that, from the moment equations were found that do not admit roots other than those of the form $a + ib$, these "fictive quantities" have been introduced "into all of analysis." "By what right I will not discuss in this place," he limited himself to saying (1799, p. 4). In any case, he declared, "I will complete my demonstration without any help of imaginary quantities." The theorem itself was stated in the form that mathematicians had traditionally used since the time of D'Alembert: "Every algebraic entire function can be split into factors of the first or second degree."

The demonstration was preceded by a detailed and precise critique of the earlier demonstrations, from D'Alembert to Lagrange[1] (see §1.2). In all these cases, Gauss observed, instead of a demonstration we find a *petito principii*, in that in the course of the demonstration the existence of the roots is tacitly assumed. This was so in Euler's version as well as in the improved version given by Lagrange (1774), who had sought to give a new demonstration of the factorization of a polynomial $P(x) = P_1(x)P_2(x)$ in order to fill the holes in Euler's demonstration. But, Gauss observed, "[Lagrange's] entire investigation is also based on the assumption that every equation of the mth degree really has m roots (1799, p. 20).

Since the preceding attempts did not escape objections and errors, Gauss preferred to let himself be guided "by entirely different principles," even though he recognized the validity and the "genuine strength of the demonstration" that had been given by D'Alembert. "I believe that ... a rigorous demonstration of our theorem can be constructed [on the basis of D'Alembert's]," Gauss declared. However, D'Alembert, like all other mathematicians, had not doubted "the *existence* of the values of x to which the given values of X correspond, but supposed it and only investigated the *form* of these values" (1799, p. 9).

The fundamental novelty of Gauss' demonstration lies entirely in his awareness of the need for an *existence proof* of a root, even though his own proof, as we will see, is not free from criticism. From his demonstration it follows implicitly that such a root is of the form $a + ib$; it would be futile and incomprehensible to imagine a kind of hierarchy of imaginaries beginning from the complex number $a + ib$. Gauss called such numbers "a true shadow of a shadow," conjectured by some eighteenth-century mathematician (1799, p. 14). The field $\mathbb{C}$ is algebraically closed, as Euler and D'Alembert had already maintained (see §1.2).

Gauss began his demonstration with a proof of the following lemma.

"If the quantity r and the angle ϕ are determined in such a way as to give the equations

$$(4.1.1) \quad T = r^m\cos m\phi + Ar^{m-1}\cos(m-1)\phi + Br^{m-2}\cos(m-2)\phi$$
$$+ \cdots + Lr \cos \phi + M = 0$$

$$(4.1.2) \quad U = r^m\sin m\phi + Ar^{m-1}\sin(m-1)\phi + Br^{m-2}\sin(m-2)\phi$$
$$+ \cdots + Lr \sin \phi = 0,$$

the function $x^m + Ax^{m-1} + Bx^{m-2} + \cdots + Lx + M = X$ will be divisible by the factor $x^2 - 2 \cos \phi + r^2$ if $r \sin \phi \neq 0$, and by the factor $x - r \cos \phi$ if $r \sin \phi = 0$" (1799, p. 21).

The demonstration of the actual theorem is consequently reduced to determining r and ϕ in such a way as to verify (4.1.1) and (4.1.2).

If one thinks of a [monic] polynomial $P_m(z)$ for $z \in \mathbb{C}$ and $z = r(\cos \phi + i \sin \phi)$, then $P_m(z) = T + iU$ and Gauss' requirement signifies nothing other than the existence of a z such that $T = 0$ and $U = 0$, that is, the existence of a root of the polynomial.

The foundation of Gauss' demonstration consists of supposing the continuity of the functions T and U, which he implicitly assumes. It was on this point that Bolzano's criticisms were based, as we have seen above (§3.3).

Gauss in fact considers a reference plane in polar coordinates (r,ϕ). As r and ϕ vary, the functions $T(r,\phi)$ and $U(r,\phi)$ define two surfaces, each of which is partly above and partly below the plane. For r sufficiently large, the signs of T and U are given by the first terms of $r^m \cos m\phi$ and $r^m \sin m\phi$ respectively, which change sign as ϕ varies. "Wherefore the fixed plane is necessarily cut by the first surface," Gauss observes, and similarly for the surface U. The curves of intersection $T = 0$ and $U = 0$ thus obtained are composed of different branches, each of which is nevertheless given by a *continuous line*. The demonstration thus reduces to showing that "there is at least one point in the plane where any branch of the first line is cut by a branch of the second line" (1799, p. 23).

In order to show this Gauss considers a circle in the plane (r,ϕ) with a radius r sufficiently large and center at the origin C. "At an infinite distance from the point C", Gauss observes, the curve of the equation $U = 0$ "coincides" with the line of the equation $\sin m\phi = 0$. Now the equation $\sin m\phi = 0$ defines a system of m lines through the origin, which form the angles

$$0, \frac{\pi}{n}, \frac{2\pi}{n}, \dots, \frac{(n-1)}{n}\pi$$

with the x axis. In other words, they form the asymptotes of the system of $2m$ branches of the curve $U = 0$. The same thing happens with the curve $T = 0$, which has the lines of the equation $\cos m\phi = 0$ for asymptotes, forming the angles

$$\frac{\pi}{2m}, \frac{3\pi}{2m}, \dots, \frac{(2m-1)}{2m}\pi$$

with the axis.

The lines of a system are consequently the bisectors of the angles formed by the lines of the other (Figure 7). Gauss then demonstrates that "from the center C one can describe a circle on whose circumference there are $2m$ points at which $T = 0$ and just as many at which $U = 0$, in such a way that one of the latter lies between two of the former" (1799, pp. 24-5).

Now, since the branches of the curve $U = 0$ alternate with those of the curve $T = 0$ outside a circle of sufficiently large radius, inside the circle there must be at least one point (in fact m points) at which they intersect. This is the core of Gauss' demonstration, which appeals to continuity and the topological behavior of algebraic curves. "Now it is well known from higher geometry that

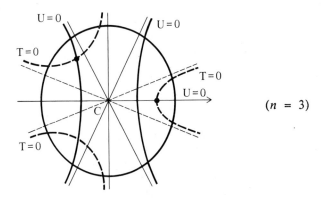

Figure 7

an algebraic curve (or one part of any algebraic curve if it is composed of many) either returns to itself or extends to infinity on both sides, and so if any branch of an algebraic curve enters a definite space, it must necessarily again somewhere exit from this space" (1799, p. 27).

Such a curve, he adds in a note, cannot in fact "suddenly break off somewhere," as happens for example with the transcendent curve $y = 1/\log x$, nor can it "almost lose itself in some point" like the logarithmic spiral.

Until now no one has advanced any doubts about this, Gauss asserts, "but still, if someone requests, I will be ready to provide a demonstration free of all doubts on some other occasion" (*ibid.*). The arguments on which he based his reasoning, he finally observes, derive "from the principles of the geometry of place [topology]..., which are no less valid than the principles of the geometry of magnitudes"[2] (1799, p. 28).

If, on a circle of radius r, we consider the $2m$ points at which $T = 0$ and the $2m$ points, alternating with the first, at which $U = 0$, and if we connect the points of the first system two-by-two with continuous lines inside the circle, we cannot similarly connect the points of the second system without the new continuous lines intersecting the first. At such points of intersection, therefore, $U = 0$ and $T = 0$, which concludes the demonstration.

Although at first sight this appears to be different from that of D'Alembert, Gauss observes at this point, "if you look at the essentials" it is in fact the same. This is in fact how Gauss translated D'Alembert's "strength of showing" [*nervus probandi*]: we consider any variable point M on the curve $T = 0$ and, on it, $|U|$. It is necessary to show that as M varies one arrives at a point at which $|U| = 0$. Indeed, if at a certain point of the curve $T = 0$ one has for $|U|$ a minimum $\neq 0$, then Gauss asserts that such a point will be multiple for the curve and the module of U can also diminish by passing to another branch of $T = 0$. By repeating the reasoning as necessary, Gauss concludes, we will in the end arrive at a point at which $|U| = 0$. But in so doing he also makes recourse to an intuitive argument of continuity -- it is not in fact evident that $|U|$ must necessarily have a minimum.[3]

Although he said he had demonstrated the theorem "with all rigor," Gauss does not seem to have been completely satisfied in demonstrating a theorem of algebra by appealing to arguments from higher geometry.[4] In fact, this theorem remained a favored topic of Gauss' research. In the course of his life he published four different demonstrations of it, the last in 1849.

Two short proofs appeared in 1816. The first (1816a) was of a purely algebraic character, inspired by the one that had been proposed by Euler and revived by Lagrange.[5] This was immediately followed by a different proof (1816b) that was purely analytical in nature.

As in (1799), in this second proof Gauss considered a polynomial $X = T + iU$ where the functions T and U are given by the first members of (4.1.1) and (4.1.2) respectively. Setting $T = t$ and $U = u$, Gauss then considered the derivatives

$$r\,\frac{\partial t}{\partial r} = t', \qquad r\,\frac{\partial^2 u}{\partial r\,\partial\phi} = t'',$$

(4.1.3)   and

$$r\,\frac{\partial u}{\partial r} = u', \qquad -r\,\frac{\partial^2 t}{\partial r\,\partial\phi} = u''.$$

Gauss did not here explicitly remark on the fact that

(4.1.4) $\dfrac{\partial t}{\partial r} = \dfrac{1}{r}\,\dfrac{\partial u}{\partial\phi}$ and $\dfrac{1}{r}\,\dfrac{\partial t}{\partial\phi} = -\dfrac{\partial u}{\partial r}$

are valid, and express the Cauchy-Riemann conditions in polar coordinates for the function $X = t + iu$. These equations are nevertheless implicit in his calculation of the functions t', t'', u', u''.

Again in polar coordinates, $X = t + iu$ can be written as $X = P(\cos \Phi + i \sin \Phi)$, where $P^2 = t^2 + u^2$ and $\Phi = \Phi(r,\phi) = \arctan^{u/t}$.

Gauss then considers

(4.1.5) $y = \dfrac{\partial^2 \Phi}{\partial r\,\partial\phi}$,

which, by means of (4.1.3) and (4.1.4), can be expressed as a rational function $y = y(r, t, u, t', u', t'', u'')$ with the denominator $rP^4 = r(t^2 + u^2)^2$.

Taking a circle with its center at the origin and a radius R sufficiently large, Gauss demonstrates that for $r = R$, one has

(4.1.6) $tt' + uu' > 0$,   where $\dfrac{\partial \Phi}{\partial \phi} = \dfrac{tt' + uu'}{t^2 + u^2}$.

But $t^2 + u^2 > 0$. Consequently, outside of a circle that is sufficiently large it is impossible that t and u simultaneously equal 0.

The theorem hence reduces to demonstrating that within a circle of radius R "there must exist such values of the indeterminates r, ϕ for which simultaneously $t = 0$ and $u = 0$" (Gauss, 1816a, p. 61).

Gauss reasons by *reductio ad absurdum*. Let us suppose that the theorem is not true. Then $t^2 + u^2 > 0$ within the circle and the value of y given by (4.1.5) must always be finite. Let us now consider the double integral

$$\Omega = \iint\limits_{|z| \leqslant R} y \, dr \, d\phi \quad (z = r(\cos \phi + i \sin \phi)),$$

which will give the same finite and determinate value, whether by first integrating with respect to r and then with respect to ϕ, or vice versa. But

(4.1.7) $\quad \int y \, d\phi = \dfrac{tu' - ut'}{r(t^2 + u^2)} = \dfrac{\partial \Phi}{\partial r}$

is zero for $\phi = 0$ and $\phi = 2\pi$, and therefore $\Omega = 0$.

On the other hand,

$$\int y \, dr = \frac{tt' + uu'}{t^2 + u^2} = \frac{\partial \Phi}{\partial \phi}$$

is positive for (4.1.6) and hence $\Omega > 0$. These are contradictory and therefore the theorem is demonstrated.

In a note Gauss adds that

$$\int_0^{2\pi} \frac{\partial \Phi}{\partial \phi} \, d\phi = 2m\pi,$$

something that "can be demonstrated in another place" and depends on the fact that arctan u/t is a "multiform" function. Inverting the order of integration in calculating the integral Ω consequently leads to different results. This is an observation that is of great interest, which Gauss takes up again in the last part of his work so as to better clarify the nature of the contradiction on which he bases his demonstration of the theorem.

Since the supposition that t and u do not become zero simultaneously within the circle is contradictory, it is necessary to conclude that for at least one value of r and ϕ one has simultaneously $t = 0$ and $u = 0$. But, "in such a case it is not possible to calculate the integrals $\iint y \, dr \, d\phi$," Gauss observes, since y becomes infinite[6] and consequently "analytical operations applied to nonsensities by blind calculation lead to absurdities" (Gauss, 1816b, p. 63).

Besides, Gauss adds, let us consider any real function $\eta = \eta(\xi)$ and the definite integral $\int \eta \, d\xi$ which, "generally speaking," expresses the area included between the ξ axis and the curve $\eta = \eta(\xi)$. If "we treat it according to the usual rules and *neglect continuity*, [my emphasis] we are frequently entangled in contradictions" (1816b, p. 64), as happens in the case of $\eta = 1/\xi^2$ when we take an interval of integration that includes the origin $\xi = 0$.

How one must treat these and other similar "paradoxes of analysis we will more fully pursue on another occasion," Gauss promises. However, he never returned to this question or to the problem of the order of integration in a double integral. As we will see in the next section, this instead became Cauchy's object of study and the point of departure for his more profound discoveries in the field of complex analysis.

4.2. Cauchy's *Memoire* on Definite Integrals

In 1814 the 25-year-old Cauchy presented the Institut with a paper on the calculus of "definite" integrals, or more precisely, on the evaluation of improper (real) integrals when one or both of the limits of integration are infinite. (With enormous delay, this work was not printed until 1827.) Improper integrals had been a topic of discussion among mathematicians since the time of Euler. Euler himself, Laplace in his research on approximation connected with the theory of probability, Poisson, and Legendre in his *Exercices de calcul intégral* (1811), had all been using "a type of induction based on the passage from the real to the imaginary," Cauchy wrote (1827a, p. 329). Laplace had held the view that such procedures could be considered "as a means of discovery similar to the induction which the geometers have long used. But these methods, even when employed with much care and restraint, always leave the demonstrations and their results wanting" (In: Cauchy, 1827a, pp. 329-30).

This then is Cauchy's objective: "to establish the passage from the real to the imaginary by a direct and rigorous analysis" (*ibid.*, p. 330). In his work we find many of the concepts that constitute the foundations of Cauchy's theory of functions of a complex variable, from the conditions of monogeneity to the first formulation of Cauchy's "integral theorem."

Cauchy begins by considering the integral

$$\int f(y)dy,$$

and supposing that $y = g(x,z)$, with x and z new independent variables. There then follows the equation between the "differential coefficients,"

$$(4.2.1) \qquad \frac{\partial}{\partial z}\left[f(y)\,\frac{\partial y}{\partial x}\right] = \frac{\partial}{\partial x}\left[f(y)\,\frac{\partial y}{\partial z}\right].$$

"One can verify this equation directly only by means of differentiation," he says (1827a, p. 337).

Equation (4.2.1), Cauchy says, is also valid when y is "in part real, in part imaginary": $y = M + iN$ (where M and N are real functions of

x and z) and $f(y) = P' + iP''$. Then for (4.2.1) we can substitute

(4.2.2) $\dfrac{\partial S}{\partial z} = \dfrac{\partial U}{\partial x}, \qquad \dfrac{\partial T}{\partial z} = \dfrac{\partial V}{\partial x}$

where

(4.2.3)

$$S = P'\frac{\partial M}{\partial x} - P''\frac{\partial N}{\partial x}, \qquad U = P'\frac{\partial M}{\partial z} - P''\frac{\partial N}{\partial z}$$

$$V = P'\frac{\partial N}{\partial z} + P''\frac{\partial M}{\partial z}, \qquad T = P'\frac{\partial N}{\partial x} + P''\frac{\partial M}{\partial x}.$$

Equations (4.2.2), Cauchy asserts, contain "the entire theory of the passage from the real to the imaginary" (1827a, p. 338). The first part of the paper, on "The equations that enable the passage from the real to the imaginary," develops the consequences that such equations entail for the calculation of "definite" integrals.

By integrating (4.2.2), Cauchy obtains,

(4.2.4)

$$\int (S'' - S')dx = \int (U'' - U')dz;$$

$$\int (T'' - T')dx = \int (V'' - V')dz$$

(where the primes indicate the values assumed by the functions at the limits of integration), assuming that the functions S, U, T, V always have a well determined value within the limits of integration. From this he obtains a number of applications to particular cases. In the first place, by setting $M(x,z) = x$ and $N(x,z) = z$, (4.2.2) become the "Cauchy-Riemann equations,"

$$\frac{\partial P'}{\partial x} = \frac{\partial P''}{\partial z}; \qquad \frac{\partial P'}{\partial z} = -\frac{\partial P''}{\partial x},$$

which in fact Cauchy never even wrote. He instead rewrote (4.2.4) by taking $[0,x]$ and $[0,z]$ as the intervals of integration, thus:

(4.2.5)

$$\int_0^x P\,dx - \int_0^x p\,dx = \int_0^z p''dz - \int_0^z P''\,dz$$

$$\int_0^x P''\,dx \qquad\qquad = \int_0^z P'\,dz - \int_0^z p'\,dz$$

where $f(x) = p$ and $f(iz) = p' + ip''$.

Cauchy then illustrates with an example how one can use (4.2.5).

By setting $f(x) = e^{-x^2}$ and letting the upper limit of integration become infinite, by means of a few transformations he finds the integrals

$$\int_0^\infty e^{-x^2} \cos 2xz\,dx = e^{-z^2}\int_0^\infty e^{-x^2}\,dx = \frac{1}{2}\sqrt{\pi}\,e^{-z^2}$$

$$\int_0^\infty e^{-x^2} \sin 2xz\, dx = e^{-z^2} \int_0^z e^{z^2}\, dz,$$

which had already been known to Laplace.

The paper continues by showing how one can calculate like integrals by setting other values of y and $f(y)$ in (4.2.2) and (4.2.3).

The applications to the calculation of improper integrals represent Cauchy's primary aim, and the first part of the paper did not seem to Legendre, the referee appointed by the Institute, to have a particularly suitable title. "It seems to me to have an objective completely different from that announced in the title," he wrote in his report (In: Cauchy, 1827a, p. 321). In fact, in his view, "the use that Mr. Cauchy makes of imaginaries ... can only conform to the ordinary rules of analysis, and are not subject to any difficulty" (*ibid.*).

Cauchy discussed the implications of equations (4.2.2 - 5), which had escaped Legendre, in a series of footnotes that were added to the paper in 1825, when it was finally presented to the Secretary of the Academy for publication. He here pointed out, for instance (1827a, p. 340), that equations (4.2.5) can be replaced by the single formula

$$(4.2.6) \quad \int_0^x f(x + iz)dx - \int_0^x f(x)dx = i\left[\int_0^z f(x + iz)dx - \int_0^z f(iz)dz\right].$$

But by this time, as he elaborated the intuitions and ideas that had been present in this paper, Cauchy was already developing the theory of integration in the complex field (see §4.4c below).

In the second part of the paper Cauchy discussed the possibility of inverting the order of integration of a double integral, and it was at this point that, in Legendre's view, the young mathematician obtained his most original results.

The problem, Cauchy said, is already present when one considers the definite integral of a function of one variable. The equation

$$(4.2.7) \quad \int_a^b \phi'(x)dx = \phi(b) - \phi(a),$$

ceases to be true when the function is discontinuous between the limits of integration. As he had already pointed out in the introduction,

if, when one allows the variable to increase by insensible degrees, the function is found to pass abruptly from one value to another, the variable always being included between the limits of integration, the difference of these two values must be subtracted from the definite integral taken in the ordinary way, and each of the abrupt jumps that the resulting function can make will necessitate a correction of the same type (1827a, p. 332).

In other words, as Cauchy shows in the course of his article, if "the function $\phi(z)$ passes abruptly from a determinate value to another value sensibly different from the first, in such a way that, designating by ξ a very small quantity, we have

$$\phi(Z + \xi) - \phi(Z - \xi) = \Delta,$$

then the ordinary value of the definite integral (4.2.7) must be diminished by the quantity Δ" (*ibid.*, p. 403).

In a footnote added to the printed edition Cauchy pointed out that this is the idea that he called the "principal value" of the definite integral in his *Résumé* (1823) (see §4.4a below), which, in modern terms, is written as

$$(4.2.8) \qquad p \cdot v \int_a^b \phi'(x)dx = \lim_{\xi \to 0} \left[\int_a^{Z-\xi} \phi'(x)dx + \int_{Z+\xi}^b \phi'(x)dx \right],$$

where Z is an infinity point of $\phi(x)$ in $[a,b]$.

In fact, Cauchy says, if we apply this procedure to the two integrals of the second member of (4.2.8) and calculate the integrals "by the ordinary method," we obtain $\phi(Z - \xi) - \phi(a)$ and $\phi(b) - \phi(Z + \xi)$ respectively. Their sum is then $\phi(b) - \phi(a) - \Delta$ (1827a, p. 404).

As an example, he gives the following

$$\int_{-2}^4 \frac{dz}{z} = \log(4) - \log(-2) - \Delta,$$

where $\Delta = -\log(-1)$ (1827a, pp. 404-5).

What then must one do in a double integral when, after the first integration, the function under the integral sign becomes infinite or indeterminate for certain values of the variable within the interval of integration?

In this case the double integral that appears in (4.2.4), which can also be written as

$$\int_{x_1}^{x_2} \int_{z_1}^{z_2} \frac{\partial S}{\partial z} dx \, dz = \int_{z_1}^{z_2} \int_{x_1}^{x_2} \frac{\partial U}{\partial x} dx \, dz,$$

$$(4.2.9)$$

$$\int_{x_1}^{x_2} \int_{z_1}^{z_2} \frac{\partial T}{\partial z} dx \, dz = \int_{z_1}^{z_2} \int_{x_1}^{x_2} \frac{\partial V}{\partial x} dx \, dz,$$

is indeterminate. Moreover, the two successive integrations can give completely different values. Contrary to what happens for functions of a single variable, Cauchy observes, if a function $f(x,z)$ assumes the form $0/0$ for $x = a$, $z = b$, it goes to different limits according to how we let $z \to b$ and $x \to a$. Then, when we calculate a double integral of a function that presents such an indeterminate form, we obtain two distinct and well determined values according to whether we first integrate with respect to x and then with z or vice versa.

"It is easy to see," Cauchy notes, "that the error produced by this reversal rests entirely with the part of the double integral that corresponds to systems very close" to the values $x = a$ and $z = b$

(1827a, p. 334). In this way Cauchy was led to consider "a peculiar kind of definite integral in which the limits relative to each variable are infinitely close to each other without the integrals being zero." Cauchy gives integrals of this type the name of "singular integrals," and a large part of his paper is devoted to their study.

Cauchy first supposes that the indeterminate point of the integrand is in a corner, then along a side of the rectangle of integration, and finally inside it. As he shows for the case of a simple integral (4.2.7), it is necessary to separate the part of the domain of integration which does not create problems from that which includes critical points. An example is the calculation of the double integral $\int_{a'}^{a''}\int_{b'}^{b''}(\partial K/\partial z)dx\,dz$, where $K = \phi(x,z)$ becomes indeterminate for $x = a'$ and $z = b'$. We have

$$\int_{a'}^{a''}\int_{b'}^{b''}\frac{\partial K}{\partial z}\,dx\,dz = \int_{a'}^{a''}\int_{b'}^{b''}\frac{\partial K}{\partial z}\,dz\,dx + A,$$

where A is the "singular" integral

(4.2.10) $A = -\int_0^\varepsilon \phi(a' + \xi, b' + \zeta)d\xi,$

ε is a very small quantity, and ζ must be equal to zero after the integration.

Cauchy illustrates his result with an example that to our eyes resembles the one given by Gauss at the end of (1816b).

Let $K = \phi(x,z) = z/(x^2 + z^2)$. Now consider the integral

$$\int_0^1\int_0^1\frac{\partial K}{\partial z}\,dx\,dz.$$

If we integrate with respect to z and then with respect to x, we obtain

$$\int_0^1\int_0^1\frac{\partial K}{\partial z}\,dx\,dz = \int_0^1\frac{dx}{1 + x^2} = \frac{\pi}{4}.$$

But, if we invert the order it is necessary to add a quantity A that, according to (4.2.10), is given by

$$A = -\int_0^\varepsilon \phi(\xi,\zeta)d\xi = -\int_0^\varepsilon \frac{\zeta}{\xi^2 + \zeta^2}d\xi = -\arctan\frac{\varepsilon}{\zeta}.$$

But, for $\zeta = 0$, arctan $\varepsilon/\zeta = \pi/2$, and therefore

$$\int_0^1\int_0^1\frac{\partial K}{\partial z}dx\,dz = \frac{\pi}{4} + A = -\frac{\pi}{4}.$$

Cauchy then shows that for the general case, if the indeterminate point is (X,Z), we have

$$A = \int_0^\varepsilon [\phi(X - \xi, Z - \zeta) + \phi(X + \xi, Z + \zeta)$$

$$- \phi(X - \xi, Z + \zeta) - \phi(X + \xi, Z - \zeta)]d\xi.$$

He then applies this result to the calculation of many improper
integrals, thus completing the analysis he had already given in the
first part of the paper.

The calculation with "singular" integrals attracted Legendre's
attention. In his report he noted that some of the integrals
calculated by Cauchy

> present several cases where the law of continuity is violated.
> One of these formulas, ... the integral

$$(4.2.11) \qquad \int_0^\infty \frac{x \cos ax}{\sin bx} \frac{dx}{1 + x^2}$$

> increases or decreases abruptly by $\pi/2$ when the ratio a/b, which
> at first is taken to be equal to a whole number, decreases or
> increases by an infinitely small quantity (In: Cauchy, 1827a, p.
> 326).

Legendre added that he had verified the correctness of Cauchy's
results for $a < b$ and $a > b$ "by methods peculiar to me," but that he
had found "his formula for the case $a = b$ incorrect" (In: Cauchy,
1827a, p. 326).

For $a = b$ Cauchy in fact gave for the integral the two values

$$(4.2.12) \qquad \frac{\pi}{2} \frac{e^b + e^{-b}}{e^b - e^{-b}}, \quad \frac{\pi}{2} \frac{3e^{-b} - e^b}{e^b - e^{-b}},$$

which in reality represent the left and right limits of the function
represented by the integral at the point of discontinuity $a = b$. They
differ by $\pi/2$ from the value that Legendre had found in the second
volume of his *Exercises de calcul intégral*, which was written almost
at the same time as Cauchy's paper and published in 1817. Legendre
had obtained the value by using a substitution procedure of the type
that Abel later criticized[7] (see §3.5).

In the *Exercises*, (Part IV, p. 124) Legendre had in fact obtained

$$(4.2.13) \qquad \int_0^\infty \frac{m \cot ax}{m^2 + x^2} dx = \frac{\pi}{e^{2am} - 1},$$

which gives the arithmetic mean of the values (4.2.12) for $m = 1$ and
$a = b$.

In a supplement attached to his paper (1827a, pp. 493-506), Cauchy
gave what seemed to Legendre to be "the true solution to this
problem and others like it" (1827a, p. 326). Cauchy asserted that the
contradiction between the two values of (4.2.12) and that given by
(4.2.13) "is only apparent" and showed how the first ones could be
reduced to the other by setting $a = b - \alpha$ or $a = b + \alpha$ respectively
(for infinitely small α), taking into account that

$$\int_0^\infty \sin \alpha x \frac{x}{1 + x^2} dx = \frac{\pi}{2} e^{-\alpha}.$$

In fact, by setting "α very small," we have, in the first case,

$$(4.2.14) \quad \int_0^\infty \frac{x \cos(b - \alpha)x}{\sin bx} \frac{dx}{1 + x^2} = \frac{\pi}{2} + \int_0^\infty \frac{x \cos bx}{\sin bx} \frac{dx}{1 + x^2} .$$

If, in place of the integral of the second member we set the value of (4.2.13), as found by Legendre, (4.2.14) will give the first value in (4.2.12).   Similarly, by setting $a = b + \alpha$, we can reduce (4.2.13) to the second value in (4.2.12).

The unusual conclusion that Cauchy reaches at this point is that the integral (4.2.14) has "two values essentially different from each other according to whether one supposes α to be zero or very small." In Legendre's view it was this trick of infinitesimals that represented the "true solution"! But it is difficult to say how convinced Cauchy was of this since, like Gauss (1816b), in this paper he showed an understanding of continuity considerably superior to that of his contemporaries. He was already in possession of the idea of the continuity of a function that he would publish in the *Cours d'analyse*. And it was precisely on this point, as we have seen, that he based the second part of his paper.

In particular, in order to study the indeterminate cases of a double integral, he put the function

$$f(M \pm iN) = P' \pm iP''$$

in the form

$$f(x) = \frac{\overline{F}(x)}{F(x)},$$

where $\overline{F}(M + iN) = Q' \pm iQ''$ and $F(M \pm iN) = R' \pm iR''$, from which he obtains,

$$(4.2.15) \quad P' = \frac{Q'R' + Q''R''}{R'^2 + R''^2}, \qquad P'' = \frac{Q''R' - Q'R''}{R'^2 + R''^2} .$$

The "critical" points were given by the zeros of $F(x)$. Equations (4.2.15) show why for such values of x one has an indeterminate form for $f(x)$ and not an infinity point.

The calculation of improper integrals by means of double integrals provides a glimpse of the developments that these studies would lead to a decade later, with Cauchy's definition of an integral in a complex domain and the creation of the theory of residues (see §4.4c below).

4.3. Complex Numbers and Complex Variables

The paper that Cauchy presented to the Academy in 1814 did not contain any explicit theory of complex variables. Since it was a work of research, Cauchy made use of complex quantities by taking them to be things already known, at least according to then common mathematical practices.   Nor did he say anything about the geometrical interpretation of complex numbers, or the possibility of interpreting his calculus with double integrals as the evaluation of an integral $\int_{ABCD} f(z)dz$ along the boundaries of a rectangle $ABCD$ in the complex plane. Nevertheless, the geometrical interpretation of complex numbers had been a topic of lively discussion among the Parisian mathematicians during these years.

According to Cauchy's own account (1847c, pp. 175) during his stay in Cherbourg he had discovered that as early as 1786 a certain Henri-Dominique Truel had found a way of representing complex numbers in a plane and had communicated his discovery to Augustin Normand, a shipbuilder at Le Havre. The discovery, however, had remained unpublished.

The Danish cartographer Gaspard Wessel (1745-1818) had also had the same idea, and in 1797 had presented a paper on it to the Danish Academy of Sciences, where it was published. But Wessel's article went completely unnoticed until it was rediscovered a hundred years later and republished in a French translation (Wessel, 1897).

Other roughly contemporary publications had also gone unnoticed, for example, by abbé M. Buée (1748-1826) in 1806, and that of the Swiss dilettante mathematician R. Argand (1768-1822), who published an anonymous pamphlet on the geometric interpretation of complex numbers at his own expense.

After he had moved to Paris, Argand discussed his discovery with Legendre, who wrote about it to J. F. Francais (1775-1833), a graduate of the École Polytechnique who was then a professor at Metz, although without mentioning Argand's name. Francais in turn published the essential elements of the idea in an article that appeared in Gergonne's *Annales* in 1813, whereupon Argand promptly claimed credit for his theory. In the following years different writers took differing positions on Argand's proposal that complex numbers be interpreted as vectors in a plane, until Argand finally published a definitive paper in which he explained his idea in detail and, as an example of its utility, provided a new demonstration of the fundamental theorem of algebra.[8]

But Argand's demonstration was not generally accepted and was soon forgotten, along with his geometrical theory of complex numbers. When Cauchy gave a demonstration of the same theorem in 1817 which in large part reused Argand's argument, he did not mention the discussion that Argand's work had generated, but claimed he had been inspired by a demonstration published in Legendre's *Théorie des nombres* (1808).

It is likely that Cauchy had found Argand's argument to be insufficiently rigorous, or had been completely ignorant of the entire affair, as Petrova has maintained (1974, p. 259). But in any case, when he set out to explain the foundations of the theory of complex numbers and of series of complex terms in the *Cours d'analyse*, he chose to take a completely different approach.

Here complex quantities are introduced in a formal manner. Cauchy first defines a "symbolic expression" as "any combination of algebraic signs that do not signify anything in themselves or to which one attributes a value different from that which it naturally has"[9] (1821, p. 153).

This expression has the aim of simplifying the calculation. "One calls *symbolic equations* all those which, taken to the letter and interpreted according to generally established conventions, are inexact or have no meaning, but from which one can deduce exact results," he writes (*ibid.*).

Cauchy concludes that among the symbolic expressions that have some interest in analysis, we must above all isolate those called imaginary.

He now considers the two symbolic expressions,

$$\cos a + \sqrt{-1} \sin a \quad \text{and} \quad \cos b + \sqrt{-1} \sin b,$$

and their product by operating with the usual rule "as if $\sqrt{-1}$ was a real quantity whose square was equal to -1" (1821, p. 154). In this way he obtains

(4.3.1)
$$\cos(a + b) + \sqrt{-1} \sin(a + b)$$
$$= (\cos a + \sqrt{-1} \sin a)(\cos b + \sqrt{-1} \sin b).$$

The quantities that appear in this last equality "cannot be interpreted according to generally established conventions, and do not repesent anything real," Cauchy observes at this point. "Taken literally, (4.3.1) is inexact and has no meaning" (*ibid.*). In order to obtain exact results, Cauchy adds, it is necessary to calculate the product of the second member and then set the real and imaginary parts equal, thus obtaining the usual form:

$$\cos(a + b) = \cos a \cos b - \sin a \sin b$$

$$\sin(a + b) = \sin a \cos b + \cos a \cos b.$$

Introducing in this way the new entities $a + ib$ $(a,b \in \mathbb{R})$, Cauchy defines with scrupulous precision the operations with such "imaginary expressions," their properties, etc. The definition of function that he gives for the case of a complex variable does not present any new elements; instead, a function of a complex variable is defined simply as an "imaginary expression" in which the real term and the coefficient of the imaginary term are real functions of x and y.

Cauchy then concerns himself with what is meant by the various usual functions, trigonometric and logarithmic, when we pass from the real field to the "imaginary."

There is an evident correspondence between the various chapters of the first and second part of the *Cours*, where Cauchy "translates" the concepts and terms established for real variables to the case of complex variables. This is done for the concept of infinitesimal and of the continuity of a function as well as for series and convergence criteria, which are transformed into conditions on modules. He finally explains the series of powers $\Sigma a_n z^n$, which are shown to converge inside the circle $|z| < 1/A$ where $A = \lim \sup|a_n|^{1/n}$.

In Chapter X Cauchy takes up the demonstration of the fundamental theorem of algebra that he had already given in 1817.

Whatever the real or imaginary values of the constants a_0, ..., a_n are, the equation

$$(4.3.2) \quad f(x) = a_0 x^n + a_1 x^{n-1} + \cdots + a_{n-1} x + a_n = 0,$$

in which n designates a whole number equal to or greater than one, always has real or imaginary roots" (1821, p. 331).

Setting $x = u + iv$ and $f(x) = \Phi(u,v) + i\chi(u,v)$, it is necessary to show that there exists at least one $x_0 = u_0 + iv_0$ such that

$$(4.3.3) \quad 0 = F(u_0,v_0) = |f(x_0)|^2 = \Phi^2(u_0,v_0) + \chi^2(u_0,v_0).$$

Like Gauss (see §4.1), Cauchy bases his demonstration on an argument of continuity that he takes to be evident: given $F(u,v) > 0$, continuous and increasing without limit for increasing values of $|x|$, it "will one or more times arrive at a certain lower limit which it will never surpass" (1821, p. 154). Let A be such a minimum and let $x_0 = u_0 + iv_0$ such that $F(u_0,v_0) = A$. Then, if h, $k \in \mathbb{R}$ and α is an infinitesimal,

$$(4.3.4) \quad \Delta F(u_0,v_0) = F(u_0 + \alpha h, v_0 + \alpha k) - F(u_0,v_0) \geqslant 0.$$

At this point Cauchy expands (4.3.4) according to the power of the increment $\alpha(h + ik) = \alpha\rho(\cos\theta + i\sin\theta)$. The sign of $\Delta F(u_0,v_0)$ is then given by the sign of the term of the expansion containing the power of the lowest degree of the infinitesimal α. Let m be such a term. Then

$$\Delta F(u_0,v_0) = KA^{1/2}\alpha^m \rho^m \cos(\phi + m\theta) + \cdots$$

where $K > 0$ and ϕ is a certain determinate angle in the expansion of (4.3.4) in polar coordinates.

If $A \neq 0$, θ being completely indeterminate, $\cos(\phi + m\theta)$ can assume negative values for suitable values of θ, and $\Delta F(u_0,v_0)$ will be negative, which is impossible. It necessarily follows that $A = F(u_0,v_0) = 0$ and x_0 is a root of (4.3.2).

Cauchy can then show that (4.3.2) can be split into the product of a_0 for n linear factors of the type $x - a - ib$, which are not necessarily all distinct, and finally that an algebraic equation of degree n has n roots in $\mathbb{C}$ when each is taken with its own respective multiplicity. In the next chapter Cauchy applies these results to splitting the rational functions $f(x)/F(x)$ into simple fractions. This is the same problem preliminary to the calculation of integrals of rational functions with which D'Alembert had begun his demonstration of the fundamental theorem of algebra (see §1.2).

But, as with the differential calculus, so also the integral calculus was not included among the topics treated in the "analyse algébrique." The *Cours d'analyse* instead concluded with a final chapter on the transformation of rational functions into recurrent series and left the problem of integration to the *Résumé* of 1823.

4.4. Cauchy's Theory of Integration

In 1823 Cauchy published a paper on the integration of linear differential equations with constant coefficients (1823b). In the *Observations générales et additions* he wrote,

> In the paper you have just read, I consider every definite integral taken between two given limits as being nothing other than the sum of the infinitely small values of the differential expression placed under the $\int$ sign, which corresponds to the various values of the variable enclosed within the limits in question. When we adopt this manner of conceiving definite integrals, we easily demonstrate that a similar integral has a unique and finite value whenever, the two limits of the variable being finite quantities, the function under the $\int$ sign itself remains finite and continuous in the entire interval included between these limits (1823b, pp. 333-4).

In a postscript to this paper he further observed,

> We are naturally led by the theory of quadratures to consider each definite integral which is taken between two real limits as being nothing other than the sum of the infinitely small values of the differential expression placed under the $\int$ sign which correspond to the various real values of the variable which are included between the limits in question. Now, it seems to me that this manner of conceiving a definite integral ought to be adopted in preference, as I have done, because it is equally suitable to all cases, even to those in which we cannot pass generally from the function placed under the $\int$ sign to the primitive function. In addition, it has the advantage of always giving real values for the integrals which correspond to real functions. Finally, it allows us to easily separate each imaginary equation into two real equations. All that would no longer be so if we considered a definite integral taken between two real limits as necessarily equivalent to the difference of the extreme values of a discontinuous primitive function, or if we made the variable pass from one limit to another by a series of imaginary values (*ibid.*, p. 354).

This latter was for example the method used by Poisson (1820a) to calculate the integral

$$(4.4.1) \qquad \int_{-1}^{1} \frac{dx}{x}$$

by setting $x = -(\cos z + i \sin z)$ taken from $z = 0$ to $z = (2n + 1)\pi$.

By taking x from -1 to $+1$, we avoid passing through the origin (where the integrand is not defined). After then changing the variable, we have

$$dx = -i(\cos z + i \sin z)dz,$$

and

$$(4.4.2) \qquad \int_{-1}^{1} \frac{dx}{x} = \int_{0}^{(2n+1)\pi} i \, dz = +i(2n + 1)\pi.$$

It was now necessary to make these ideas rigorous and to establish a systematic theory of integration, both real and complex. This was the objective Cauchy pursued at the beginning of the 1820s.

a) The Definite Integral

Having set out the foundation of the differential calculus in the first part of the *Résumé* of 1823 (see §3.6), in the second part Cauchy turned to the integral calculus. Here we find the definition of the integral as a limit of sums that has become "classic."

Cauchy considers a function of a real variable $y = f(x)$ continuous in the interval $[x_0, X]$. Given $n - 1$ values of x, where

$$x_0 < x_1 < x_2 < \cdots < x_{n-1} < X,$$

he considers the sum S of the products

$$(4.4.3) \qquad S = \sum_{i=1}^{n} (x_i - x_{i-1})f(x_{i-1}), \qquad x_n = X.$$

Then S depends: (a) on the number n of the intervals (x_i, x_{i+1}); (b) on the way of dividing $[x_0, X]$ that is adopted.

"Now it is important to note," Cauchy writes, "that if the numerical values of the elements $[X_{i+1} - x_i]$ become very small and the number n very large, the method of division will have only an insensible influence on the value of S" (Cauchy, 1823a, pp. 122-3).

In order to demonstrate this, Cauchy notes that for a finer subdivision of the interval one has

$$(4.4.4) \qquad S = \sum_{i=1}^{n} (x_i - x_{i-1})f(x_{i-1} + \theta_{i-1}(x_i - x_{i-1})), \qquad |\theta_{i-1}| < 1.$$

By then setting

$$f(x_{i-1} + \theta_{i-1}(x_i - x_{i-1})) = f(x_{i-1}) + \epsilon_i, \qquad (\epsilon_i \text{ small})$$

(4.4.4) becomes

$$S = \sum_{i=1}^{n} (x_i - x_{i-1}) f(x_{i-1}) + \sum_{i=1}^{n} \epsilon_i (x_i - x_{i-1}).$$

Now, Cauchy observes, if the differences $(x_i - x_{i-1})$ are "very small," because $f(x)$ is continuous (in fact, uniformly continuous), every ϵ_i will "differ very little" from zero. By repeating the reasoning and implicitly reverting to the 'Cauchy criterion' that he had demonstrated in the *Cours d'analyse* (see §3.4), Cauchy concludes,

> if we let the numerical values of these elements decrease indefinitely by increasing their number, the value of S will end by being sensibly constant or, in other words, it will end by attaining a certain limit that will depend uniquely on the form of the function $f(x)$ and the extreme values x_0, X attributed to the variable x. This limit is what we call a *definite integral* (Cauchy, 1823a, p. 125).

In the cases in which the extremes of the interval become infinite or $f(x)$ does not remain finite and continuous between x_0 and X, "we no longer see what meaning should be attached to the notation that serves to represent the limit of S in general" (*ibid.*, p. 141). In order to eliminate all uncertainty, Cauchy says, "it is sufficient to extend by analogy the equations

(4.4.5) $\displaystyle\int_{x_0}^{X} f(x)dx = \int_{x_0}^{x_1} f(x)dx + \int_{x_1}^{x_2} f(x)dx + \cdots + \int_{x_{n-1}}^{X} f(x)dx$

and

(4.4.6) $\displaystyle\int_{x_0}^{X} f(x)dx = \lim_{\substack{\xi \to X \\ \xi_0 \to x_0}} \int_{\xi_0}^{\xi} f(x)dx$

to those cases where they cannot be rigorously demonstrated" (*ibid.*).

If $f(x)$ becomes infinite at the points $x_i \in [x_0, X]$ $(i = 1, ..., m)$, Cauchy asserts, we consider an infinitesimal number ϵ and the positive constants v_i, μ_i. Then equation (4.4.5) becomes

(4.4.7) $\displaystyle\int_{x_0}^{X} f(x)dx = \lim_{\epsilon \to 0} \left[\int_{x_0}^{x_1 - \epsilon \mu_1} f(x)dx + \int_{x_1 + \epsilon v_1}^{x_2 - \epsilon \mu_2} f(x)dx \right.$

$$\left. + \cdots + \int_{x_m + \epsilon v_m}^{X} f(x)dx \right],$$

and if the extremes of integration become infinite, (4.4.6) becomes

$$\int_{-\infty}^{+\infty} f(x)dx = \lim_{\varepsilon \to 0}\left[\int_{-1/\ \varepsilon\mu}^{x_1 - \ \varepsilon\mu_1} f(x)dx + \int_{x_1 + \ \varepsilon\ v_1}^{x_2 - \ \varepsilon\mu_2} f(x)dx\right.$$

$$\left. + \cdots + \int_{x_m + \ \varepsilon v_m}^{1/\ \varepsilon v} f(x)dx\right],$$

for v, μ positive constants

Using the terminology that he had introduced in (1827a), Cauchy calls integrals with one or more points of infinity in the interval of integration or those with infinite limits of integration "singulars." The principal value of such an integral is given by setting $v_i = \mu_i = 1$.

In lesson 25 of the *Résumé* Cauchy sets out the theory of such integrals and in the following lesson introduces the concept of an indefinite integral.

By rendering Leibniz' original conception of the integral as the sum of infinitesimal elements rigorous, Cauchy clearly distances himself from the then common practice of *assuming* the existence of the indefinite integral *in the first place* and deriving the definite integral from this according to the classic formula

$$\int_a^b f(x)dx = F(b) - F(a),$$

where $F'(x) = f(x)$.

In the *Avertissement* to the *Résumé* Cauchy had written, "it seemed to me necessary to demonstrate in general the existence of the *integrals* or *primitive functions* before making known their various properties. To arrive at this end, it was first necessary to establish the idea of *integrals taken between given limits* or *definite integrals*" (1823a, p. 10).

In fact, by setting

$$\mathcal{F}(x) = \int_{x_0}^x f(x)dx,$$

we have

$$\mathcal{F}(x) = (x - x_0)f(x_0 + \theta(x - x_0)), \qquad |\theta| < 1.$$

Then

(4.4.9) $$\int_{x_0}^{x+\alpha} f(x)dx - \int_{x_0}^x f(x)dx = \alpha f(x + \theta\alpha)$$

or

$$\mathcal{F}(x + \alpha) - \mathcal{F}(x) = \alpha f(x + \theta\alpha).$$

If $f(x)$ is finite and continuous in the neighborhood of a point x, the same will be true for $\mathcal{F}(x)$, and by passing to the limit as $\alpha \to 0$, (4.4.9) becomes

(4.4.10) $\mathcal{F}'(x) = f(x)$.

Cauchy then shows that, if $\omega'(x) = 0$, it follows that $\omega = $ constant and hence that "the general value of y," the solution of the equation

$$dy = f(x)dx,$$

is given by

$$y = \int_{x_0}^{x} f(x)dx + \text{constant}.$$

This is what he calls the indefinite integral.

Naturally things become complicated when $f(x)$ is no longer continuous in $[x_0, X]$ in an infinite number of points, a problem that will present all of its difficulties in the study of trigonometric series. However, the fundamental point which it is necessary to emphasize here is that Cauchy marks the turning point, from considering integration as the inverse operation of derivation to a modern theory of measure, where the primary object of interest is the integral.

In addition, Cauchy's approach to integration also explains the central interest he had in the questions of the *existence* of solutions for differential equations.

In the final lessons of the *Résumé* Cauchy discusses the possibility of derivation under the integral sign (lesson 33), where he asserts, "it is sufficient to differentiate under the $\int$ sign" in order to obtain

$$\frac{\partial}{\partial y} \int_{x_0}^{x} f(x,y)dx = \int_{x_0}^{x} \frac{\partial f(x,y)}{\partial y} dx.$$

In lesson 34 he takes up the principal theme of (1827a) by treating the "comparison of two kinds of simple integrals that result in certain cases from a double integral." He finally points out, for the case of simple poles, the first developments in the field of complex integration; while the last lessons (36-38) are devoted to the discussion of the Taylor series (in one or more variables) with the remainder in the form of definite integral (see §3.6).

b) The Integration of Ordinary Differential Equations

The theory of ordinary differential equations was the subject of the lessons that Cauchy gave to students in their second year at the École Polytechnique. While the *Cours d'analyse*, the *Résumé*, and the *Leçons sur l'application du calcul infinitesimal à la géometrie* contain the subjects that Cauchy dealt with in the first year of study, the lack of documentation has long hindered any attempt to form a precise idea of his second-year lectures. The recent discovery of proof sheets for these lectures (Cauchy, 1981) now allows a fairly complete reconstruction of Cauchy's entire course on analysis.

In keeping with the official program required by the school's Council on Instruction, Cauchy's initial lectures for the second-year course presented the "classical" results of "exact" integration, by means of which one can "in certain cases" integrate a differential equation of the first order,

(4.4.11) $P(x,y)dx + Q(x,y)dy = 0.$

For example, if

$$\frac{\partial P}{\partial y} = \frac{\partial Q}{\partial x}$$

then (4.4.11) is the exact differential of a $u(x,y)$ and, by means of "immediate integration," the integral being sought is $u(x,y) =$ constant. This is still the case when the first member of (4.4.11) can be reduced to an exact differential by means of an integrating factor or by a change of variables.

At this point, Cauchy defines the concepts of general, particular, and singular integrals of ordinary differential equations, and applies these methods to the integration of the linear equation

$$y'\Phi(x) + y\chi(x) + \Psi(x) = 0,$$

and of the homogeneous equation of the first order.

The integration of the equation

(4.4.12) $y = f(x,y'),$

which Cauchy presents in lesson 3, can also be reduced to a generalization of the method of substitution. In fact, by assuming y' to be a new variable and differentiating, we obtain

(4.4.13) $y'dx = \Phi(x,y')dx + \chi(x,y')dy',$

where $\Phi(x,y') = [\partial f(x,y')/\partial x]$ and $\chi(x,y') = [\partial f(x,y')/\partial y']$.

"Having established this," Cauchy observes at this point, "if by any method one is able to find the general integral and the singular integrals of equation (4.4.13), it is only necessary to eliminate y' between these integrals and equation (4.4.12) in order to find its general integral and its singular integrals" (Cauchy, 1981, p. 15).

Cauchy illustrates this method with Clairaut's equation,

$$y = xy' + f(y'),$$

whose general integral is

$$y = Cx + f(C), \quad C \text{ constant},$$

and whose singular integrals are obtained by eliminating y' from

$$y = xy' + f(y')$$

$$0 = x + \frac{\partial f(y')}{\partial x} .$$

Cauchy returns to this example in lesson 5, where he shows that by eliminating the constant C from the equations $F(x,y,C) = 0$ and $\partial F/\partial C = 0$ one obtains the singular integrals derived from the general integral.

In the first lessons of the course Cauchy limits himself to explaining in a clear and concise manner the traditional methods for determining the general and singular integrals of differential equations without discussing their domain of existence. But, beginning with lesson 6, the course takes on an original character; here Cauchy begins to present the problem of finding an integral of $y' = f(x,y)$ that satisfies the condition of assuming a given value y_0 for an assigned value x_0 of x. This is the so-called 'Cauchy Problem.' In the next lesson he shows that there exists a general integral of the differential equation if there exists $y = \bar{F}(x)$ that resolves the 'Cauchy Problem'. With this, as Gilain has observed, Cauchy "reverses the ordinary procedure which consisted of finding the general integral before anything else" (In: Cauchy, 1981, p. xxxiv). Moreover, Cauchy's overall problem becomes that of demonstrating the existence of a solution to the 'Cauchy Problem.' Until then mathematicians had sought to find the expression of a similar integral whose existence was admitted without question.

Cauchy gives the existence proof by means of a method that he calls "approximation" in contrast to the "exact" integration that he had set out in the preceding lessons. Cauchy's technique thus makes rigorous an idea that had already been present since Euler, that of considering a polygon that approximates the integral curve as closely as desired.

In fact, if $y = \bar{F}(x)$ is a solution of the 'Cauchy Problem' for the equation $y' = f(x,y)$, then $\bar{F}(X)$ for $x = X$ "will differ very little" from the value Y found by eliminating y_i from the equations

$$y_1 - y_0 = (x_1 - x_0)f(x_0,y_0)$$

$$\vdots \qquad \vdots \qquad \vdots$$

$$y_i - y_{i-1} = (x_i - x_{i-1})f(x_{i-1},y_{i-1})$$

$$\vdots \qquad \vdots \qquad \vdots$$

$$Y - y_{n-1} = (X - x_{n-1})f(x_{n-1},y_{n-1}),$$

where $(x_i - x_{i-1})$ is very small.

Setting $Y = \bar{F}(x_0, ..., x_{n-1},X,y_0)$ as the value thus determined, Cauchy demonstrates that if the function $f(x,y)$ is continuous[10] and $|f(x,y)| < A$ on $[x_0,X]$, then Y can be given in the form

$$Y = y_0 + (X - x_0)f(x_0 + \theta(X - x_0), y_0 + \theta_1 A(X - x_0)),$$

where $|\theta| < 1$, $|\theta_1| < 1$.

Under the additional hypothesis that both $\partial f(x,y)/\partial y$ and $f(x,y)$ are finite, continuous, and bounded in the same region of the x,y plane, and by utilizing the theorem of finite increments, Cauchy further shows that a new subdivision of the interval $[x_0,X]$ into subintervals of a smaller size will give a new value Y' that differs "very little" from Y.

He finally shows[11] that "if one makes the numerical values of the elements of the difference $X - x_0$ decrease to infinity, the value of Y determined by the equation $Y = \mathcal{F}(x_0, ...,X,y_0)$ will converge to a limit that will depend uniquely on the three quantities x_0, X, and y_0" (Cauchy, 1981, p. 47).

On the basis of these results, Cauchy proves in the following lesson the theorem on the existence of the solution to the 'Cauchy Problem' in the interval $(x_0,x_0 + a)$ under the hypothesis that both $f(x,y)$ and $\partial f(x,y)/\partial y$ are continuous and bounded for $x_0 < x < x_0 + a$ and $y_0 - Aa \leqslant y < y_0 + Aa$, A being a majorant of $|f(x,y)|$ in the interval.

This is a theorem of local existence,[12] which poses to Cauchy the problem of the global existence or nonexistence of the solution. To this end he studies the conditions of the continuation of such a solution to an interval $x_0 + a_1$, with $a_1 > a$, where the conditions of regularity required by the theorem exist.

By repeating his previous reasoning, Cauchy considers the sequence

(4.4.14) $x_0 + a$, $x_0 + a_1$, $x_0 + a_2$, ...,

which will diverge or converge to a certain limit Ξ. In the first case X can increase in absolute value beyond every given limit, while in the second case X "can indefinitely approach the limit" Ξ.

Similarly for $y = \mathcal{F}(x)$, "it will be necessary either that the quantity $\mathcal{F}(\Xi)$ be infinite, that one of the functions

$$f(x,\mathcal{F}(x)), \quad \frac{\partial f(x,\mathcal{F}(x))}{\partial y}$$

becomes infinite for the particular value $x = \Xi$, or finally that one of these functions becomes discontinuous in the neighborhood of the particular value in question"[13] (Cauchy, 1981, p. 63).

On the other hand, Cauchy shows that the solution cannot be extended to the entire real axis by producing the counterexample of the equation $dy = dx/(x + y)$ with the initial conditions $x_0 = 1$, $y_0 = 0$.

As for the uniqueness of the solution to the 'Cauchy Problem,' Cauchy limits himself to emphasizing in lesson 10 the global nonuniqueness character of his result. "Moreover, one can," he says, "even under the given hypothesis, conceive of different functions of x which, being similarly able to fulfill the stated conditions, coincide in the neighborhood of the particular value $x = x_0$ and diverge for certain values of x sensibly different from x_0."

But, as Gilain has rightly observed, the example that Cauchy gives, the equation

$$(x + 1)dy - (y + 1)dx = 0,$$

with the conditions $x_0 = 0$, $y_0 = 0$, shows Cauchy's ambiguity and imprecision in this regard.

After devoting two additional lessons to a discussion of "the distinctive character" of singular integrals and the various methods that can be used in their numerical calculation, in lesson 13 Cauchy finally turns to the problem of the local existence of the solution of the 'Cauchy Problem' for a system of linear differential equations. At this point, however, the printed text stops.

Why? Gilain's suggestion seems probable enough: the methods and contents of Cauchy's lessons received strong opposition, not only from the students but also from the Council of Instruction (Cauchy, 1981, p. xix).

This was not the first time Cauchy had been reproached by the scholastic authorities. Already in 1820 the Directors of the school had told him to conform to the program and not waste the time that should be devoted to teaching the applications of the calculus (which was more important for future engineers) in discussing abstruse questions of rigor, "a luxury of analysis undoubtedly suitable for papers to be read at the Institute, but superfluous for teaching the students of the École" (In: Belhoste, 1984, p. 36).

The first pages of Cauchy's second-year lessons revived such criticisms, and Cauchy's reaction was immediate: "M. Cauchy announces that, in order to conform to the wishes of the Council, he will no longer give, has he has until the present, completely rigorous demonstrations" (Cauchy, 1981, p. xix).

This may explain why Cauchy lost interest in publishing his lectures, which could no longer adequately reflect his conceptions of rigor. What today appears as the first step towards the teaching of modern rigor in mathematics, was criticized by Cauchy's contemporaries as a "luxury of analysis" or even an unrecommended "lack of clarity," if not counterproductive for the students of the École Polytechnique.

The pages of Cauchy's second-year lectures make evident the profound conceptual unity that stood at the foundation of his course of analysis. In particular, the methods for integrating differential equations of the first order directly recall the theory of integration presented in the *Résumé*.

In both cases, Cauchy emphasizes the importance of overturning the traditional approaches and above all of demonstrating the existence of the definite integral or of the solution of the 'Cauchy Problem' in order to then return to the concept of the indefinite integral or of the general integral of a differential equation. In both cases this translated into a demonstration of the existence of a limit, which Cauchy established in both cases by reverting to the (implicit) concept of uniform continuity and to the use of "Cauchy's criterion."

In this way, the rejection of the method of series that had been anticipated in the *Cours d'analyse* was transformed into an exposition of the principles of the differential and integral calculus without recourse to series and, similarly, into a rejection of the method of series for the integration of differential equations that was in widespread use at the time. In place of series, for Cauchy the concept of limit became the central concept in all of analysis. Even the integral $\int_{z_0}^{z_1} f(z)dz$, $z \in \mathbb{C}$, could be based on this concept, as Cauchy showed in 1825 in a paper that marked the beginning of his dominating interest in complex analysis.

Significantly, after the publication of his second-year lectures had been interrupted, Cauchy limited himself to recalling his method of integration "by approximation" in the initial pages of his Prague paper (1835), which was primarily devoted to a presentation of a second method of existence proofs and was related to his studies of complex functions[14] (see §4.5 below).

c) Complex Integration

Cauchy's studies on integration, which were begun in the 1814 paper and subsequently expanded in his lessons at the École Polytechnique, culminated in a pamphlet that in many ways represents his masterwork (Cauchy, 1825). Its object was the definite integral whose limits of integration are complex numbers. In it he writes,

In order to fix generally the meaning of the notation

$$\int_a^c f(x)dx,$$

where a and c designate real limits and $f(x)$ a real or imaginary function of the variable x,[15] it is sufficient to consider the definite integral represented by this notation as equivalent to the limit or to one of the limits towards which the sum

$$(x_1 - a)f(a) + (x_2 - x_1)f(x_1) + \cdots + (c - x_{n-1})f(x_{n-1})$$

converges when the elements of the difference $c - a$, that is to say

$$x_1 - a, \ x_2 - x_1, \ ..., \ c - x_{n-1},$$

these quantities having the same sign as the difference, receive ever smaller numerical values. Therefore, in order to include in the same definition integrals between real limits and integrals taken between imaginary limits, it is necessary to represent by the notation

(4.4.15) $\displaystyle\int_{a+ib}^{c+id} f(z)dz$

the limit or one of the limits towards which the sum of the products of the form

$$[(x_1 - a) + i(y_1 - b)]f(a + ib)$$

$$[(x_2 - x_1) + i(y_2 - y_1)]f(x_1 + iy_1)$$

. . .

$$[(c - x_{n-1}) + i(d - y_{n-1})]f(x_{n-1} + iy_{n-1})$$

converge, when, in each of the two series,

$$a, x_1, x_2, ..., x_{n-1}, c$$

$$b, y_1, y_2, ..., y_{n-1}, d$$

these being composed of terms that always go on increasing or decreasing from the first to the last, these same terms indefinitely approach each other, and their number increases continuously (Cauchy, 1825, pp. 42-3).

One can think of obtaining the values x_i and y_i by means of two continuous monotonically increasing or decreasing functions of the same variable t,

$$x = \phi(t), \qquad y = \psi(t) \qquad t \in [\alpha, \beta]$$

such that

$$\phi(\alpha) = a \qquad \psi(\alpha) = b; \qquad \phi(\beta) = c \qquad \psi(\beta) = d.$$

If we indicate the integral (4.4.15) by $A + iB$ and substitute for x and y the functions of t defined above, we find for this integral the form,

(4.4.16) $A + iB = \displaystyle\int_{\alpha}^{\beta} [\phi'(t) + i\psi'(t)]f[\phi(t) + i\psi(t)]dt,$

which, by setting

$$\phi'(t) = x', \qquad \psi'(t) = y',$$

can be written

(4.4.17) $A + iB = \displaystyle\int_{\alpha}^{\beta} (x' + iy')f(x + iy)dt.$

We now suppose that the function $f(x + iy)$ remains finite and continuous whenever x remains between the limits a and c and y between the limits b and d. In this particular case, we can easily prove that *the value of the integral* (4.4.15), *that is to say, the imaginary expression $A + iB$, is independent of the nature of the functions $x = \phi(t)$ and $y = \psi(t)$* (*ibid.*, p. 44).

This is the form in which Cauchy states his 'integral theorem.' The usual formulation can be readily obtained if we observe that the functions $x = \phi(t)$ and $y = \psi(t)$ can be thought of as the parametrical equations of a curve in the (x,y) plane joining the points $(a + ib)$ and $(c + id)$.

Cauchy then establishes that the integral (4.4.15) is independent of the path of integration. The hypotheses made for $f(z)$ -- that it is finite and continuous -- are however insufficient. In the demonstration Cauchy in fact tacitly assumes and implicitly uses both the existence and the continuity of $f'(z)$.

In order to demonstrate the theorem Cauchy resorts to the method of variations; he considers the increment of two functions $x(t)$ and $y(t)$ given by

(4.4.18) $\varepsilon u, \quad \varepsilon v \qquad u = u(t), \, v = v(t),$

where   is a first-order infinitesimal and $u(\alpha) = u(\beta) = v(\alpha) = v(\beta) = 0$.

By expanding the corresponding variation of the integral (4.4.17) "in ascending powers of ε,"[16] Cauchy obtains a series whose first-order infinitesimal part with respect to ε is given by

(4.4.19) $\varepsilon \int_{\alpha}^{\beta} [(u+iv)(x'+iy)f'(x+iy) + (u'+iv')f(x+iy)]dt + 0(\varepsilon^2).$

This integral is zero, as we can see by integrating by parts, and hence the increment of the integral (4.4.17) "is an infinitesimal of a second or higher order" (*ibid.*, p. 44). We can arrive at the same result in a direct way, Cauchy adds, observing that (4.4.19) is "simply the total variation of the integral (4.4.17)" (*ibid.*, p. 45). This variation is zero because the integrand is an exact differential. If the conditions of regularity of $f(x + iy)$ are no longer satisfied and $z = a + ib$ is a simple pole of $f(x + iy)$, Cauchy considers the limit

(4.4.20) $c = \lim_{\substack{x \to a \\ y \to b}} [(x - a) + i(y - b)]f(x + iy)$

which, "without noticeable error," is given by

$c = \varepsilon f(a + ib + \varepsilon).$

By first considering the variation of x and y given by εu, εv and the corresponding variation of the integral (4.4.17), Cauchy obtains

$$A' + iB' - (A + iB) = \int_\alpha^\beta [x' + \epsilon u' + i(y' + \epsilon v')] \cdot$$
(4.4.21)

$$\cdot f[(x + \epsilon u + i(y + \epsilon v)]dt - \int_\alpha^\beta (x' + iy')f(x + iy)dt,$$

which is always "negligible" except in the neighborhood of the pole $a + ib$.

The theory of singular integrals developed in the *Résumé* (1823) provided Cauchy with a suitable tool for evaluating (4.4.21).

By setting $t = \tau$ as the value for which $x(\tau) = a$ and $y(\tau) = b$, Cauchy considers $t = \tau + \epsilon w$, for infinitesimal w. From this he obtains

(4.4.22) $(x' + iy')f(x + iy) = c \dfrac{x + iy}{x - a + i(y - b)} = \dfrac{c}{\epsilon} \dfrac{1}{w}$

and

(4.4.23) $[x' + \epsilon u' + i(y' + \epsilon v')]f[x + \epsilon u + i(y + \epsilon v)] = \dfrac{c}{\epsilon} \dfrac{1}{w + \lambda + i\mu}$

with

$$\lambda + i\mu = \dfrac{\gamma + i\delta}{\alpha + i\beta},$$

where $\alpha = x'(\tau)$, $\beta = y'(\tau)$, $\gamma = u(\tau)$, and $\delta = v(\tau)$. Equation (4.4.21) is thus transformed into

(4.4.24) $A' + iB' - (A + iB) = c\displaystyle\int_{-1/\sqrt{\epsilon}}^{1/\sqrt{\epsilon}} \dfrac{dw}{w + \lambda + i\mu} - c\displaystyle\int_{-1/\sqrt{\epsilon}}^{1/\sqrt{\epsilon}} \dfrac{dw}{w}.$

Since in the second integral the integrand becomes infinite when $w = 0$, if we pass to the principal value of the integral (which is zero) and then set $\epsilon = 0$, we finally obtain

(4.4.25) $A' + iB' - (A + iB) = -ic\displaystyle\int_{-\infty}^{+\infty} \dfrac{dw}{(w + \lambda)^2 + \mu^2} = \pm\pi ci,$

where the sign depends on the sign of $\alpha\delta - \beta\gamma$.

In this way there appears the concept of the residue of a function in a simple pole (4.4.20) and, for it, the value of the integral along a closed path around the pole.

In the subsequent paragraphs Cauchy defines in a similar manner the residue for the case of multiple poles of order m in $x = x_1$ and then shows how to apply (4.4.25) to calculating improper integrals, just as he had done in (Cauchy, 1827a).

Cauchy's guiding idea was to extend the interval of integration (for a real integral) to a closed path in the complex field within which the function is holomorphic except at isolated points, and then to evaluate the integral by means of the theorem of residues.

In this way he succeeded in showing, for example, that

$$\int_0^\infty \frac{\sin x}{x} = \frac{\pi}{2},$$

or in calculating Poisson (or Fourier) integrals, such as

$$\int_{-\infty}^{+\infty} g(x)e^{iax}dx, \qquad a \in \mathbb{R}$$

The calculus of residues thus becomes, in Cauchy's eyes, "a new kind of calculation similar to the infinitesimal calculus." This was in fact what Cauchy wrote in the title of a note that appeared in the first issue of the *Exercices de mathématiques* (1826a), a sort of 'personal review,' which began to appear in this year.

In this article Cauchy for the first time introduced the term "residue," preceded by a characteristic symbol, $\mathcal{E}$, that indicates "the extraction of the residue" of a function, while with "integral residue taken between the given limits" $\overset{X}{\underset{0}{\mathcal{E}}}\overset{Y}{\underset{Y}{\mathcal{E}}}$ he indicated the sum of the ^{0} residues of a function at the poles (the only singularities known to him) within a given rectangular domain. He finally defines the property of the "operator" $\mathcal{E}$.

If $x = x_1$ is a simple pole of $f(x)$, Cauchy defines the "residue of the function $f(x)$ for $x = x_1$" as

(4.4.26) $\quad \mathcal{E} f(x_1 + \varepsilon)\big|_{\varepsilon =0}$

and, for a pole of order m, similarly

(4.4.27) $\quad \dfrac{1}{(m-1)!} \dfrac{d^{m-1}[\varepsilon^m f(x_1 + \varepsilon)]}{d\varepsilon^{m-1}}\bigg|_{\varepsilon =0}$

which can immediately be translated into modern terms for a function $f(z) = g(z)/(z - c)^n$ with $g(c) \neq 0$, respectively, as

(4.4.26') $\quad \mathrm{Res}(f,c) = \lim\limits_{z \to 0} (z - c)f(z) = g(c)$, for a simple pole,

and

(4.4.27') $\quad \mathrm{Res}(f,c) = \dfrac{1}{(m-1)!} g^{(m-1)}(c)$, for a pole of order m.

The calculus of residues was applicable to many types of questions, Cauchy asserted. "For example, one immediately derives with the calculus of residues Lagrange's interpolation formula, the splitting of rational fractions in the case of equal or unequal roots, formulas suitable for determining the values of definite integrals, the sum of a multitude of series and particularly of periodic series, the integration of linear equations with finite or infinitely small differences and constant coefficients, with or without a variable last term, Lagrange's series and other series of the same kind, the resolution of algebraic or transcendental equations, etc. ..." (Cauchy, 1826a, p. 24).

This program was realized by Cauchy in a series of articles that appeared in the *Exercices*. In addition to an imposing number of new results, Cauchy also published some refinements to the theory of residues. In particular, he showed that

$$\int f(z)dz = 2\pi i \, \mathcal{E}f(z),$$

first for the case of a rectangle (1826b), and then for a circle in the complex plane.[17]

Thus by the end of the 1820s Cauchy had progressed well beyond his contemporaries in the study of complex analysis. Nevertheless, he could not imagine that the 'integral theorem' -- to our eyes the most important result as regards complex integration--had already been known to Gauss. The latter, in fact, in a letter he had written to his friend Bessel (1784-1846) in 1811, had said,

> Now what should one think of $\int \phi x \, dx$ for $x = a + bi$? Obviously, if we want to begin from clear concepts, we must assume that x passes through infinitely small increments (each of the form $\alpha + \beta i$) from the value for which the integral is 0 to $x = a + bi$, and then sum all the $\phi x \, dx$. In this way the meaning is completely established. But the passage can occur in infinitely many ways: just as one can think of the entire domain of all real magnitudes as an infinite straight line, so one can make the entire domain of all magnitudes, real and imaginary, meaningful as an infinite plane, wherein each point determined by abscissa $= a$ and ordinate $= b$ represents the magnitude $a + bi$ as it were. The continuous passage from one value of x to another $a + bi$ accordingly occurs along a line and is consequently possible in infinitely many ways. I now assert that the integral $\int \phi x \, dx$ always maintains a single value after two different passages, if ϕx nowhere $= \infty$ within the region enclosed between the lines representing the two passages. This is a very beautiful theorem,[18] for which I will give a not difficult proof at a suitable opportunity (Gauss, 1811, pp. 90-1).

But it is well known that Gauss, far from the competitive climate of the Parisian mathematicians, published his works with great parsimony and only when these had attained the perfection he desired, following the motto "*pauca sed matura*" (few but ripe) that he imposed on himself.

4.5. The "Calcul des limites" and Differential Equations

Following the July revolution of 1830, which lead to the fall of the
Bourbon dynasty and the rise to power of Philippe d'Orleans,
Cauchy began to think about leaving Paris. He appeared rarely at
the meetings of the Academy and at the École Polytechnique, whose
students had participated actively in the Revolution. He eventually
left France under conditions of self-exile and stayed for a time in
Switzerland, where he considered founding an academy and college
founded on monarchic and religious principles at Freibourg.

In order to seek support for this project from Catholic sovereigns
and people close to the Bourbons, Cauchy traveled to Italy. But
instead of receiving aid for his projected academy, he was instead
offered a chair of "higher physics" by the king of Piedmont and
Sardinia. Cauchy accepted this offer and taught at Turin from the
beginning of 1832 until the summer of the following year.

The Italian mathematical community was at this time largely
dominated by Lagrange's influence and Cauchy felt the contrast
with his new ideas of analysis very sharply.

On his arrival in Italy Cauchy presented a long article that did not
contain any particularly new results, but which nevertheless sounded
completely new to the Italian mathematicians (Cauchy, 1830).
Cauchy in fact presented a succinct summary of his ideas as they
had been set out in his courses at the Ecole, which included an open
disagreement with Lagrange's ideas.[19]

One of the primary things Cauchy insisted on was of course rigor.
It was the lack of adequate rigor in the methods commonly used in
astronomy and celestial mechanics, together with the extraordinary
length and difficulty of the calculations, that motivated the first
paper he presented to the Academy of Science of Turin in 1831.[20] "I
consequently think that the geometers and astronomers will attach
some value to my work," Cauchy here asserted, "when they
understand that I have been able to establish general principles and
a simple method for the expansion of functions, whether explicit or
implicit, with the aid of which one can not only rigorously
demonstrate the formulas and show the conditions of their existence,
but also fix the limits of the errors that one commits by neglecting
the remainders that must complete the series" (Cauchy, 1841a, p. 51).
(Hence the name, "calculus of limits," given to it by Cauchy.)

But Cauchy's faith in the acceptance of his work was misplaced, as
we read in the report that appeared in the Mémoires of the
Academy. "The reading of the paper was interrupted by several
verbal discussions between the author and Mr. Plana," it says
(Terracini, 1957, p. 186). Plana (1781-1864), who had been a student
of Laplace and Monge at the École Polytechnique and was a
convinced follower of their methods, certainly did not agree when
Cauchy accused the author of the *Mécanique céleste* of a lack of
rigor.[21]

Cauchy's work found a friendler reception in Milan. After he had finished editing it between 1832 and 1833, it was translated into Italian and published in 1834 in the *Opuscoli matematici e fisici*, a new journal that had been founded by G. Piola (1791-1850).

Cauchy began by showing that, when $\bar{x} = Xe^{ip}$, for $-\pi \leqslant p \leqslant \pi$, if we take a continuous and finite function $f(x)$ for $|x| < X$, together with its derivative, then

(4.5.1) $\int_{-\pi}^{\pi} f(\bar{x})dp = 2\pi f(0).$

Under the further hypothesis that the successive derivatives, $f'(x)$, $f''(x)$, ..., $f^{(n)}(x)$ are finite and continuous for $|x| < X$,[22] Cauchy finds, by integrating

$$\int_{-\pi}^{\pi} \frac{f(\bar{x})}{\bar{x}^n}dp$$

by parts, that is was also

(4.5.2) $\int_{-\pi}^{\pi} \frac{f(\bar{x})}{\bar{x}^n}dp = \frac{1}{n} \int_{-\pi}^{\pi} \frac{f'(\bar{x})}{\bar{x}^{n-1}}dp = \cdots = \frac{1}{n!} \int_{-\pi}^{\pi} f^{(n)}(\bar{x})dp,$

and hence, from (4.5.1), that

$$\frac{1}{2\pi} \int_{-\pi}^{\pi} \frac{f(\bar{x})}{\bar{x}^n}dp = \frac{1}{n!} f^{(n)}(0).$$

In particular, if $f(0) = 0$, (4.5.1) becomes simply

(4.5.3) $\int_{-\pi}^{\pi} f(\bar{x})dp = 0.$

"From these formulas one can easily deduce, as we will see, those that serve to expand an explicit or implicit function of the variable x in a series ordered according to the ascending powers of this variable," Cauchy observes at this point (1841a, p. 60).

In fact, by substituting for $f(\bar{x})$ in (4.5.3) the product

$$\bar{x} \frac{f(x) - f(x)}{\bar{x} - x} \qquad |x| < X, x \neq \bar{x},$$

and expanding the "Cauchy kernel" $\bar{x}/(\bar{x} - x)$ in a geometric series, we have

(4.5.4) $f(x) = \frac{1}{2\pi} \int_{-\pi}^{\pi} \frac{\bar{x}f(\bar{x})}{\bar{x} - x}dp$

which is the famous Cauchy 'integral formula.'[23] It follows that "$f(x)$ will be expandable in a series ordered according to the ascending powers of x if the module of the real or imaginary variable x retains a value below that for which the function $f(x)$ (or its first derivative) ceases to be finite and continuous" (Cauchy, 1841a, p. 61).

The generic term of the expansion is then given by

$$(4.5.5) \qquad \frac{1}{2\pi} \int_{-\pi}^{\pi} \frac{x^n}{\overline{x}^n} f(\overline{x}) dp = \frac{x^n}{n!} f^{(n)}(0).$$

Now if $|x| = \xi$ and $\Lambda f(\overline{x})$ indicates "the largest value" of $|f(\overline{x})|$ as p varies ($-\pi \leqslant p \leqslant \pi$), then one can majorize the first member of (4.5.5) by

$$(4.5.6) \qquad \left| \frac{1}{2\pi} \int_{-\pi}^{\pi} \frac{x^n}{\overline{x}^n} f(\overline{x}) dp \right| \leqslant \left(\frac{\xi}{X} \right)^n \Lambda f(\overline{x}).$$

In a similar fashion, the remainder of the Maclaurin series, which gives the expansion of $f(x)$, will be majorized by

$$(4.5.7) \qquad |R_n| = \left| \frac{1}{2\pi} \int_{-\pi}^{\pi} \frac{x^n}{\overline{x}^{n-1}(\overline{x} - x)} f(\overline{x}) dp \right| \leqslant \frac{\xi^n}{X^{n-1}(X - \xi)} \Lambda f(\overline{x}).$$

Cauchy then extends this 'method of majorants' to functions of many variables and to the "implicit" functions defined by an equation of the type $f(x,y) = 0$. He shows with specific examples how one can determine "the limits" of the errors committed when one disregards the terms of the series after the first n.

He finally announces that the same method can be applied to the case of functions defined by differential equations, both ordinary and partial, but he does not develop this idea any further at this point. This was instead the subject of a paper that he published in Prague in 1835, where he had gone from Turin in 1833 in order to follow Charles X Bourbon and tutor his son.

In the introduction to this paper Cauchy resumed his criticism of the method of series that was then commonly used for the integration of differential equations. "The integration of differential equations was consequently illusory," he wrote, "so long as one did not provide any means of assuring that the series obtained were convergent and that their sums were functions able to verify the given equations; in such a way that it was necessary either to find such a method or to seek another method by means of which one could generally establish the existence of functions able to verify the differential equations and to calculate values indefinitely close to these same functions" (1835, p. 400).

The only known method suitable for this purpose, Cauchy adds, is that which he himself had presented in his second-year lessons at the École Polytechnique (see §4.4b). He then gave a rapid description of it, concluding that "the advantages which this method offers will be joined by many others still in what I am now going to explain" (1835, p. 404).

While in 1831 celestial mechanics had provided the occasion for his "calculus of limits," Cauchy now found his initial stimulus in a "wonderful paper where Mr. Hamilton lets the integration of the differential equations encountered in dynamics depend on the determination of one function represented by a definite integral that satisfies two partial differential equations of the second order"[24] (*ibid.*).

Thus Cauchy had thought that "there would perhaps be some advantage" in reducing the integration of a system of differential equations to the integration of a single partial differential equation of the first order.

After having shown how one can associate such an equation with the given system,[25] Cauchy obtains the general integrals of the system by setting certain integrals of the characteristic equation equal to arbitrary constants. He then shows that these particular integrals can be written as series, and it is at this point that the "calculus of limits" is able not only to assure the convergence of such series, but also to majorize the remainder by evaluating the error that arises when one stops with the first n terms of the expansion.

For a function $f(x)$ continuous in a circle of center x and radius r, (for which "one will evidently have"

$$\frac{\partial f(x + re^{ip})}{\partial r} = \frac{1}{ir} \frac{\partial f(x + re^{ip})}{\partial p}$$

Cauchy adds with some significance!) the inequality (4.5.6) is transformed into

(4.5.8) $|f^{(n)}(x)| \leqslant n! r^{-n} \Lambda f(x + \bar{x})$

where $\bar{x} = re^{ip}$, and $\Lambda f(x + \bar{x})$ indicates the maximum of $|f(x + \bar{x})|$ on the circumference of the circle.[26]

The inequalities (4.5.8) play a decisive role in demonstrating the convergence of the series that represent the integrals of the differential equation associated with the system. Cauchy immediately illustrates his "methods of majorants" for the simple case of a single differential equation,

(4.5.9) $dx = \mathcal{F}(x,t)dt,$

where $x = x(t)$. By calling ζ the value of x corresponding to the value $t = \tau$, Cauchy demonstrates the expandability of ζ in a power series of $(\tau - t)$, and consequently the existence of an integral of (4.5.9) for a suitably small value of $|t - \tau|$, "the value of the module r of $\bar{x}$ being subjected only to the condition that the function

$$\mathcal{F}[(x + \bar{x}, t + \theta(\tau - t)]   (|\theta| < 1)$$

remains finite and continuous, whatever the angle p, for this value of r and for a smaller value" (Cauchy, 1835, p. 446).

By means of (4.5.8) he then obtains a majorization of the module of the derivatives of $\mathcal{F}(x,t)$ with respect to x that appear in the terms of the series. As Gilain has observed, "if one looks at the demonstration, one recognizes that Cauchy utilizes for every θ a majorization in module of the partial derivatives of $\mathcal{F}(x,t)$ with respect to x by means of $\Lambda \mathcal{F}[(x + \bar{x}, t + (\theta - t)]$, then a uniform majorization with respect to the second variable by introducing a value $\ominus$ of θ such that the preceding module is the largest possible. Cauchy's demonstration in fact utilizes the continuity of $\mathcal{F}$ with respect to the pair (x,t) and, for each t, its analyticity with respect to x in a neighborhood of the initial values of x and t (In: Cauchy, 1981, p. xliii).

Nevertheless Cauchy illustrates his theorem with the example

$$dx = (x + t)^i dt,$$

where both x and t as well as $\mathcal{F}(x,t)$ are real quantities and i is "any positive quantity." Cauchy does not seem to recognize the conceptual difference between the existence proof demonstrated in his second-year lessons at the École (see §4.4b) and these new results. The ambiguity between the real and complex runs through Cauchy's entire paper. In fact, after having shown how to extend the demonstrated theorem "to any system of differential equations between an independent variable t and the functions x, y, z, ... of these same variables" (1835, p. 450), he writes in his final observations, "We add that these new theorems, like those we have set out here, can be easily extended to the case where the variables and the functions included in the given differential equations will become imaginary" (1835, p. 463).

It was precisely this latter motive that privileged in Cauchy's eyes the second existence proof and the "calculus of limits," which had transformed the integration of differential equations by series into a "completely rigorous" theory. After his return to Paris,[27] he rushed to republish his Turin and Prague papers in the reborn *Exercices d'analyse et physique mathématique*.[28]

Cauchy again took up the argument in various notes presented with weekly (!) frequency to the Academy between June and July 1842 (*Oeuvres* (1) **6**, pp. 461-470; 7, pp. 5-83). He there reformulated his second existence proof for a system of ordinary differential equations and then extended it to the case of first-order linear partial differential equations of the form

$$\frac{\partial u_i}{\partial t} = F_i(t, x_1, ..., x_n; u_1, ..., u_m, \frac{\partial u_1}{\partial x_1}, ..., \frac{\partial u_m}{\partial x_n})$$

with the initial conditions given by

$$u_i(0, x_1, ..., x_n) = w_i(x_1, ..., x_n) \quad (i = 1, ..., m).$$

If the F_i are analytic functions in the neighborhood of a point (for example the origin) and linear with respect to the $\partial u_k/\partial x_n$ and the w_i are also analytic in the neighborhood of the point, then Cauchy shows that there exists a unique solution of the 'Cauchy Problem,' expandable in a locally convergent power series.[29]

Thus, thanks to the "calculus of limits," the theory of differential equations became for Cauchy a part, fully relevant and meaningful in itself, of the general theory of complex functions, to which he dedicated a large part of his research in the last period of his life.

4.6. The Emergence of Cauchy's Theory of Complex Functions

In the 1840s Cauchy, by then already more than fifty years old, threw himself with renewed enthusiasm into exploiting the opportunities that had been opened by the systematic introduction of imaginaries into analysis. He presented a large number of papers to the Academy in which he clarified, reformulated, and extended the results he had earlier obtained, expanded the work of his former colleagues and students on this subject, set out new ideas, and altered old concepts.

The results were published in the brief *Comptes rendus* of the Academy (sometimes no more than the enunciations were given), and were then generally republished in a more ample version in the *Exercices d'analyse et physique mathématique*,[30] although at times these only repeated word-for-word what had already been printed in the *Comptes rendus*.

Cauchy's studies joined with those of Liouville, Laurent, Puiseux, and Hermite to create an efflorescence of research that soon led from the instrumental and heuristic use of imaginaries to the formation of a complete theory of "holomorphic" functions, to use the terminology that was later introduced by Briot and Bouquet (1875) and is still in use today.

One of the first contributions to the theory of functions was that of the engineer P. A. Laurent (1813-1854), who was inspired by Cauchy's work on the "calculus of limits" and expansions in series. In 1843 he presented the Academy with a paper on the representability by a power series of a holomorphic function within an annulus, that is, a ring-shaped domain bounded by two concentric circles of center c, when the series is generalized in such a way as to also admit negative powers of $z - c$.

In his report on the paper to the Academy, Cauchy first recalled his own theorem on this topic (see §4.5) and then stated Laurent's theorem as follows:[31]

If x designates a real or imaginary variable, then a real or imaginary function of x can be represented by the sum of two ordered, convergent series, the one with integral, increasing powers, the other with integral, decreasing powers of x, in so far as the module of x retains a value between two limits where the function or its derivative remains finite and continuous" (Cauchy, 1843a, p. 116).

In fact, by generalizing Cauchy's integral theorem, Laurent shows that if a function $f(z)$ is holomorphic in an annulus of center c, then it is representable in a unique manner by the series

$$\sum_{-\infty}^{\infty} a_k (z-c)^k \quad k \in \mathbb{Z},$$

where the coefficients are given by

$$a_k = \frac{1}{2\pi i} \int_{\ell} (z-c)^{-k-1} f(z) dz,$$

ℓ being a closed loop around c within the annulus.[32]

Even though Cauchy admits that, "The extension given to the theorem on the convergence of series by Mr. Laurent appears worthy of notice" (*ibid.*, p. 117), his results "are included, as a special case," in a formula that appeared in the *Exercices de mathématique* in 1826. In addition, it "can be immediately derived" from the integral theorem for an annulus that Cauchy had published in the *Exercices d'analyse* in 1840 (*ibid.*, p. 116). In any case, Cauchy repeated the contents in a short note that immediately followed his report (Cauchy, 1843b). He here asserted that the simplest way of arriving at Laurent's theorem was to utilize the mean value formula,

$$f(c) = \frac{1}{2\pi} \int_{-\pi}^{\pi} f(c + re^{ip}) dp ,$$

for a function which is finite and continuous, together with its derivative, within the annulus.

Together with the "calculus of limits" and the 'integral theorem,' the theory of residues seemed in Cauchy's eyes to promise the widest and most fruitful applications.

In the same year Cauchy presented the Academy with a number of short papers that showed how the theory of residues could also provide results for elliptic functions. The theory of elliptic functions had recently enjoyed a resurgence of interest after Jacobi (1835) had shown that a meromorphic function of a complex variable cannot have more than two distinct periods whose ratio is necessarily imaginary.[33]

This result seemed to open new research opportunities. Was it possible to find all the functions that, like the elliptic functions, were doubly periodic?

Liouville began to study the problem in the summer of 1844. In December of that year he took advantage of the opportunity provided by a note that Chasles (1793-1880) had written on geometrical constructions of the amplitudes of elliptic functions to inform the Academy of the objective of his current research. "We know the most general expression for every periodic function, namely, $\sum_{i=0}^{\infty} A_i \cos i\pi x + B_i \sin i\pi x$. [I want] to likewise determine the most general conditions under which it acquires a second period" (Liouville, 1844, p. 1261).

He then presented a sketch of his ideas on the subject. He set out the following "general principle" that to his mind seemed "to impress on the study of elliptic functions an uncommon character of unity and simplicity."

Let z be any real or imaginary variable, and let $U(z)$ be a well-defined function of z, that is, a function which, for every value $x + iy$ of z, takes a unique value which is always the same when x and y become the same. If such a function is doubly periodic, and if one recognizes that it never becomes infinite, one can, from this alone, affirm that it reduces to a constant" (Liouville, 1844, p. 1262).

Cauchy immediately recognized the deep analogy between Liouville's theorem for elliptic functions and his own results on functions of a complex variable obtained through the theory of residues. At the very next meeting of the Academy he presented a paper (Cauchy, 1844b) in which, after having mentioned his first works on the subject, he set out "as a special case" obtainable from it the theorem:

If for every real or imaginary value of the variable z, the function $f(z)$ always has a unique and determinate value, and if it also reduces to a determinate constant F for every infinite value of z, then it will also reduce to the same constant when the variable z has any finite value" (1844b, p. 367).

Yet again Cauchy recalled his old paper of 1814 where he had established "a fundamental principle" from which the various above cited theorems follow: the fact that the difference between the results obtained from the two different orders of integration in a double integral is expressed "by a singular definite integral" (see §4.2).

It follows, among other things, that "in the theorem relative to the expansion of functions in series one can rigorously dispense with the use of derivatives [*fonctions derivées*]" (1844b, p. 368).

This was also the view expressed in a "judicious observation" that Liouville had recently sent him, Cauchy wrote. On the other hand, he pointed out that, "as far as the expansion of functions in series is concerned, it seems to me that the consideration of derivatives should not be completely abandoned" (1844b, p. 369). He had in fact done this himself by adding the condition on the continuity of the derivative to the 1841 republication of his 1831 Turin paper on the "calculus of limits."

Going on to demonstrate a few of the consequences obtainable with the calculus of residues, Cauchy showed how, from the formula,

$$\mathcal{E} f(z) = 0,$$

one can "directly" obtain the theorem that "if a function $f(z)$ of the real or imaginary variable z is everywhere continuous and consequently everywhere finite, it will simply reduce to a constant" (1844b, p. 372).

Thus Liouville's theorem could be obtained in full generality within the theory of functions set out by Cauchy. However, Liouville deserves the credit for having first recognized the theoretical importance of this "general principle."

Liouville never published a demonstration, nor did he ever follow out his announced research on doubly periodic functions. He limited himself to setting out his mature ideas on the subject a few years later in several lessons prepared for the German mathematicians Borchardt (1817-1880) and Joachimsthal (1818-1861) while they were visiting Paris in 1847.[34]

For his part, although Cauchy published five different demonstrations of Liouville's theorem obtained from his previous results in less than a year, he seemed much more interested in developing the latest refinements of his integral theorem as well as of the theory of residues. In his view this constituted the foundation of his entire theory.

He gave the first general formulation of the integral theorem in 1846. Here he considered "an area S measured on a given plane or on a given surface" bounded by a "unique" closed curve s[35] (1846a, p. 72). Then calling K a function of any variables x, y, z, ... that determine a point of S and of their derivatives with respect to s, Cauchy considers the integral

$$(4.6.1) \quad \int_{\partial s} K \, ds$$

and shows that if $K = P \, dx + Q \, dy + Z \, dz + \cdots$ is an exact differential, then the value of the integral is independent of the form of the curve s, provided that K remains finite and continuous.

Passing then to the special case of a plane with Cartesian coordinates x, y, if P and Q are finite and continuous functions of x and y, the integral

$$(4.6.2) \quad \int_{\partial s} P \, dx + Q \, dy,$$

taken along the boundary of S, reduces to the double integral

$$(4.6.3) \quad \iint_S \left[\frac{\partial Q}{\partial x} - \frac{\partial P}{\partial y} \right] dx \, dy,$$

extended to all the points of the domain S. Moreover, Cauchy added, if $P \, dx + Q \, dy$ is an exact differential, the integral (4.6.2) is zero.[36]

The theorem of residues can also be extended to the case of a plane domain bounded by a closed curve s, Cauchy observed in a note to the Academy a little more than a month later (Cauchy, 1846b). If P', P'', P''', ... are poles of a function $f(z)$ in the domain S, then S can be divided into parts A, B, C, ..., each of which contains a single pole, and

$$\int_S f(z)dz = \int_{\partial A} f(z)dz + \int_{\partial B} f(z)dz + \int_{\partial C} f(z)dz + \cdots .$$

Hence,

$$(4.6.4) \quad \int_S f(z)dz = 2\pi i \, \mathcal{E}f(z).$$

"The sign $\mathcal{E}$ indicates the sum extending to the only roots of the equation $1/f(z) = 0$ which correspond to the points situated in the interior of S" (1846b, p. 138).

In reflecting on the conditions of validity for (4.6.4), Cauchy was quick to recognize that $f(z)$ must assume "precisely the same value when, after having traced the entire curve, we return to the point of departure" (1846c, p. 154). In other words, $f(z)$ must be monodromic, using the term that Cauchy himself later introduced.

"But nothing prevents" other things from happening, as for example with the integral of a function that "contains roots of algebraic or transcendental equations" (*ibid.*). In the first part of his paper Cauchy discusses the case of single-valued functions, as for example $f(z) = 1/z$. In this case the integral $\int f(z)dz$ is multi-valued, as Gauss had already seen, and the values depend on the path of integration. He then goes on to treat the case of multi-valued functions under the integral sign. Here the periodicity moduli (*indices de periodicité*) of the integral "will no longer be generally represented by the residues" (1846c, p. 165). But the absence of a clear understanding of the nature of the branch points of algebraic functions thus kept Cauchy from forming a precise idea of the integrals of multi-valued functions.

In facing these problems, Cauchy was naturally lead to consider paths in the complex plane and curvilinear integrals, which clearly rendered inadequate the conception of complex variables as simple "symbolic expressions" that he had given in the *Cours* and which he apparently never abandoned. It is therefore not by chance that at this time Cauchy began to reflect on the foundations of complex analysis, beginning with the definition of a complex number and a complex function.

Strange as it may seem, after the publication of Argand's paper and the discussion that followed (see §4.3), the geometrical interpretation of complex numbers seems to have been quickly forgotten. It was taken up again by a certain C. V. Mourey in a pamphlet that appeared in Paris in 1828. The geometrical theory was dedicated by Mourey "to the friends of evidence" so as to show how "the truth" about the nature of complex numbers could "escape the domain of the chimerae."[37]

Not until 1831 did the geometrical theory of complex numbers become fully acceptable, on the authority of Gauss, when the latter finally published the ideas that he had already formed 30 years earlier. In a notice of his 1832 paper that appeared in the *Göttingische Gelehrte Anzeige* in 1831, he wrote, "The author calls a quantity $a + ib$, where a and b are real numbers and i is the symbol written for $\sqrt{-1}$, a complex number, when both a and b are integers. Complex quantities are consequently not opposed to the real, but include these as a special case when $b = 0$" (Gauss, 1831, p. 171).

These are Gauss' complex integers which enter into the theory of biquadratic residues. He continues,

> The transfer of the theory of biquadratic residues to the domain of complex numbers could perhaps appear shocking and unnatural to those who have less confidence in the nature of imaginary quantities and are prejudiced by a wrong understanding of them, and give rise to the belief that this research may thereby be set in the air, as it were, receive precarious support, and completely disappear from view. Nothing would be less well founded than this belief" (1831, p. 174).

In order to convince his readers, Gauss then repeated what he had written in the full paper, where he showed how one could interpret complex numbers as the points of a plane. Complex numbers, which had long been scarcely tolerated, been called impossible numbers by many and held by others to be simply a game with symbols, now received an intuitive significance. "More is not needed to admit these quantities into the domain of arithmetic," Gauss concluded.

> That one has until now seen this subject from a false point of view and consequently found a secretive darkness, is in large part a result of its inappropriate nomenclature. If one had not called $+1$, -1, $\sqrt{-1}$ positive, negative, and imaginary (or even impossible) units, but instead called them direct, inverse, and lateral units, one could never have spoken of such a darkness" (1831, pp. 177-8).

Gauss finally concluded his note with the promise of a demonstration of why commutative algebraic extensions of $\mathbb{C}$ different from $\mathbb{C}$ itself do not exist.[38]

But it was not only a question of nomenclature, as Gauss seemed to believe, that hindered the complete acceptance of complex numbers for so many years. In fact, ten years after Hamilton (1837) had shown how one can define complex numbers as ordered pairs of real numbers,[39] Cauchy returned to the topic by writing that his conception of imaginaries as "symbolic expressions" had eliminated "the need to torture the spirit to seek to find what the symbol $\sqrt{-1}$ can represent, that for which the German mathematicians substitute the letter i" (Cauchy, 1847a, p. 313). There was no need to strain oneself to find an interpretation: such a symbol, Cauchy said, is no more than "a tool, an instrument of calculation whose introduction into formulas permits a faster arrival at the very real solution of the question one has posed" (ibid.).

One can then gain clarity and put algebraic theory "within the reach of all minds" by simply completely banishing imaginaries, "reducing the letter i to no more than a real quantity" (ibid.).

While this may seem "improbable and even impossible at first sight," one can attain this end by utilizing the idea of equivalence introduced by Gauss and Kummer to study classes of quadratic forms. Let us call two real polynomials equivalent, Cauchy says, if, when they are divided by $x^2 + 1$, they give the same remainder,

$$\phi(x) \equiv \chi(x) \quad (\text{mod } x^2 + 1).$$

Cauchy shows how one can interpret calculations with complex numbers as a calculation (mod $i^2 + 1$) in the ring $\mathbb{R}[i]$, where i "is a real but indeterminate quantity."

In the expanded version of the same work which was published as usual in the *Exercices* (1847b), Cauchy asserted that the concept of algebraic equivalence introduced in this way can naturally be extended to the case in which the functions are "the sum of a convergent series." Thus, for example, given the exponential series

$$e^x = \sum_{}^{\infty} \frac{x^n}{n!}$$

and setting ix in place of x, one has

$$e^{ix} = \sum_{}^{\infty} \frac{(ix)^n}{n!} \equiv \sum_{}^{\infty} (-)^n \frac{x^{2n}}{(2n)!} + i \sum_{}^{\infty} (-)^n \frac{x^{2n+1}}{(2n + 1)!} (\text{mod } i^2+1)$$

From this one obtains Euler's relation

$$e^{ix} \equiv \cos x + i \sin x \quad (\text{mod } i^2 + 1).$$

In a similar manner, Cauchy interprets the usual arithmetic operations on these "algebraic equivalences."

Not until several years later, "after new and mature reflections," did Cauchy decide to abandon the concept of imaginaries that he had been "content to show could be made rigorous" (!) in the *Cours d'analyse* and finally adhere to the geometrical theory that had been proposed by Argand in 1806, against which, in Cauchy's view, "specious objections" had been advanced (Cauchy, 1849a, p. 152).

After remembering the names of many who had contributed to the theory (but not Hamilton or Gauss!), Cauchy asserted that he had been led to this "new" conception by profiting from the ideas that had been presented by Saint Venant (1797-1886) in a note on geometric sums which had appeared in the *Comptes rendus* of the Academy in 1845.[40]

The theory of "quantities that I will call *geometrical*," Cauchy wrote, was nothing other than the usual interpretation of complex numbers as vectors in the plane (1849a, p. 153).

After having introduced in a natural manner the concept of an entire function Z of a "geometrical quantity" z, Cauchy reformulated the fundamental theorem of algebra, which he proved by using the method of minimum for the modulo of Z. As usual, Cauchy then presented his full work in the *Exercices d'analyse* (1847c). This was followed by no less than 15 papers in which he redefined the fundamental concepts, ranging over such topics as the reduction of a "geometrical quantity" to the form $x + iy$; the definitions of functions, of continuity, and of the derivability of functions of such quantities; and the study of the properties of particular functions such as the exponential, logarithmic, and trigonometric functions.

In introducing the concept of function, Cauchy wrote,

> When ... we substitute *geometrical quantities* for *imaginary expressions*, the *imaginary variables* are nothing other than *variable geometric quantities.* It remains to know how the *functions* of imaginary variables must be defined. *This last question has often embarassed the geometers, but all difficulty disappears when, letting ourselves be guided by analogy, we extend to functions of geometric quantities the definitions generally adopted for the functions of algebraic quantities* [my emphasis]. We thus arrive at conclusions that seem singular at first sight, and nevertheless very legitimate, which I will indicate in a few words.

> Two real variables or, in other words, two variable algebraic quantities, are called *functions*, the one of the other, when they vary simultaneously in such a way that the value of one determines the value of the other. If the two variables are taken to represent the abscissas of two points constrained to move on the same line, the position of one of these points will determine the position of the other, and vice versa (Cauchy, 1847d, p. 359).

Similarly, Cauchy concludes, a geometrical quantity $w = u + iv$ can be considered a function of a variable geometric quantity $z = x + iy$ whenever the value of z determines the value of w or, in geometrical terms, when the position of the point z determines the position of the point w. To this end u and v will be determinate functions of x and y.

In commenting on Cauchy's definition, Casorati (1826-1890) wrote, "This definition can be understood as implying purely the idea of the dependence of the value w on that of z, and not necessarily the idea of an analytic expression for this dependence," even if it seems clear that Cauchy intended such expressions always to be admitted (Casorati, 1868, p. 70).

Casorati's second observation is that,

> although wanting to embrace exclusively analytic expressions, the new definition is much more general than the first, in as much as a quantity obtainable by means of operations performed on two variables x and y will not, except in relatively particular cases, be obtainable by means of a system of operations performed on the single composite quantity $x + iy$. A generalization of this type, for which any function of two variables x and y can be put between the functions of a single

complex variable $x + iy$, is not ... suggested by a truly profound analogy; for it the theory of functions of a complex variable would simply be the theory of functions of two variables ... It is thus that Cauchy himself, while conserving the given definition of a function of a complex variable $x + iy$, found it necessary to introduce an epithet to designate, among all the functions understood in the definition, only those which, like all the functions resulting from ordinary systems of operations performed on the combination $x + iy$, enjoy the property of having for every value of $x + iy$ a derivative independent of the value of dy/dx (1868, pp. 70-71).

These functions were later called *monogenic* by Cauchy, where the condition of monogenicity is expressed by the equation

$$(4.6.5) \quad i\frac{\partial w}{\partial x} = \frac{\partial w}{\partial y} \quad \text{or} \quad \frac{\partial u}{\partial x} = \frac{\partial v}{\partial y}; \frac{\partial u}{\partial y} = -\frac{\partial v}{\partial x},$$

from which the following derive:

$$(4.6.6) \quad \frac{\partial^2 u}{\partial x^2} + \frac{\partial^2 u}{\partial y^2} = 0; \quad \frac{\partial^2 v}{\partial x^2} + \frac{\partial^2 v}{\partial y^2} = 0.$$

Cauchy also proposed a number of special terms to indicate particular properties of functions. Some of these terms have remained in use while others have been completely forgotten. He defines, for example, as *monodromic* a (single-valued) function defined in a finite portion S of the complex plane when the function always takes the same value at one and the same point P in S, whatever path z may take within S to arrive at P.

Synectic are for Cauchy finite, continuous functions that are monodromic and monogenic in a portion S of the plane.

An explicit definition of the concept of the monogenicity of a function now permitted Cauchy to specify the classes of functions to which he applied the theory and to completely relieve his old uncertainties about the necessity of requiring the condition of derivability in his theorems on expansions in series. "The principles that I have just set out," he wrote with respect to (4.6.5) and (4.6.6), "confirm what I have said elsewhere about the need to specify the derivative of a function of z in the theorem" on the expansion in power series of a function with the "calculus of limits" (Cauchy, 1851a, p. 304).

For monogenic and monodromic functions, he then reformulates the theorem of residues, to which he returned many times afterwards. In 1855 he introduced the idea of a logarithmic indicator (*compteur logarithmique*) by considering the integral

$$(4.6.7) \quad \frac{1}{2\pi i} \int_{\partial S} \frac{Z'}{Z} \, dz,$$

where $Z(z)$ is a monodromic and monogenic function in a domain S except at isolated points (poles).

He then showed the remarkable formula

$$(4.6.8) \quad \frac{1}{2\pi i} \int_{\partial S} \frac{Z'}{Z} \, dz = N - P,$$

where N is the number of zeros and P the number of poles of Z within S[41] (1855a, pp. 289-90). This allowed him to immediately find both another demonstration of the fundamental theorem of algebra and a demonstation of Liouville's theorem that the number of zeros of a doubly periodic function within the parallelogram of periods equals the number of poles, provided that none of them lie on the sides of the parallelogram.

Cauchy returned to the theory of residues yet another time in a final paper that was presented to the Academy only a few months before his death (1857a). Here, after pointing out the curious parallels that had guided him in his first paper of 1826, where the residue of a function was defined as the coefficient of ϵ in the expansion of the variation of $f(x)$ in a power series (see §4.4c), in the same way in which Lagrange had defined the derivative as the coefficient of the first term of the expansion of $f(x)$ in a Taylor series, he asserted that the idea of residue must be established without considering series.

In order to do this he began directly from the integral $\int_{\partial S} Z \, dz$ of a function $Z(z)$ extended to the boundary of any area S and asserted that if Z was monogenic and monodromic everywhere in S, then $\int_{\partial S} Z \, dz = 0$, while if Z has isolated singularities z_k in S, $\int_{\partial S} Z \, dz = 2\pi i \, \text{Res}(Z, z_k)$.[42]

"We recognize here," Cauchy wrote, "how useful it is to clearly define the functions of geometric quantities, or in other words, the functions of imaginary variables, by not only distinguishing monodromic and nonmonodromic functions, but also monogenic and nonmonogenic functions" (1857a, p. 437).

"Neatly" defining the concept of complex functions allowed Cauchy, among other things, to confront in a more satisfactory manner the problem of the multi-valuedness of the integrals of algebraic functions.

Resuming his research of 1846 in the *Mémoire sur les fonctions irrationelles* (1851a), Cauchy considered a function u defined by the algebraic equations $U(u,z) = 0$ which takes the values $u_1, u_2, u_3, \ldots$ for a given value of z. He then considered the points c_i of the plane C where a function u_g "becomes infinite or equivalent to another term u_h of the series $u_1, u_2, u_3, \ldots$" (1851a, p. 293). The points c_i are then the poles or the branch points of the function u. If z describes a closed curve such that the points c_i are all outside of it, the u_k branches of the function will remain continuous functions and each will reacquire its initial value when z returns to its initial position. This will not occur if any point c_i is within the path followed by z.

Cauchy then considers the lines $C_i D_i$, the indefinite extensions of the "ray vectors" traced by a fixed point 0 to the points c_i, the "*points d'arrêt.*" By means of these "*lignes d'arrêt,*" which "can be compared to the obstacles at which the moving point stops, without ever crossing them" (1851a, p. 294), Cauchy divided the complex plane into regions where the u_k are continuous functions. He could then set the problem of evaluating the integral $\int u\, dz$ along a curve that can either meet the *lignes d'arret* or not.

This research intertwined with that which had already been published by Puiseux (1850). In a later session of the Academy Cauchy limited himself to announcing that he had succeeded in determining the number of periods (*indices de periodicité*) of the integral of a function u defined by an algebraic equation $f(u,z) = 0$ "for the most general case" (*Oeuvres* (1) 11, pp. 300-1, 304-5).

Algebraic functions and their integrals were in fact the arguments put forward by Victor Puiseux (1820-1883) in a paper that appeared in Liouville's *Journal* in 1850, which has since become famous.

Beginning from Cauchy's studies on the continuity of the solutions of an algebraic equation $f(u,z) = 0$, where f is a polynomial of the complex variables u and z, Puiseux first of all introduced a distinction between the poles and the branch points of a function. For transcendental functions, he introduced the idea of a "pole of infinite order" (or an essential singularity, as we say today) as for example the point $z = 0$ for the function $e^{1/z}$.

Puiseux then demonstrated that, if u_1 is a solution of $f(u,z) = 0$ and if b_1 is the value of u_1 corresponding to $z = c$, then the function u_1 assumes the same value b_1 when z returns to c after having followed a closed path γ in the plane that does not include poles or branch points. In other words, along such a path, u_1 is single-valued and $\int_\gamma u_1\, dz = 0$.

If, on the other hand, $z = a$ is a branch point of order p and if we consider a closed path around a which does not include any other critical points, Puiseux shows that the p functions $u_1, \ldots, u_p$ can be subdivided into a certain number of "circular systems," or, in modern terms, the orbits of the subgroups of the Galois group of the equation $f(u,z) = 0$, for which Puiseux calculates the order.

He further demonstrates the fundamental theorem that if a function u_i belongs to a circular system of order q, then u_i can be expanded in the neighborhood of the point according to the power

$$\sum_{n=m} a_n(z - a)^{n/q} \quad (m \in \mathbb{Z}).^{43}$$

In the last part of his paper Puiseux finally faces the problem of determining the periods of the integral $u_1 dz$ along any path from a point c to a point k in the complex plane. He in fact proposes to: "1⁰ find all the *distinct* periods which belong to a value of $\int_c^k u_1\, dz...$; 2⁰ to discover if each period p belongs to all the values of the integral $\int_c^k u_1\, dz$ or only to a part of them; 3⁰ to determine the values of $\int_c^k u_1\, dz$ which remain distinct when one abstracts the whole multiples of the periods" (Puiseux, 1850, p. 439).

He therefore limits himself to considering a few particular cases, such as the so-called hyper-elliptic integrals and the case $\int u \, dz$ when u satisfies the binomial equation $\phi(z)u^m - H^m = 0$, where ϕ and H are polynomials without any common factor.

In the course of this research, Puiseux demonstrates that a single-valued algebraic function is always rational and that if f is an irreducible polynomial, from a root u_i of $f(u,z) = 0$ one can follow a suitable path in the complex plane to any other u_j which has the same starting value as u_i. This theorem, which was among the implicit assumptions needed to validate Puiseux's results on the periods of the integrals, also gave him a useful criterion for deciding, with a finite number of operations, whether an algebraic equation is irreducible or not.

Like Puiseux, Cauchy's works on complex functions also inspired Charles Hermite (1822-1901), who became one of the leading French mathematicians of the second half of the century.

In 1851 the young Hermite presented the Academy with a note on one of his favorite subjects, the theory of elliptic functions.[44] Utilizing the idea of a parallelogram P of the periods and that of residue, Hermite showed that the sum of the residues of a doubly periodic function, monogenic and monodromic, is zero with respect to P. He further proved that every such function can be rationally expressed by means of elliptic functions. Both results had been known to Liouville.

Liouville himself, on the occasion of Cauchy's report, recalled his old paper to the Academy (Liouville, 1844) and the lessons he had given to his German friends Borchardt and Joachimsthal. He then gave the Academy Borchardt's manuscript which contained the essential part of his work. In addition to Hermite's results and Liouville's theorem, it included the demonstration "of the equally important proposition that the number of roots that satisfy the [doubly periodic] function is always precisely equal to the number of infinites of this function and that, moreover, the sums of the values of the variable, relative to these two circumstances of a null or infinite function, are always equal to each other up to the multiples of the periods" (*Oeuvres* (1) 11, pp. 373-4).

Liouville's observations provoked an immediate response from Cauchy in which he claimed for his own calculus of residues the priority "for the fundamental principle invoked by Liouville for doubly periodic functions" and for the consequences he had derived from this principle, thus renewing a quarrel that had taken place a few years earlier. For his part Liouville decided to devote the second semester of his course at the Collège de France for 1850-51 to doubly periodic functions.

Among the auditors of Liouville's lectures were Briot (1817-1882) and Bouquet (1819-1885). Liouville's "beautiful theory" inspired both the notes they presented to the Academy on doubly periodic functions and their 1859 book on this subject. In this volume they explicitly said that, "The learned lectures of the illustrious geometer [Liouville] and the wonderful works of M. Hermite on the same subject have been the point of departure for our own research"[45] (Briot and Bouquet, 1859, preface).

In the first part of their book Briot and Bouquet gave a systematic exposition of the theory of functions of a complex variable as it had been set out by Cauchy. This was the text that Cauchy himself, perhaps because he had never been required to write a handbook as he had for his courses at the École Polytechnique, had never written.

Even though the enunciations of the theorems and demonstrations in Briot's and Bouquet's volume are sometimes incomplete and unsatisfactory, their work played a decisive role in spreading Cauchy's point of view and his results to European mathematicians.

From an instrument for the discovery of new theorems and the most diverse applications, as it had long been for Cauchy, the "passage from the real to the imaginary" acquired with Briot and Bouquet's treatise the character of a systematic theory. Their book remained for a long time the standard text of the "French school." It was translated into German in 1862, and reprinted in French in 1875.

On the other hand, in Germany, with Riemann and Weierstrass, the theory of complex functions followed an autonomous path of development that went far beyond what had been set out by Cauchy. The same can be said for real analysis. It is consequently the latter subject to which we must devote our attention before we turn to the works of the analysts of Göttingen and Berlin.

Notes to Chapter 4

[1]At the time Gauss had no knowledge of the demonstration suggested by Laplace in his mathematical lectures at the École Normale in 1795, which were not published until 1812 in the *Journal de l'Ecole Polytechnique* (In: *Oeuvres* **14**, pp. 10-111). According to Lagrange (1797-8, pp. 200-201), Laplace's demonstration "leaves nothing to be desired as a simple demonstration," even though it appeared to him to be impossible to realize because of the difficulty of the calculations. Laplace in fact began with a completely different idea than Euler and Lagrange. He utilized an inductive argument applied to concepts like symmetric functions and the discriminants of a polynomial. For a modern explanation of Laplace's demonstration, see Remmert (1983).

[2]Ostrowski showed how one can render Gauss' demonstration rigorous, although it requires refined arguments of a topological nature that "are present neither in the dissertation itself nor in the pre-Gaussian literature" (1927, p. 1).

[3]The Artin-Schreier theory of ordered fields has definitively clarified why a continuity argument appears, under different forms, in demonstrations of the fundamental theorem of algebra.

[4]Among other things, he implicitly reverts to Bezout's theorem when he asserts that the circumference and the algebraic curves $T = 0$ and $U = 0$ respectively have $2m$ points of intersection.

[5]An explanation in modern terms can be found in Dieudonné (1978, I, pp. 70-71).

[6]Actually y becomes indeterminate since t and u also appear homogenously in the numerator of y.

[7]A detailed explanation of the technical particulars of the controversy between Legendre and Cauchy is given in Grattan-Guinness (1970b, pp. 36-40).

[8]See Petrova (1974). Argand's articles and those of the mathematicians who participated in the discussion in the *Annales* have been published in Argand (1874).

[9]In commenting on this definition, Hankel could do no more than say, "I do not believe that I am exaggerating when I call this a scandalous wordplay, unsuitable for mathematics, which is and should be proud of the clarity and evidence of its concepts" (1867, p. 14).

[10]For the continuity of a function of many variables Cauchy implicitly recalls the *Cours d'analyse*, where he had affirmed that a function $f(x_1, ..., x_n)$ is continuous at a point $(a_1, ..., a_n)$ if it is continuous with respect to every variable x_i taken one at a time. This nevertheless is not sufficient (Heine, 1870, p. 361), as we see from the example $f(x,y) = xy/(x^2 + y^2)$ for $(x,y) \neq (0,0)$, $f(0,0) = 0$. The function is in fact discontinuous at the origin, notwithstanding that $f(x,0) = f(0,y) = 0$ are both continuous functions at the origin (Pringsheim, 1899, p. 48).

[11]As Gilain has observed (1981, p. xxvi), by returning "in an intuitive manner" to the 'Cauchy criterion' for the convergence of a sequence set out in the *Cours d'analyse*.

[12]In 1876 Lipschitz weakened the conditions of the theorem, requiring that $f(x,y)$ satisfy the "Lipschitz condition" instead of continuity (see §7.2b). The existence theorem is today called that of 'Cauchy-Lipschitz.'

[13]Contrary to what Cauchy seemed to believe, it is not always true that $\mathcal{F}(x)$ has a determinate limit for $x \to \Xi$.

[14]On numerous occasions, however, Cauchy referred more or less explicitly to the contents of his second-year lectures, before abbé Moigno had published them in Volume II of his *Lecons* (1840-44).

There were based on those of Cauchy, but in a form considerably less coherent and rigorous. See in particular, Cauchy (1830), (1835), and (1841b).

[15]A complex function of a real variable x is a quantity expressible by $f(x) + i\phi(x)$, where f and ϕ are real functions of x, which should not be confused with a function of a complex variable.

[16]Cauchy here uses a technique that was widely adopted at the time but which he had nevertheless not hesitated to criticize from the point of view of rigor. Furthermore, one must remark that he writes as if the analyticity of the integral were stated, even though it is not among the hypotheses of the theorem.

[17]As we will see, (§4.6) until the end of his life Cauchy continued to work on the theory of residues, to which he always seems to have attributed much more importance than to his "integral theorem."

[18]At this point Gauss adds the footnote, "In reality it can here be assumed that ϕx is itself a uniform function of x, or at least that, for those values within every complete region of the surface, only one system of values can be taken without a break of continuity" (Gauss, 1811, p. 91 n.).

[19]On Cauchy's stay in Italy see Terracini (1957). For the contrast between the Lagrangian tradition in Italy and the "modern analysis" of Cauchy, see Bottazzini (1981).

[20]This was first published in 1834, and then reprinted in part as (1841a), which version I use here.

[21]Plana was for a long time the director of the Observatory at Turin and subsquently the secretary of the Academy of Science. He also acquired considerable fame in Europe for his works on the theory of the moon.

[22]In reality it is sufficient to suppose the existence and the continuity of $f'(x)$.

[23]From (4.5.4) we immediately obtain the usual form

$$f(z) = \frac{1}{2\pi i} \int_c \frac{f(\zeta)}{\zeta - z} \, d\zeta$$

by describing the circumference c of the circle of convergence with $\zeta = re^{i\phi}$. In fact,

$$\frac{1}{2\pi} \int_{-\pi}^{\pi} \frac{\zeta f(\zeta)}{\zeta - z} \, d\phi = \frac{1}{2\pi i} \int_{-\pi}^{\pi} \frac{f(\zeta)}{\zeta - z} i r e^{i\phi} d\phi = \frac{1}{2\pi i} \int_c \frac{f(\zeta)}{\zeta - z} \, d\zeta.$$

In a similar manner we obtain the integral formulas for the

derivatives from (4.5.5).

[24]This refers to the first part of Hamilton (1834), in which the Irish mathematician introduced the "canonical" (so named by Jacobi) equations by reformulating Lagrange's dynamical equations,

$$\frac{dq_i}{dt} = \frac{\partial H}{\partial p_i}; \quad \frac{dp_i}{dt} = \frac{\partial H}{\partial q_i},$$

where the Hamiltonian function $H(p_i, q_i, t)$, has the physical meaning of total energy.

Hamilton's ideas were taken up by Jacobi and further developed in his *Vorlesungen über Dynamik*, which he gave at Königsberg in 1842-3. They were published in 1866.

[25]Subsequently called the "characteristic equation" by Cauchy.

[26]In order to arrive at this result Cauchy utilizes the mean value formula,

$$f(x) = \frac{1}{2\pi} \int_{-\pi}^{\pi} f(x + re^{ip}) dp,$$

which Poisson had earlier found (1823c, p. 498).

[27]Cauchy returned to Paris in 1838. He regained his position in the Academy, but not his teaching positions, because of his refusal to take the required oath of allegiance to Louis Philippe. When the latter was subsequently overthrown in 1848, Cauchy finally regained his chair at the Sorbonne and retained it until his death in 1857.

[28]At this time the theory of differential equations was the subject of intense research in France, particularly on the part of Sturm and Liouville. Initially inspired by the works of Fourier on the propagation of heat, in a short time they arrived at the so-called "Sturm-Liouville theory." An accurate analysis of this topic can be found in Lützen (1983).

In the shadow of Cauchy's publications, Weierstrass had also obtained the same result in a work on the integration of a system of ordinary differential equations, which was not printed until 1894. Weierstrass (1842) there shows that, given the system

$$\frac{dx_i}{dt} - G_i(x_1, ..., x_n) \quad (i = 1, ..., n),$$

where $x_i = x_i(t)$ and G_i are polynomials in $x_1, ..., x_n$, then "first, n power series,

$$P_1(t), ..., P_n(t),$$

can be determined which converge (ordinarily) in a certain neighborhood of the point ($t = 0$), and which, when substituted for

x_1, ..., x_n, satisfy the preceding differential equation and simultaneously take any predetermined values a_1, ..., a_n for $t = 0$" (Weierstrass, 1842, p. 75).

[29]Without apparently being aware of it, Cauchy's theorem was rediscovered and generalized to any analytic function F_i by Sonia Kowalewski (1850-1891) in a thesis written under Weierstrass' direction, which was published the following year (Kowalewski, 1875).

The demonstrations of Cauchy and Kowalewski were then simplified by the works of several mathematicians, among them Darboux and Goursat. For a discussion of the 'Cauchy Problem' and the methods of solution developed at the time, see Hadamard (1923). For more recent developments see Dieudonné (1978, II, pp. 119-20). For a scientific biography of Kowalewski and a full discussion of her contributions to mathematics, see Cooke (1984).

[30]The individual issues of the *Exercices* appeared with increasing delay, until the last number of Volume 4 carried the date 1847 but was actually published between 1850 and 1851. This has sometimes confused the dating of Cauchy's extensive (but often repetitive) work during his later years. In order to form a correct chronology of these works it is necessary to look at the notices that appeared in the *Comptes rendus* of the Academy.

[31]Laurent's paper was not published at this time. Its basic contents appeared in a posthumous article (Laurent, 1863). The same results had already been obtained, but not published, by Weierstrass (1841a).

[32]As it is well known, Laurent's theorem allows an elegant classification of isolated singularities. If

$$\sum_{-\infty}^{\infty} a_k(z - c)^k$$

is the expansion in a Laurent series in the neighborhood of an isolated singularity of a function $f(z)$, holomorphic in a domain $D \backslash c$, then c is an eliminable singularity, a pole, or an essential singularity according to whether $a_k = 0$ for $k < 0$, $a_k = 0$ for $k < -m$ and $a_{-m} \neq 0$, or $a_k \neq 0$ for infinite $k < 0$ and vice versa. For a demonstration see Remmert (1984, p. 252).

[33]Jacobi's theorem provided the initial stimulus to Casorati's research on multi-valued functions with more than two periods. On this argument see Bottazzini (1977b).

[34]The text of Liouville's lectures, edited by Borchardt, was not published until 1880. For a discussion of Liouville's demonstration of this theorem, as given in his manuscript, see Pfeiffer (1983).

[35]The reference to curved surfaces has led Kline (1972, p. 640) to hypothesize that Cauchy perhaps reformulated his theorem after "learning of Green's work in 1828." However, there is no explicit reference to the English mathematician in Cauchy's 1846a paper.

[36]Cauchy limits himself to requiring continuity in S for P and Q and says nothing about their derivability nor the continuity of their derivatives. Besides, as we have seen Cauchy himself say with respect to Laurent's theorem, he himself frequently changed his mind about the hypotheses required for $f(z)$, both for his integral theorem and for expansion in series by the method of majorants. About 40 years later E. Goursat (1884) demonstrated Cauchy's theorem $\int_\gamma f(z)dz = 0$ by assuming that $f(z)$ was derivable with continuity in a domain bounded by the closed curve γ. He divided the domain into small squares to which he then added the residual parts near the contour γ and calculated the integral $\int_\gamma f(z)dz$ as the sum of the integrals taken along the boundary of these subdomains.

A few years later Goursat wrote that his 1884 demonstration did not really require the continuity of $f'(z)$ and concluded his second demonstration of Cauchy's theorem with the observation, "We see that by assuming Cauchy's point of view it is *sufficient* to build the theory of analytic functions, to suppose the *continuity* of $f(z)$ and the *existence* of the derivative" (Goursat, 1900, p. 16). But to do this he was forced to demonstrate a subtle and difficult lemma on the uniform approximation of the difference quotient by the derivative.

In the following year Pringsheim took up Goursat's demonstration, observing that "the true *heart* of this integral theorem lies in its validity for a *special* region of the *simplest* form, for example, a *triangle*. ... The possibility of transferring it to a region bounded by a *curved* line unfortunately depends on *continuity* conditions which belong to the integrals of *every continuous* function" (1901, p. 418). Pringsheim consequently first demonstrated Cauchy's theorem for a triangle and then went on to the case of a closed curve γ by approximating it with polygons. Pringsheim's demonstration has become standard in modern texts (see Hille, 1963, I, pp. 163-167; as well as Remmert, 1984, pp. 136-141). Cartan (1961, pp. 70-1) demonstrates Cauchy's theorem for any rectangle within the domain and then, on the hypothesis that $f'(z)$ is continuous, gives a second demonstration using the Green-Riemann formula. For the case of a rectangle R with boundary γ, we have for $f(z) = u + iv$,

$$\int_\gamma f(z)dz = \int_\gamma u\,dx - v\,dy + i\int_\gamma v\,dx + u\,dy$$

$$= -\iint_R \left[\frac{\partial u}{\partial x} + \frac{\partial u}{\partial y}\right]dx\,dy + i\iint_R \left[\frac{\partial u}{\partial x} - \frac{\partial v}{\partial y}\right]dx\,dy.$$

But in the double integrals the integrands are zero for the Cauchy-Riemann conditions. The demonstration is generalizable to any simply connected domain.

The converse of Cauchy's integral theorem was demonstrated in 1886 by G. Morera (1856-1909).

[37]Ten years later Mourey's pamphlet came to Liouville's attention. He found there the clue to his own demonstration of the fundamental theorem of algebra (Liouville, 1839-40).

[38]In his 1863 Berlin lectures Weierstrass demonstrated that $\mathbb{C}$ is, apart from isomorphisms, the only commutative algebraic extension of $\mathbb{R}$ An analogous demonstration was published by Hankel, who concluded with the observation, "Thus the question has been answered, whose solution Gauss had promised in 1831 but not given" (1867, p. 107).

[39]Hamilton's paper of 1837 was an attempt to construct algebra "as the Science of Pure Time." According to Hamilton, algebra should be a science "deduced by valid reasonings from its own intuitive principles, and thus not less an object of *a priori* contemplation than Geometry" (1837, p. 5). While the basis of geometry is the idea of space, the basis of Hamilton's concept of algebra (when it is taken to include analysis) is the intuitive idea of time. After having explained his philosophical ideas in the introductory remarks, in the first part of his essay Hamilton introduced the integers, rational, and real numbers. Although his construction of the real number system was unsatisfactory, it nevertheless represents an important step towards the arithmetization of analysis. Hamilton then went on to illustrate his "theory of conjugate functions, or algebraic couples" in the second part of his article. Using a peculiar and today obsolete terminology, he introduced the following operations for ordered pairs of real numbers (a_1, a_2).

$$(b_1, b_2) + (a_1, a_2) = (b_1 + a_1,\ b_2 + a_2)$$

$$(b_1, b_2) - (a_1, a_2) = (b_1 - a_1,\ b_2 - a_2)$$

$$(b_1, b_2)(a_1, a_2) = (b_1 a_1 - b_2 a_2,\ b_2 a_1 + b_1 a_2)$$

$$\frac{(b_1, b_2)}{(a_1, a_2)} = \left[\frac{b_1 a_1 + b_2 a_2}{a_1^2 + a_2^2},\ \frac{b_2 a_1 - b_1 a_2}{a_1^2 + a_2^2} \right],$$

and defined the "couples" (1,0) and (0,1) as the "primary unit," equivalent to the number 1, and the "secondary unit," respectively. He further defined the commutativity of the addition and multiplication of couples, their distributivity, and the relation $(a,0) = a$ for any arbitrary real number a. He finally defined the power of number couples by considering the case $(-1,0)^{1/2} = (0,1)$. The field of complex numbers can be then considered as a model of algebraic couples so defined. For Hamilton the theory of conjugate couples represented a first step towards the elaboration of quaternions. A detailed examination of Hamilton's construction of the system of real numbers is given in Mathew (1978). For an evaluation of his theory within the development of English algebra see Nový (1973).

[40]It is not unreasonable to suppose that Cauchy's attention had been directed to Saint Venant (1845) by Hermann Grassman (1809-1877). The latter had learned in 1847 that Saint Venant had set out a vector system similar to his own and, not knowing his address, sent a letter to him in care of Cauchy, together with two copies of his *Ausdehnungslehre* of 1844. Saint Venant replied sometime later that he had never received the copy of Grassman's book.

Another priority question arose several years later between Cauchy and Grassman in regard to an article, "Sur les clefs algébriques," which appeared in the *Comptes rendus* in 1853 (*Oeuvres* (1) 11, pp. 439-445; (2) 14, pp. 417-66). In this article Cauchy presented a calculus that was very close to the algebraic methods developed by Grassman in the *Ausdehnungslehre*. For an analysis of Saint Venant's article and the priority dispute between Grassman and Cauchy, see Crowe (1967, pp. 81-85).

[41]If S is a domain limited by a simple closed rectifiable oriented curve and $f(z)$ is meromorphic in $S \cup \partial S$, then (4.6.8) shows that when z describes ∂S, the argument of $f(z)$ increases by a multiple of 2π given by $N - P$. For this reason Cauchy's theorem on the *compteur logarithmique* is sometimes called the "Argument principle" (e.g. see Hille, 1963, I, pp. 252-254).

[42]The general formulas of the theorem of residues for any closed curve requires the introduction of a notion that translates the intuitive idea of the number of times that the curve turns around each singularity into analytic terms. In order to do this one uses the idea of the index of a point with respect to a curve, which is a special case of the *compteur logarithmique*. In fact, if γ is a closed path in $\mathbb{C}$ and z is a point not on γ, then one defines the "index" of z with respect to γ by the integer

$$\mathrm{Ind}(\gamma,z) = \frac{1}{2\pi i} \int_\gamma \frac{d\zeta}{\zeta - z},$$

and the theorem of residues for a holomorphic function in a domain $D\backslash z_k$ ($z_k \in D$, $k \in N$) limited by a curve γ homologous to zero is given by

$$\frac{1}{2\pi i} \int_\gamma f(z)dz = \sum_k \mathrm{Ind}(\gamma,z_k)\mathrm{Res}(f,z_k),$$

on the assumption that no z_k is on γ.

If γ is a simple closed path, $\mathrm{Ind}(\gamma,z) = 1$ for every z within γ. For the case of a compact set the fundamental theorem can be stated more simply in the following manner.

"Let D be any open region of the Riemann sphere S_2 and let f be a function holomorphic in D except perhaps for isolated points that

are singular for f.   Let Γ be the oriented boundary of a compact
topological space A contained in D and let us suppose that Γ does not
contain any singular point of f nor the point at infinity.   The
singular points z_k contained in A are then finite in number and we
have the relation:

$$\int_\Gamma f(z)dz = 2\pi i \left(\sum_k \text{Res}(f,z_k) \right),$$

where $\text{Res}(f,z_k)$ designates the residue of the function f at the point
z_k; the sum includes all the singular points $z_k \in A$, including
eventually the point at infinity" (H. Cartan, 1961, p. 93).

The Riemann sphere is the sphere $x^2 + y^2 + u^2 = 1$ of the space $\mathbb{R}^3$
provided with the topology induced from $\mathbb{R}^3$.   If $P \equiv (0,0,1)$, then the
mapping $(x,y,u) \rightarrow z$ where $z = x+iy/1-u$ is a homeomorphism of $S_2 - P$ onto $\mathbb{C}$.

[43]A similar expansion in series of a branch of a real algebraic
curve $f(z,y) = 0$ in the neighborhood of a branch point is already
found in Newton, who also had the idea of determining the
exponents n/q by means of a graphic procedure (the polygon of
Newton). See Brill and Noether (1894, pp. 116-123).

Puiseux's theorem can be interpreted as the first important result
of the modern theory of uniformization: up to a linear change of
coordinates, u can be locally written as a holomorphic function of t
$= (z - a)^{1/q}$ $(q > 0)$. t is also a local uniformizing parameter for the
curve $f(u,z) = 0$.

[44]Hermite's article was not included in his *Oeuvres*. For an idea of
its contents it is consequently necessary to refer to the report written
by Cauchy (1851d) for the Academy.

[45]To be truthful, Liouville did not really appreciate Biot and
Bouquet's initiative. In the year their book was published, 1859, he
reasserted his priority by again choosing the theory of doubly
periodic functions as the subject of his lectures at the Collège de
France. Many years later among his unpublished manuscripts he
gave vent to his bitterness by writing, "MM. Briot and Bouquet,
cowardly thieves, but the most worthy Jesuits. Elected as thieves by
the Academy!!!!!" (In: Belhoste and Lützen, 1984, p. 28).

Chapter 5
THE CONVERGENCE OF FOURIER SERIES

5.1. The "Demonstrations" of Cauchy and Poisson

At the beginning of the last century, the study of partial differential equations was a topic of the greatest interest among European mathematicians. Fourier had provided a striking example by integrating the equation for the propagation of heat, while the mathematical treatment of many other physical problems which were then under investigation led to equations of the same type. Chief among these were problems involving the propagation of plane waves and potential theory. (For a further discussion of the latter, see the Appendix).

Laplace had investigated the first problem already in 1778. Shortly thereafter, in 1781, Lagrange had studied the propagation of surface waves in a thin layer of water, but had died before he could revise the section on hydrodynamics in the second edition of his *Mécanique analytique* (1811-15).

As a result, at the beginning of the nineteenth century, the theory of hydrodynamics had not advanced far beyond the point where the work of the eighteenth-century mathematicians had left it. In 1815 the Institute therefore decided to propose the following question: "A heavy fluid mass, initially at rest and of an indefinite depth, is set in motion by a given cause. Describe, at the end of a determinate period of time, the form of the outer surface of the fluid and the velocity of each of the molecules situated on this surface" (In: Cauchy, 1827c, p. 5).

The prize was won by a paper written by Cauchy[1] (1827c). In it, he arrived at the equation

$$(5.1.1) \qquad \frac{\partial^2 q_0}{\partial a^2} + \frac{\partial^2 q_0}{\partial b^2} = 0,$$

where q_0 indicates the velocity potential. He integrated this in the form

$$q_0 = \Sigma \int_0^\infty \cos(am) \cdot e^{bm} \cdot f(m)dm$$

$$+ \Sigma \int_0^\infty \cos(am) \cdot e^{-bm} \cdot g(m)dm,$$

though without indicating how he arrived at the solution, and then demonstrated in note *ix* that the solution is "general" (*ibid.*, pp. 146-9).

Fourier had also integrated Laplace's equation (5.1.1) for the case of a lamina, but Cauchy did not mention this. It appears that the information he had about Fourier's paper was limited to Poisson's brief summary of 1808, and that this remained the case until 1818.

In fact, in a note (*xix*) that was later added to the paper, where Cauchy explained the "reciprocal" formulas

$$f(x) = \sqrt{(2/\pi)} \int_0^\infty \phi(\mu) \cos \mu x \, d\mu$$

$$\phi(x) = \sqrt{(2/\pi)} \int_0^\infty f(\mu) \cos \mu x \, d\mu$$

that he had found, but which were also known to Fourier, he claimed that he had found them on the basis of his own research and of Poisson's work on the theory of waves. After Fourier had drawn his attention to the results of his studies of 1807 and 1811, Cauchy wrote, "I discovered the same formulas there, and I hastened to render them the justice that is due him in this regard in a second note published with the date of December, 1818" (Cauchy, 1827c, p. 301).

A similar thing happened with the notation for the definite integral that is in common use today. Fourier had first introduced it, and after Cauchy began to use it in the 1820s, he always gave explicit credit to Fourier, writing, "If we write, as does M. Fourier ..."

In the same year that Cauchy's paper won the prize, and even before the prize had been awarded, Poisson (who had not been able to participate in the competition because he was already a member of the Academy) presented his own paper on the subject (Poisson, 1816). In it he integrated the equation

$$\frac{\partial^2 \phi}{\partial x^2} + \frac{\partial^2 \phi}{\partial y^2} + \frac{\partial^2 \phi}{\partial z^2} = 0,$$

(where ϕ has the same meaning as q_0 in (5.1.1)), and presented the solution by means of trigonometric series and Fourier integrals.[2]

Thus Cauchy, Poisson, and Fourier were all interested in studying functions obtained from the integration of partial differential equations. In these years all three were working on the same types of problems and often had heated disagreements over their findings. For example, once after Poisson had read a paper to the Academy, Fourier raised a series of objections to his results. Poisson's reply led to a heated discussion, in which Cauchy also joined, over the proper method of integrating partial differential equations for the propagation of waves and vibrations of a plate.[3]

The latter problem involved the extension to two dimensions of the old problem of the vibrating string. In 1808 the German physicist E. Chladni (1756-1827) had attracted the attention of the members of the Institute and even of Napoleon himself with a series of ingenious experiments on the vibrations of glass and metal plates, in which he succeeded in showing the nodal lines during the vibration. Chladni's interesting experiments described a phenomenon that, in order to be fully explained, required the determination of differential equations.

This problem was the object of a "*prix extraordinaire*" offered by the Institute in the following year which was renewed again in 1813 and 1816. The winner was a paper submitted by Sophie Germain (1776-1831), but, as usual, the publication was delayed for many years.[4]

In 1818 Fourier also presented a paper on the same subject, but in the *Bulletin de la Société philomatique* Poisson called this as "a question of pure curiosity," of little relevance from the mathematical point of view, which did not add anything to the studies Poisson had published himself. In reply, Fourier pointed out that both his own analysis and that of Poisson on the propagation of plane waves could be derived "from the principles that have been used to determine the analytical laws of the motion of heat," that is, Fourier's memoirs of 1807 and 1811, which at that time were still unpublished.

Fourier's paper was followed a short time later by a long article by Poisson on the propagation of waves in a three-dimensional fluid and by two additional works of Cauchy on the same subject which were published in the *Journal de l'Ecole polytechnique*. These studies anticipated one of the most widely discussed subjects of the 1820s, the theory of elasticity which had been initiated by the works of Claude Navier (1785-1836) and Cauchy.[5]

From the mathematical point of view, the integration of the partial differential equations to which these studies generally led rendered more pressing than ever the question that Fourier had proposed regarding the representability of functions by convergent trigonometric series. Indeed, beginning in the 1820s, Cauchy and Poisson frequently concerned themselves with the problem of the convergence of such series, although they used different techniques and had different intentions.

One of the first works was Poisson (1820b). At the beginning of this paper he wrote,

> When one applies analysis to questions of physics or mechanics, or even to the simple problems of geometry, one must sometimes express certain functions by a series of sines or cosines of arcs proportional to the variable. In certain cases, these functions must therefore be represented for all the real values of the variable, from negative infinity to positive

infinity; the series of which we are speaking then change into
definite integrals. ... At other times, the functions being
considered are only given in a limited range of the values of the
variable; it is only for these values that the functions must be
reduced to series of periodic quantities, or, if you wish, there
are only parts of the functions for which it is then necessary to
give this form (Poisson, 1820b, p. 417).

Curiously enough, Poisson began from an old work of Lagrange
that had been written in the course of a discussion on vibrating
strings (see §1.3). He consequently assumed that $t = 0$ in equation
(1.3.17) and wrote it as

$$(5.1.2) \qquad f(x) = \frac{2}{\ell} \int \left[\sum_i \sin \frac{i\pi x}{\ell} \sin \frac{i\pi\alpha}{\ell} \right] f(\alpha)d\alpha.$$

This formula represents, Poisson said, all the values of $f(x)$ for x
lying between 0 and ℓ (which are also the limits of integration of
the integral). At this point Poisson made the important observation
that, "In general, an infinite series of periodic quantities, like that
contained in the preceding formula, can only have a clear and
precise meaning when one treats it as the limit of a convergent
series" (1820b, pp. 421-2).
   He then multiplied the general term of the series in (5.1.2) by the
exponential e^{-ki} ($k > 0$) to obtain

$$(5.1.2') \qquad f(x) = \frac{2}{\ell} \int \left[\sum_i e^{-ki} \sin \frac{i\pi x}{\ell} \sin \frac{i\pi\alpha}{\ell} \right] f(\alpha)d\alpha$$

"in which one must make k infinitely small or zero, after having
performed the calculation. The introduction of this exponential,"
Poisson writes, "by rendering the series convergent, dissipates the
difficulties that Lagrange's formula (5.1.2) presented; and, in this
form, one will see that it becomes susceptible to a direct and
rigorous demonstration" (1820b, p. 422).
   In order to do this, he assumes the result (known from the theory
of series) that

$$\sum_i e^{-ki} \cos i\theta = \frac{e^k - e^{-k}}{2(e^k - 2\cos\theta + e^{-k})} - \frac{1}{2}$$

in which $(x - \alpha)\pi/\ell$ is substituted in place of θ. Then, after
multiplying both members by $f(\alpha)(d\alpha/\ell)$ and integrating, he obtains

$$\int \left[\sum_i e^{-ki} \cos \frac{i\pi(x - \alpha)}{\ell} \right] f(\alpha) \frac{d\alpha}{\ell} + \frac{1}{2\ell} \int f(\alpha)d\alpha$$

$$(5.1.3) \qquad = \int \frac{(e^k - e^{-k})f(\alpha)d\alpha}{2\ell[e^k - 2\cos((x-\alpha)\pi/\ell) + e^{-k}]} .$$

Taking the integral from 0 to ℓ and supposing that x is included
between the same values and that $f(x)$ is not infinite, we have

$$(5.1.4) \qquad f(x) = \int \left[\sum_i e^{-ki} \cos \frac{i\pi(x - \alpha)}{\ell} \right] f(\alpha) \, \frac{d\alpha}{\ell} + \frac{1}{2\ell} \int f(\alpha) d\alpha$$

"provided that one makes $k = 0$ after the integration" (*ibid.*).

Equation (5.1.4), after interchanging the signs of integration and series, becomes the complete Fourier series for $f(x)$, while the integral of the second member of (5.1.3) is the so-called 'Poisson integral,' which first appeared in Poisson (1811).

Poisson reasons as follows: let us suppose that k is an infinitesimal. Then it will also be the integrand of the second member of (5.1.3) for all values of α, except those that render $\cos i\pi(x - \alpha)/\ell$ infinitely close to 1.   (In this case the integrand assumes the indeterminate form 0/0.)

In order to avoid the indeterminacy, Poisson considers $\alpha = x + u$ (with u an infinitesimal), and takes the integral from $u = -\beta$ to $u = \beta$, where β is a positive infinitesimal quantity. In such an interval, Poisson says, $f(\alpha)$ can be considered to be constant and equal to $f(x)$. Moreover, disregarding the third order infinitesimals with respect to k and u, we have

$$e^k - e^{-k} = 2k, \; e^k - 2 \cos \frac{(x - \alpha)\pi}{\ell} + e^{-k} = k^2 + \frac{u^2\pi^2}{\ell^2}$$

values which, when substituted into the integral of the second member of (5.1.3) give

$$(5.1.5) \qquad f(x) \int \frac{\ell k \, du}{k^2\ell^2 + \pi^2 u^2} = \frac{f(x)}{\pi} \arctan \frac{\pi u}{k\ell} + \text{constant.}$$

This, after passing to the definite integral between $-\beta$ and β, becomes

$$\frac{2f(x)}{\pi} \arctan \frac{\pi\beta}{k\ell} \; .$$

Finally, setting $k = 0$, we obtain

$$(5.1.6) \qquad \frac{2f(x)}{\pi} \arctan \infty = f(x),$$

which completes the demonstration.

But from (5.1.5) on, Poisson uses a technique that, as we have seen, Abel questioned in 1826; the substitution $k = 0$ is legitimate if the series that appears in the integral (5.1.3) is convergent, something that Poisson instead intends to prove.[6]

After giving his "demonstration," Poisson uses (5.1.3) to obtain the expansion in series under particular hypotheses (such as $f(x)$ even, etc.) and concludes, "There exist many other formulas of the same type as the preceding, which also have the property of expressing the values of any function whatsoever between the given limits, and which differ from each other by the different conditions that they fulfill at these limits."   After which he adds, "The studies of M. Fourier on the propagation of heat in solid bodies and my paper on this same subject[7] contain many formulas of this kind" (1820b, p.

425). Poisson, repeated this demonstration much later, almost unaltered (1823a and b), and it was taken up again much later still by H. A. Schwarz (1843-1921), who avoided the recognizable error.[8]

In opposition to the method Fourier had used for finding the coefficients, Poisson raised objections that seemed to be widely shared. "It has seemed to me that the formula for the expansion of functions in series has not in fact been demonstrated in a precise and rigorous manner," he wrote in the paper mentioned above (1823a, p. 46). And later, "the determination of the coefficients essentially supposes that one knows, in addition to the form of the functions, that they are expandable" (1835, p. 186). This insightful observation was seconded by Sturm when he wrote: "Fourier and other geometers seem to have misunderstood the importance and the difficulty of this problem [that of demonstrating the possibility of an expansion], which they have confused with that of determining the coefficients" (1836, p. 400).

This was the situation when, in 1827, Cauchy confronted the question of the expansion of functions in periodic series. He was unconvinced by Poisson's demonstration, which he felt was founded on intuitive and nonrigorous reasoning, and wrote that:

> In series of this type, the coefficients of the different terms are ordinarily definite integrals that include sines and cosines; and when the integrations can be made, by reason of the particular form attributed to the function that it is necessary to expand, it is easily seen that the series obtained are convergent. Nevertheless it is always desirable that this convergence be demonstrated in a general manner, independently of the values of the functions (1827b, p. 12).

Cauchy's paper is interesting in many ways. It in fact provides a mixture of "modern analysis" and old conceptions of special significance, in addition to a notable example of his way of using the methods of real and complex analysis (including his theory of residues).

Therefore, given the series

$$(5.1.7) \qquad f(x) = \frac{1}{a} \left\{ \int_0^a f(\mu)d\mu + 2 \sum_{n=1}^\infty \int_0^a f(\mu)\cos \frac{2n\pi(x - \mu)}{a} d\mu \right\}$$

he writes, "it is now necessary to demonstrate its convergence" between 0 and a. In order to do this, he makes use of a formula that he had found in the course of his own studies on the propagation of waves (1827c, p. 236) and on complex integration (1825).

Indeed, Cauchy says that "it is sufficient to remember" that for a function $\phi(\mu + i\nu)$, if one has

$$(5.1.8) \qquad \lim_{\nu \to +\infty} \phi(\mu + i\nu) = 0, \quad \lim_{\nu \to -\infty} \phi(\mu + i\nu) = 0,$$

then there follow respectively,

$$\int_0^a \phi(\mu)d\mu = -i \int_0^\infty [\phi(a + i\nu) - \phi(i\nu)]d\nu + 2\pi i \, {}_0^a\mathcal{E}_0^\infty \, \phi(z),$$

(5.1.9)

$$\int_0^a \phi(\mu)d\mu = i \int_0^\infty [\phi(a - i\nu) - \phi(-i\nu)]d\nu - 2\pi i \, {}_0^a\mathcal{E}_{-\infty}^0 \phi(z),$$

where $\mathcal{E} \, \phi(z)$ indicates the sum of the residues of the function in the given domain (see §4.4c).

By first setting $\phi(\mu) = e^{bi\mu}f(\mu)$ and then $\phi(\mu) = e^{-bi\mu}f(\mu)$, where $b > 0$ and $f(\mu)$ is bounded, from the integrals (5.1.9) one can obtain,

$$\int_0^a e^{bi\mu}f(\mu)d\mu = -i \int_0^\infty [e^{abi}f(a + i\nu) - f(i\nu)]e^{-b\nu}d\nu,$$

(5.1.10)

$$\int_0^a e^{-bi\mu}f(\mu)d\mu = i \int_0^\infty [e^{-abi}f(a - i\nu) - f(-i\nu)]e^{-b\nu}d\nu.$$

Rewriting (5.1.7) as

$$f(x) = \frac{1}{a}\int_0^a f(\mu)d\mu + \frac{1}{a} \sum_n \int_0^a e^{\frac{2n\pi}{a}(x-\mu)i} f(\mu)d\mu$$

(5.1.11)

$$+ \frac{1}{a} \sum_n \int_0^a e^{-\frac{2n\pi}{a}(x-\mu)i} f(\mu)d\mu,$$

from (5.1.10) Cauchy finally obtains

$$f(x) = \frac{1}{a}\int_0^a f(\mu)d\mu - \frac{i}{a}\sum_n e^{-\frac{2\pi n}{a}xi} \int_0^\infty e^{-\frac{2\pi n}{a}\nu}[f(a+i\nu) - f(i\nu)]d\nu$$

(5.1.12)

$$+ \frac{i}{a}\sum_n e^{\frac{2\pi n}{a}xi} \int_0^\infty e^{-\frac{2\pi n}{a}\nu}[f(a + i\nu) - f(-i\nu)]d\nu.$$

By now setting in the generic term v_n of the series (5.1.12) $(2n\pi/a)\nu = z$, as n grows to infinity, v_n reduces to

(5.1.13) $w_n = -\dfrac{1}{2n\pi} [f(a) - f(0)]\sin\dfrac{2n\pi x}{a}$.

"Now, it is clear that the series which has the expression $[w_n]$ for a general term will be a convergent series," Cauchy writes, and therefore the given series $\Sigma \, v_n$ will also be convergent, which completes the proof.

In other words, Cauchy does not try to give a direct demonstration of the convergence of the Fourier series, but, so to say, "transforms" the convergence into that of a series which is easier to study by means of a convergence "criterion" which can be formulated as follows: "If the series $\Sigma \, w_n$ is convergent and the general term $w_n \to v_n$ as n tends to infinity, then the series $\Sigma \, v_n$ is also convergent."

The fact that this criterion can be easily refuted by Dirichlet's counterexample (see §5.2 below) should not lead us to conclude that Cauchy's error here was particularly clumsy. Instead, the error is extremeley indicative of the attitude of a "working mathematician" towards his own science -- a person who was firmly pledged to "do mathematics" in a new way, to try unexplored paths. Cauchy sets out propositions, considers them, returns to the statements, and takes up questions apparently resolved in order to give them new solutions with new perspectives.

Mathematics here seems more like an experimental science than the hypothetico-deductive framework to which we are accustomed. But then, logical structures appear more often at the time the theory is being reorganized than at the actual moment of discovery. This was particularly true in the case of Cauchy, as becomes clear when we compare his lectures at the Ecole Polytechnique with his research papers.

However, according to Dirichlet, Cauchy himself was not completely satisfied with his work. "The author of this work himself admits that his proof is defective for certain functions for which the convergence is, however, incontestable," he wrote (Dirichlet, 1829, p. 119). In any case, Cauchy took up the question the following year from a completely different point of view, utilizing his new theory of residues (1827d). This work appeared in the second volume of his *Exercices de mathématique*, but it passed unnoticed at the time.[9]

These studies on the convergence of Fourier series did not provide Cauchy with an opportunity for returning to his theorem on the continuity of a function that is the sum of a series of continuous functions. Thus, in his Turin lectures, (see §4.5) which were published in the *Résumes analytiques* of 1833, we find the theorem in question still being restated in the same terms he had used in the *Cours d'analyse*.

5.2. Lejeune-Dirichlet's Memoir

The question of the convergence of Fourier series was approached from a completely different point of view, one which was later to become dominant, by Lejeune-Dirichlet. During an extensive stay in Paris, Dirichlet had belonged to the group of intellectuals that had gathered around the figure of Fourier. But Dirichlet's mathematical work was influenced not only by Fourier but by Gauss. Indeed, the latter's *Disquisitiones arithmeticae* provided a valuable stimulus to his research on number theory.[10]

The combined influence of Fourier and Gauss was also responsible for Dirichlet's research into various fields of mathematical physics, particularly potential theory. In his lectures on this subject, Dirichlet stated his famous "principle," which remained a topic of discussion and research among mathematicians for many years (see the Appendix).

The paper in which Dirichlet confronted the question appeared in the fourth volume of the *Journal für die reine und angewandte*

Mathematik,[11] the new review founded by Crelle which became one of the most prestigious mathematical journals of the century and has remained so even to the present day. Abel had helped to found it in 1826 and Crelle had offered him the editorship. After Crelle's death in 1855, its editors included men like Kronecker and Weierstrass. It would not be improper to say that, at least for its first 40 to 50 years of life, Crelle's journal published the most significant articles in mathematics, those that were destined to set the trends of mathematical research. Abel, Jacobi, Dirichlet, Kummer, Dedekind, Hermite, Cayley, Riemann, Weierstrass, and Kronecker are among the names that we encounter in its annual volumes.[12]

Dirichlet's paper begins with a few critical observations on Cauchy's article of (1827b), the only one, Dirichlet declares, that he knows on this question.

Dirichlet's central argument against Cauchy's demonstration is that, if a series has $A(\sin nx)n$ (this is how Dirichlet writes (5.1.13)) as a general term and is convergent, this does not imply that a series Σv_n is also convergent when the ratio of the terms of the two series with the same index differs from unity by as little as desired, providing the indices are taken to be suitably large.

Dirichlet proposes the following counterexample:

$$\Sigma \frac{(-1)^n}{\sqrt{n}},$$

which is a convergent series;

$$\Sigma \frac{(-1)^n}{\sqrt{n}}\left[1 + \frac{(-1)^n}{\sqrt{n}}\right],$$

which is divergent, and yet the ratio of their generic terms is $1 \pm 1/\sqrt{n}$, which converges to 1 as n increases.

The method used by Dirichlet follows directly from the analysis of the demonstration given by Fourier for (2.3.15). As Darboux has observed, "This method consists, as we have seen, in expressing the sum of the first m terms of the series by a definite integral, and then of searching for the limit of this integral" (In: Fourier, *Oeuvres* 1, p. 158). This is how Dirichlet develops his demonstration. We consider, he says, a number h with $0 < h < \pi/2$ and let $f(\beta)$ be a continuous function, positive and decreasing monotonically in $[0,h]$. We form the integral

$$(5.2.1) \qquad \int_0^h \frac{\sin i\beta}{\sin \beta} f(\beta)d\beta,$$

where i is a positive quantity. It is now necessary to study what happens to the integral as i increases. In order to do this Dirichlet divides the interval of integration $[0,h]$ into partial intervals, $[0,\pi/i]$, $[\pi/i, 2\pi/i]$, ..., $[(r - 1)\pi/i, r\pi/i]$, $[r\pi/i, h]$, where $r\pi/i$ indicates the largest multiple of $\pi/i < h$.

Now it is clear that the factor $\sin i\beta$ in the integrand allows the integral to successively change sign as β crosses the various intervals, and that consequently the $(r + 1)$ integrals into which the integral (5.2.1) is decomposed are alternately positive and negative.

Moreover, each of these is smaller in absolute value than the preceding. In fact, if we consider any two consecutive values,

(5.2.2) $\displaystyle\int_{(\nu-1)\pi/i}^{\nu\pi/i} \frac{\sin i\beta}{\sin \beta} f(\beta)d\beta$ and $\displaystyle\int_{\nu\pi/i}^{(\nu+1)\pi/i} \frac{\sin i\beta}{\sin \beta} f(\beta)d\beta$,

and substitute $\beta + \pi/i$ into the second, this latter is transformed into the integral

(5.2.3) $\displaystyle\int_{(\nu-1)\pi/i}^{\nu\pi/i} \frac{\sin(i\beta + \pi)}{\sin(\beta + \pi/i)} f(\beta + \pi/i)d\beta$

$\displaystyle = -\int_{(\nu-1)\pi/i}^{\nu\pi/i} \frac{\sin i\beta}{\sin (\beta + \pi/i)} f(\beta + \pi/i)d\beta$,

which is certainly smaller in absolute value than the first integral of (5.2.2) because

$$f(\beta + \pi/i) < f(\beta)$$

and

$$\sin(\beta + \pi/i) > \sin \beta$$

(with $\sin \beta$ increasing for $0 < \beta < h < \pi/2$).

We now consider the first of the two integrals (5.2.2). The two factors that form the integrand, $\sin i\beta/\sin \beta$ and $f(\beta)$, are both continuous functions of β within the limits of integration and the first of these always has the same sign. Then the integral can be expressed as the integral of the first factor for a quantity ρ_ν included between $f((\nu - 1)\pi/i)$ and $f(\nu\pi/i)$, or as

(5.2.4) $\displaystyle\rho_\nu K_\nu = \rho_\nu \int_{(\nu-1)\pi/i}^{\nu\pi/i} \frac{\sin i\beta}{\sin \beta} d\beta$,

a quantity that depends on ν and on i and is positive or negative as $\nu - 1$ is even or odd.

By changing the variables in the integral (5.2.4), that is, by putting γ/i in place of β, with γ variable, we find that as i increases the integral converges to the limit

(5.2.5) $\displaystyle k_\nu = \int_{(\nu-1)\pi}^{\nu\pi} \frac{\sin \gamma}{\gamma} d\gamma$

(these integrals are also alternately positive and negative). On the other hand, it is known that the integral

$$\int_0^\infty \frac{\sin \gamma}{\gamma} d\gamma$$

or, equivalently, the series of alternating terms

(5.2.6) $k_1 - k_2 + k_3 - \cdots$

derives from (5.2.5) converges to $\pi/2$.

At this point we consider the behavior of *Dirichlet's integral*, as the integral (5.2.1) is commonly called. We determine the limit, as i increases indefinitely, by dividing the infinite number of integrals that are obtained in two groups. The first,

(5.2.7) $K_1 \rho_1 - K_2 \rho_2 + K_3 \rho_3 + \cdots - K_m \rho_m$ (m even),

and the remainders,

$$\sum_{t=m+1}^{\infty} (-1)^{t+1} K_t \rho_t.$$

The integrals (5.2.7) are such that the quantities $\rho_1, \ldots, \rho_m$ tend to the limit $f(0)$ as i increases, while $K_1, \ldots, K_m$ tend to the values $k_1, \ldots, k_m$ respectively under the same hypothesis for i.

Therefore (5.2.7) converges to the limit

(5.2.8) $\sum_{n=1}^{m} (-1)^{n+1} k_n f(0) = S_m f(0).$

On the other hand, from (5.2.8) it follows that the sum in question will always be less than $K_{m+1} \rho_{m+1}$, which converges to $k_{m+1} f(0)$. When m is suitably large, k_{m+1} will be less than any given quantity and therefore, finally, the integral (5.2.1) will converge to the limit $(\pi/2) f(0)$ as i increases indefinitely.

Dirichlet also demonstrates this result for a constant function $f(\beta)$ and for the case of $f(\beta)$ increasing. In summarizing, Dirichlet concludes this first part of the proof by stating: "Whatever be the function $f(\beta)$, provided that it remains continuous between the limits 0 and h (h being positive and at the very most equal to $\pi/2$), and that it increases or decreases from the first of these limits to the second, the integral $\int_0^h (\sin i\beta/\sin \beta) f\beta \, d\beta$ will end by constantly differing from $(\pi/2) f(0)$ by a quantity smaller than any assignable number, when one makes i increase beyond every positive limit" (1829, p. 127).

In the same way Dirichlet shows that the integral $\int_0^g (\sin i\beta/\sin \beta) f(\beta) d\beta$ converges to $(\pi/2) f(0)$ when $0 < g < h < \pi/2$, and therefore that when the limits of integration are g and h (that is the integral (5.2.1) is taken between g and h), it converges to zero.

At this point Dirichlet can move on to demonstrate the convergence of Fourier series. "The steps that we are going to follow," he writes, "will lead us to establish the convergence of these series and at the same time to determine their values" (1829, p. 128). We are here facing a new approach, that is, the idea of showing that, under certain conditions, the series (2.3.14) is convergent (which was the object of the earlier demonstrations by Poisson and Cauchy) and in fact converges to $\phi(x)$.[13]

Dirichlet begins by considering the series

$$(5.2.9) \quad \frac{1}{2\pi} \int \phi(\alpha)d\alpha + \frac{1}{\pi} \sum_{1}^{\infty} \begin{Bmatrix} \cos nx \int \phi(\alpha)\cos n\alpha \, d\alpha \\ \sin nx \int \phi(\alpha)\sin n\alpha \, d\alpha \end{Bmatrix},$$

where the integrals are taken from $-\pi$ to π.

By taking the first $2n + 1$ terms of the series, that is,

$$\frac{1}{\pi} \int_{-\pi}^{\pi} \phi(\alpha)d\alpha \left[\frac{1}{2} + \sum_{i=1}^{n} \cos i(\alpha - x) \right],$$

and summing, this can be put in the form

$$(5.2.10) \quad \frac{1}{\pi} \int_{-\pi}^{\pi} \phi(\alpha) \frac{\sin(n + 1/2)(\alpha - x)}{2 \sin (1/2)(\alpha - x)} d\alpha,$$

(which is also commonly called "Dirichlet's integral").

In order to solve the problem it is necessary to study the behavior of the integral (5.2.10) as n increases, utilizing the results obtained from the similar integral (5.2.1).

In the case of (5.2.10) we divide the interval of integration into two parts, from $-\pi$ to x and from x to π, and substitute $x - 2\beta = \alpha$ into the first integral obtained and $x + 2\beta = \alpha$ into the second. The integral (5.2.10) will then be transformed, leaving aside the factor $1/\pi$, into the following:

$$(5.2.11) \quad \begin{aligned} &\int_{0}^{(\pi+x)/2} \frac{\sin(2n + 1)\beta}{\sin \beta} \phi(x - 2\beta)d\beta; \\ &\int_{0}^{(\pi-x)/2} \frac{\sin(2n + 1)\beta}{\sin \beta} \phi(x + 2\beta)d\beta. \end{aligned}$$

We now consider the second of these integrals. We suppose that $\frac{1}{2}(\pi - x) \leqslant \pi/2$ (when $x = \pi$ the integral is zero for every value of n) and let $\ell, \ell', \ell'', ..., \ell^{(\nu)}$ be the points corresponding to the values of β in which the function $\phi(x + 2\beta)$ has discontinuities or maxima and minima between $\beta = 0$ and $\beta = \frac{1}{2}(\pi - x)$. We now consider the second integral to be split into $\nu + 1$ integrals between the limits of integration $0, \ell, \ell', \ell'', ..., \ell^{(\nu)}, \frac{1}{2}(\pi - x)$.

If we suppose that $\frac{1}{2}(\pi - x) \leqslant \pi/2$, then from what has already been demonstrated, all the integrals will have zero as a limit as n increases, excluding the first which will tend to the limit $(\pi/2)\phi(x + \epsilon)$, ϵ being a given arbitrarily small positive quantity.

If $\frac{1}{2}(\pi - x) > \pi/2$, then we will again divide the interval of integration into two parts, from $\beta = 0$ to $\beta = \pi/2$ and from this value to $\beta = \frac{1}{2}(\pi - x)$. In the first case we will find the situation already discussed; in the second, by using the substitution $\beta = \pi - \gamma$, we will find a form like the preceding, the limit of the integral being zero, except for $0 = \frac{1}{2}(\pi + x)$, or $x = -\pi$, when the limit is $\phi(\pi - \epsilon)$.

We can proceed in a similar manner for the first integral of

(5.2.11), finding that it is zero for $x = -\pi$, has $(\pi/2)[\phi(\pi - \varepsilon) + \phi(-\pi + \varepsilon)]$ as a limit when $x = \pi$, and in the other cases converges to $(\pi/2)\phi(x - \varepsilon)$.

Now from this Dirichlet concludes that, "it has been proven that this series (5.2.9) is convergent, and one finds by means of the preceding results that it is equal to:

$$(5.2.12) \quad \frac{1}{2}[\phi(x + \varepsilon) + \phi(x - \varepsilon)]$$

(where ε is an infinitesimal quantity) for every value of x between $-\pi$ and π, and that for each of these extreme values, $-\pi$ and π, it is equal to

$$\frac{1}{2}[\phi(\pi - \varepsilon) + \phi(-\pi + \varepsilon)]" \quad (1829, \text{ p. } 131).$$

If x is not a point of discontinuity, (5.2.12) effectively gives $\phi(x)$ as a value of the series in x.

These are the conclusions that Dirichlet draws from the fact that the function $\phi(x)$ is continuous except at a finite number of discontinuities and that it has a finite number of maxima and minima. Then $\phi(x)$ is of bounded variation, as we say today, using Jordan's terminology.[14]

"It remains to consider the cases where the assumptions that we have made concerning the number of solutions of continuity [discontinuities] and the number of maximal and minimal values do not hold," Dirichlet adds at this point (1829, p. 131).

If the points of discontinuity are infinite in number, (5.2.10) makes sense only when the function is given in such a way that, for any two values a and b where $-\pi < a < b < \pi$, we can find two values r and s, with $a < r < s < b$, such that the function is continuous in the interval (r,s).

"One readily feels the necessity of this restriction on considering that the various terms of the series are definite integrals and on returning to the fundamental concept of an integral," Dirichlet says (1829, pp. 131-2). As an example of a function that does not satisfy this condition, he gives the celebrated 'Dirichlet function,' a function that is equal to a constant c when x is rational and a constant $d \neq c$ when x is irrational.

"The function so defined has finite and determinate values for every value of x, and yet one does not know how to substitute it in the series, seeing that the various integrals that enter into this series will loose all meaning in this case" (1829, p. 132). The restriction on the discontinuity of a function is thus directly tied to Cauchy's concept of an integral, the only one at hand. Cauchy had defined the integral for continuous functions in an interval or discontinuous functions in a finite number of points. Dirichlet, probably working from the idea that the integrability of a function is equivalent to the fact that the set of its points of discontinuity form what is today called a "nowhere dense" set, overcomes the difficulty inherent in the presence of an infinity of points of discontinuity by requiring that the function be "piecewise" continuous. This, together with the requirement that it not become infinite,[15] is the only condition that must be imposed on $\phi(x)$.

Thus Dirichlet seems to be of the opinion that a continuous function, even though it has an infinite number of maxima or minima, can always be represented by a convergent Fourier series. More than twenty years later, in 1853, when Gauss returned to the subject and wrote to him suggesting that, probably without any particular difficulty, one could extend the demonstration of the convergence of trigonometric series to the case in which the function to be expanded has an infinite number of maxima and minima, Dirichlet replied, "After a detailed consideration of the issue I find your conjecture entirely valid, provided one wishes to ignore certain entirely singular cases" (Dirichlet, *Werke* 2, p. 386).

The letter continued with a sketch of how one could demonstrate this. That Dirichlet's optimism on the power of his theorem was misplaced did not become clear until 1876, when Du Bois Reymond (1831-1889) provided a counterexample of a continuous function whose Fourier series does not converge in isolated points.[16]

In this circumstance Du Bois Reymond confirmed that he had learned from a conversation with Weierstrass "that Dirichlet appears never to have lost confidence in his proof" (1876, p. xx, n. vi).

Dirichlet himself, however, was well aware that the examination of cases excluded from his conditions of convergence involved the most delicate questions of analysis, those of the continuity and integrability of a function. He closed his paper with the observation that "the thing, in order to be done with all the clarity that one can desire, requires certain details linked to the fundamental principles of infinitesimal analysis which will be set forth in another article" (1829, p. 132).

5.3. Dirichlet's Concept of a Function

The paper foreseen in this remark never saw the light of day, even though Dirichlet published a second article on the same argument several years later in the first issue of the *Repertorium der Physik* (1837), a review that was directed by Dirichlet himself, together with Jacobi and F. Neumann, although its editor was H. W. Dove (1803-1879).

However, Dirichlet's work here does not present any substantial advance with respect to the contents of the 1829 paper. Thus it was with good reason that Kronecker (1823-1891), in a discussion he had with the Italian mathematician Casorati in Berlin in 1864, said, "the paper on continuity published by Dirichlet in Dove's *Repertorium* does not contain anything essential that is not already in Crelle's journal."[17]

In the *Repertorium* article Dirichlet limited himself to a

restatement of known results, together with the definition of a few fundamental concepts -- that of an integral (in Cauchy's sense), and that of the continuity of a function in an interval.

The notoriety that this work later acquired probably came from this latter definition. Kronecker himself left this impression by calling it simply "the paper on continuity." The real argument, however, as given by the title, was "the representation of completely arbitrary functions by means of series of sines and cosines."

This is the definition in question:

One thinks of a and b as two fixed values and of x as a variable quantity that can progressively take all values lying between a and b. Now if to every x there corresponds a single, finite y in such a way that, as x continuously passes through the interval from a to b, $y = f(x)$ also gradually changes, then y is called a continuous function of x in this interval. It is here not at all necessary that y depend on x according to the same law throughout the entire interval; indeed one does not even need to think of a dependence expressible by mathematical operations. Presented geometrically, that is with x and y thought of as the abscissa and ordinate, a continuous function appears as a connected [*zusammenhängende*] curve which for every value of the abscissa contained between a and b has only one point. This definition prescribes no common law for the individual parts of the curve; one can think of it as being put together from the most dissimilar parts or drawn entirely arbitrarily. It follows from this that a function of this kind can only be seen as completely determined in an interval when it is either given graphically for its entire range or subjected to mathematically valid laws for its individual parts. As long as one has determined the function for only a part of the interval, the manner of its extension to the rest of the interval remains completely arbitrary (Dirichlet, 1837, pp. 135-6).

In commenting on this definition in 1870, Hankel wrote that Fourier's results had revealed that the older concept of a function was insupportable as well as a tacit although essential hypothesis, that the properties of algebraic functions as regards their continuity, expandability into power series, etc., can be extended to all functions, and that therefore the requirement that any given function be representable analytically is revealed to be without significance. "Once this knot had been cut," Hankel says, the way was open to a definition of function like the following: "y is called a function of x when to every value of the variable quantity x within a certain interval there corresponds a definite value of y, no matter whether y depends on x according to the same law in the entire interval or not, or whether the dependence can be expressed by a mathematical operation or not" (Hankel, 1870, p. 67).

This is the form in which Hankel presents Dirichlet's definition.

However, he goes well beyond Dirichlet's intentions and words when he addes,

> This purely nominal definition, which in the following I will associate with the name of Dirichlet because it reverts fundamentally to his works on Fourier series which clearly demonstrated the indefensibility of all the older concepts, is however no longer sufficient for the needs of analysis, in that functions of this kind do not possess general properties, and with this all relationships between the values of the function for various values of the argument fall to the wayside (*ibid.*).

Thus, says Hankel, there is a great deal of confusion over what is meant by a function, as even a rapid survey of the best textbooks will reveal. There are those who define it according to Euler, others who define it according to Dirichlet, others who say that a functional dependence is given by a law, and still others who do not give any definition at all. "But everyone draws from his concept conclusions that are not contained in it" (*ibid.*).

There is consequently a need, Hankel continues, to give a definition of a function that is capable of establishing a theory and a calculus with functions that is consistent with the given definition, but is not so general as to become purely nominal and empty of any real content. Yet at the same time it must be able to identify the vast classes of interesting mathematical objects that are in sight. "But it is only in the most recent time [1851] that a solid foundation with a truly philosophical spirit has been newly laid by Riemann, in that he, beginning from Dirichlet's concept, established the (monogenic) function of a *complex variable*, and thus gave that empty definition a new content which approaches that of the older concept" (1870, p. 68) (see §6.2).

Now, Hankel's objective is to distinguish, in the case of functions of a *real* variable, between the functions that he calls *legitimate* and those that are *illegitimate* (like Dirichlet's function) with respect to the singularities that they can present. The instrument that he uses is an extension to the case of functions of the *principle of the permanence of formal properties* that he had enunciated for numerical systems.[18] In the latter case, the whole numbers and the properties of the operations that can be performed on them are the givens that are taken as permanent in successive extensions. In the case of functions it is the algebraic functions that offer the typical form to which all other functions must be more or less directly referred. This then, according to Hankel, is how we can formulate this principle in an attempt to overcome the "nominality" of Dirichlet's definition.

> When an expansion is given for a domain of the variable which cannot be directly extended beyond its limits because it loses its meaning outside [those limits], and when further for another domain of the variable another expansion is given which

cannot be extended into the first domain and there is no valid expansion for both domains, then both expansions will be seen to belong to one function when the function presents the same *properties* in each of the two domains (1870, p. 66).

But we are dealing with a formulation whose ambiguity Hankel himself was the first to see, since in the observations concluding his article he writes that "an unambiguous and acceptable form for this principle has however neither ever been given nor could it ever be found" (1870, p. 103).

The presence of points of discontinuity in functions of real variables, and particularly an infinite number of such points, forces us to search for other means of inquiry. We find, Hankel goes on to say, that such points can more easily be *outflanked* than *overcome*, as for example by considering *complex* variables instead of *real* ones, "in that by limiting the variability to the real values of the argument, one cannot arrive at a definition of a function satisfactory for the needs of analysis" (*ibid.*).

Beginning from theoretical premises that were radically different, Karl Weierstrass (1815-1897) arrived at similar conclusions during these same years. In his introductory lectures on the theory of analytic functions, he maintained that Dirichlet's definition, because it was too general, allowed one to "do too few things." If it has nevertheless come to be used, he says, it is because "we have tacitly transported the properties which all the functions considered possess to these general functions" (In: Dugac, 1973, p. 71).

In regard to Weierstrass' analytic work Dugac has said,

> One of the important tasks on which [Weierstrass] set his mind, which stands out clearly when we cast a synthetic look over his work, is to determine the largest class of functions for which one can give an analytic representation and which can most fully satisfy the needs of analysis. ... For Weierstrass this class was that of continuous functions, thanks to his theorem on the representation of continuous functions by uniformly convergent series of polynomials (Dugac, 1973, p. 71).

On the basis of this conception, the Taylor series becomes the most suitable foundation for analysis and for Weierstrass' theory of analytic functions.

Through a coherent analysis, Weierstrass thus arrives at a critical approach to Dirichlet's conviction that one can represent any continuous function whatsoever by a Fourier series. He first of all observes that the Fourier series of a continuous function does not always converge towards the function. In addition, an expansion of this kind does not permit a profound understanding of the properties of the function. "We cannot demonstrate that it is derivable," Weierstrass observes, "thus it lacks the most important means for knowing the properties of the variation of functions" (In:

Dugac, 1973, p. 71).

The interpretations of Dirichlet's definition of a *continuous* function that were given by Hankel and Weierstrass (but as we have seen, this specification completely disappears in their two comments!) have been wholeheartedly adopted by modern historians of mathematics. Thus E. T. Bell saw in it the first formulation of the modern concept of a correspondence between sets, writing, "Dirichlet's definition of a (numerical-valued) function of a (real, numerical-valued) variable as a table, or correspondence, or correlation between two sets of numbers hinted at a theory of equivalence of point sets" (Bell, 1945, p. 156). This demonstrates a remarkable interpretative fantasy and an ability to read into the text things that are completely absent from it.

For his part, C. B. Boyer gives this version of the story:

> Lejeune-Dirichlet ... in 1837 suggested a very broad definition of function: if a variable y is so related to a variable x that whenever a numerical value is assigned to x, there is a rule according to which a unique value of y is determined, then y is said to be a function of the independent variable x. This comes close to the modern view of a correspondence between two sets of numbers, but the concepts of "set" and "real number" had not at that time been established. To indicate the completely arbitrary nature of the rule of correspondence, Dirichlet proposed a very "badly behaved" function: when x is rational, let $y = c$, and when x is irrational, let $y = d \neq c$ (Boyer, 1968, p. 600).

This is all very suggestive, but unfortunately has nothing to do with the real story. Dirichlet neither proposed such a definition nor gave the example of the "pathological" function cited to illustrate the arbitrariness of the "rule of correspondence." Bourbaki is even more cavalier when he simply observes, "We know that it was on this occasion that Dirichlet, in clarifying the ideas of Fourier, defined the general notion of function as we understand it today" (1960, p. 247). What both authors seem to have missed is the adjective "continuous" that accompanies the term "function;" without this awareness one ends by understanding nothing of Dirichlet's ideas. Dirichlet was interested in defining in an explicit manner what Fourier had meant by an "arbitrary" function. In so doing he linked himself to a long discussion that had begun with Euler's definition of a continuous function and later been extended by Cauchy's new definition of continuity. What Dirichlet wanted to clarify was simply that a continuous function in an interval can be given either arbitrarily as a graph (where his idea of a continuous curve is largely intuitive), or as a mathematical formula which is not necessarily the same in every part of the interval. The fundamental idea is still that every continuous function, no matter how "arbitrary", can be expanded as a Fourier series.

How far Dirichlet was from what is today generally called 'Dirichlet's concept of a function' also emerges in what he wrote in the same 1837 paper while discussing the value of the Fourier series at the discontinuous points of a function. After observing that one can find isolated points of discontinuity for the function, he adds,

> The curve whose abscissa is β and whose ordinate is $\phi(\beta)$ thereupon consists of many pieces whose connectivity is interrupted at those points of the abscissa that correspond to those special values of β. For every such point on the abscissa there are exactly *two ordinates*, one of which belongs to the part of the curve that ends there while the other belongs to the part that begins there. In the following it will be important to distinguish these *two values of* $\phi(x)$ [my emphasis] and we will indicate them by $\phi(\beta - 0)$ and $\phi(\beta + 0)$ (1837, p. 156).

Having obtained formula (5.2.12), which he writes as

$$\frac{1}{2} [\phi(x + 0) + \phi(x - 0)]$$

(using the notation he introduced), Dirichlet comments on what is today called the right and left limits of $\phi(x)$ at the point x: "Where an interruption of the continuity occurs and thus the function $\phi(x)$ has *exactly two values*, [my emphasis] the series, which by its nature has a unique value for every x, represents half the sum of these values" (1837, p. 159).

Such are the "completely arbitrary" functions that already appear in the titles of the two papers. The origin of the attribution of 'Dirichlet's concept of a function' to Dirichlet probably results from his example of a "pathological" function:

$$f(x) = \begin{cases} c, & \text{if } x \text{ is rational} \\ d, & \text{if } x \text{ is irrational.} \end{cases}$$

But all of this was already present in 1755 in Euler's idea of a function as any correspondence whatever between variables. The crucial step must therefore be measured by the *use* of such functions. Dirichlet himself did not seem to show much interest in this function. He did not even include it among those "completely arbitrary" functions for which he bothered to speak of integrals and representation in Fourier series. Thus, the conceptual framework does not change until Riemann, and precisely with a new definition of the integral.

5.4. The Uniform Convergence of Series

Dirichlet's works illustrated in an unequivocal manner what Abel
had already shown by a counterexample -- that Fourier series can
represent discontinuous functions. There consequently existed an
entire class of functions that contradicted Cauchy's theorem on the
continuity of the sum of a series of continuous functions.

But the fact that a statement has been refuted does not mean that
it will be clear where the incriminating point lies. Nor, on the other
hand, did Dirichlet seem to have been aware of the contradiction
that existed between the proper demonstration of the convergence of
Fourier series and Cauchy's theorem. In his works he made no
comment about it.

On the other hand, no one at the time seems to have had any idea
of the convergence of a series different from that defined by
Cauchy in the *Cours d'analyse*. Not until in the 1840s did there
begin to emerge, with difficulty and in differing contexts, other
ways of considering the convergence of a series of functions. The
first person to explicitly denounce the patent contradiction in which
the theory of series found itself was Seidel (1821-1896), who had
been a student of Dirichlet at Berlin and then later studied under
Jacobi and F. Neumann at Königsberg. In an 1847 article devoted
to series that represent discontinuous functions, he wrote,

> One finds in Cauchy's *Cours d'analyse algébrique* ... a theorem
> which states that the sum of a convergent series whose
> individual members are functions of a quantity x and
> continuous in the vicinity of a particular value of x, is likewise
> always a continuous function of the same quantity in this
> neighborhood. It follows from this that series of this kind are
> not adapted to represent discontinuous functions in the vicinity
> of points where their values jump (Seidel, 1847, p. 35).

After briefly outlining the path of Cauchy's demonstration, Seidel
continued,

> Nevertheless, the theorem stands in contradiction to what
> Dirichlet has shown, that, for example, Fourier series also always
> converge if one forces them to represent discontinuous functions
> -- in fact, the discontinuity will frequently be found in the form
> of those series whose individual members are still continuous
> functions. ...
> When one begins from the certainty thus obtained that the
> proposition cannot be generally valid, then its proof must
> basically lie in some still hidden supposition. When this is
> subjected to a precise analysis, then it is not difficult to discover
> the hidden hypothesis. One can then reason backwards that this

cannot occur with series that represent discontinuous functions (*ibid.*, pp. 36-7).

The theorem that Seidel found to characterize these series was the following: "If one has a convergent series which represents a discontinuous function of a quantity x, whose individual members are continuous functions, then one must be able to give values of x in the immediate neighborhood of the point where the function jumps for which the series converges *arbitrarily slowly*" (Seidel, 1847, p. 37).

The demonstration Seidel gave for this can be sketched as follows: if $s(x)$ is the sum of the series and $s_n(x)$ and $r_n(x)$ are the sums of the first n terms of the series and the remainder respectively, then

(5.4.1) $s(x) = s_n(x) + r_n(x).$

Then, by taking into consideration the variation in $s(x)$ when we add an increment δ to x, we have

(5.4.2) $s(x + \delta) - s(x) = [s_n(x + \delta) - s_n(x)] + [r_n(x + \delta) - r_n(x)],$

an equation that decides on the continuity or discontinuity of $s(x)$ in the neighborhood of x. Since, in order for $s(x)$ to be continuous, $|s_n(x + \delta) - s_n(x)|$ must be $< \tau$, the entire question is reduced to studying the behavior of $|r_n(x + \delta) - r_n(x)|$.

Now, in order for the series to converge both $r_n(x + \delta)$ and $r_n(x)$ must be smaller than an arbitrary small ρ', ρ respectively, when n is suitably large. If the function $s(x)$ is continuous, then for positive ε smaller than ρ and ρ', one has $|r_n(x + \delta) - r_n(x)| < \varepsilon$ for $n > n_0(\varepsilon)$. And if, as δ tends to zero, there exists an integer N that is the largest of the successively determined n_0 such that, for $n > N$

(5.4.3) $|s(x + \delta) - s(x)| < \tau + \rho + \rho',$

then this inequality precisely expresses the continuity required for $s(x)$.

If, however, the number n_0 increases beyond every finite value when δ, beginning from an initial value η, tends to zero, then (5.4.3) ceases to be true and the convergence of the series in the neighborhood of x becomes arbitrarily slow. "However," Seidel remarks at the end of his work, "it is not impossible, barring further research, that the same thing may also happen with series whose values do not jump" (1847, p. 44).

The question that Seidel left open, whether the continuity of the sum of the series by itself implies uniform convergence, would find a negative response only much later with the construction of suitable counterexamples by Darboux (1875), Du Bois Reymond (1876), and Cantor (1880).

When Seidel's paper appeared, other mathematicians had already been studying what is today called the uniform convergence of a series of functions. Among these was Weierstrass, who was probably inspired by Gudermann (1798-1852), his teacher at the University of Münster. In an article that was published in the 1838 issue of Crelle's *Journal*, Gudermann had for the first time illuminated the property of the uniform convergence of certain infinite series that give the expansion of elliptic functions.

Weierstrass had followed Gudermann's course on elliptic functions in the Winter of 1839-40, "and it is thus that he had the opportunity to familiarize himself with this notion" (Dugac, 1973, p. 47), which he utilized in his first works on power series in 1841 and 1842. These works, however, remained unpublished until 1894.

Thus, with the demonstration that a convergent power series is uniformly convergent within the domain of convergence (see §7.3), Weierstrass also found a rigorous systematization of Abel's theorem on the continuity of a function that is the sum of a series of powers. The fundamental role played in analysis by the concept of the uniform convergence of series, however, was not explicitly emphasized by Weierstrass until the early 1860s and was subsequently developed during the course of his long career as a professor at the University of Berlin. In his lectures Weierstrass gave the definition of uniform convergence in an interval that has since become classic:

A$_1$) The series $\Sigma\, u_n(x)$ is said to be uniformly convergent in an interval $[a,b]$ when, for every arbitrarily small positive ε, there exists an $n_0(\varepsilon)$ such that $|r_n(x)| < \varepsilon$ for $n > n_0$ and for every x, $a \leqslant x \leqslant b$.

But this is not the sense with which Seidel had introduced his notion. Seidel had not been interested in the "global" properties characterizing the convergence of a series in an entire interval, but rather in the small but predetermined neighborhood of a point. This is the concept expressed by the following definition:

A$_2$) The series $\Sigma\, u_n(x)$ is uniformly convergent in the neighborhood of a point ζ of the interval $[a,b]$ if there exists a $\delta(\zeta)$ such that $|r_n(x)| < \varepsilon$ for every arbitrarily small positive ε, for $n > n_0(\zeta,\varepsilon)$ and for $\zeta - \delta(\zeta) \leqslant x \leqslant \zeta + \delta(\zeta)$.

The uniform convergence of a series in an interval as defined by A$_1$) naturally implies uniform convergence in the neighborhood of every point in the interval. The converse is also true, but the demonstration is not easy and was given by Weierstrass for the first time in 1880.[19]

Contemporary with Seidel, but as ignorant of his work as he was of Weierstrass', another young mathematical physicist was investigating the question of the mode of convergence of series, the

Englishman George Stokes (1819-1903). A descendant of the tradition begun by Peacock and Babbage, Stokes was familiar with the work of French mathematicians like Cauchy and Poisson, but was completely ignorant of what Dirichlet had done. He approached the study of series, and of Fourier series in particular, on the basis of Cauchy's theorem on the continuity of the sum of a series of continuous functions (he did not take it from the *Cours d'analyse* but rather from Moigno's *Lecons de calcul différentiel et de calcul intégral* (1840-44) which was based on Cauchy's lectures), and of Poisson's research on Fourier series. Stokes used Cauchy's definition of the continuity of a function and adopted his distinction between "convergent" and "divergent" series, but introduced an additional distinction between "essentially" and "accidentally" convergent series. This corresponds to the modern distinction between "absolute" and "conditional" convergence, which had already been introduced at the time by Dirichlet. Stokes also used Poisson's method of sums and the integral (5.1.3) in order to find the sum of a series, once he had demonstrated convergence in another way.

The most original point in Stokes' work lies in his study of Fourier series in the neighborhood of a point of discontinuity of the function. It is the same problem as Seidel's but Stokes followed a completely different approach.

In his words, he deals with the following question:

Let

(5.4.4) $u_1 + u_2 + \cdots + u_n + \cdots$

be a convergent infinite series having U for its sum. Let

(5.4.5) $v_1 + v_2 + \cdots + v_n + \cdots$

be another infinite series of which the general term v_n is a function of the positive variable h, and becomes equal to u_n when h vanishes. Suppose that for a sufficiently small value of h and all inferior values the series (5.4.5) is convergent, and has V for its sum. It might at first sight be supposed that the limit of V for $h = 0$ was necessarily equal to U. This however is not true (Stokes, 1849, p. 279).

This is the problem that Abel had treated for the special case of power series, which involves a question of double limit. In fact, if, as usual, we indicate the sum of the first n terms of the series (5.4.5) as $s_n(h)$, the question becomes that of knowing when we can exchange the operation of the limit in the two following limits:

$$\lim_{h \to 0} \lim_{n \to \infty} s_n(x) = V_0 = V(0)$$

and

$$\lim_{n \to \infty} \lim_{h \to 0} s_n(x) = U.$$

If we think that Stokes supposes the $v_n(h)$ to be continuous functions of h, then the problem is essentially that treated by Seidel relative to the continuity of the sum-function $V(h)$ when $h = 0$. Stokes' response is given by the following

"**Theorem.** *The limit of V can never differ from U, unless the convergency of the series* (5.4.5) *becomes infinitely slow when h vanishes.*

The convergency of the series is here said to become infinitely slow when, if n be the number of terms which must be taken in order to render the sum of the neglected series numerically less than a given quantity e which may be as small as we please, n increases beyond all limit as h decreases beyond all limit.

Demonstration. If the convergency do not become infinitely slow, it will be possible to find a number n_1 so great that for the value of h we begin with and for all inferior values greater than zero the sum of the neglected terms shall be numerically less than e" (Stokes, 1849, p. 281).

There is a clear distinction between what Stokes meant by an "infinitely slow" convergence and Seidel's "arbitrarily slow" convergence, as Hardy (1877-1947) has observed. "Stokes is considering an inequality satisfied for a special value of n, or at most an infinite sequence of values of n, and *not* necessarily for all values of n from a certain point onwards" (Hardy, 1918, p. 155). In other words, Stokes introduces the mode of convergence that is today called "quasi-uniform," defined as follows:

B_2) A series $\Sigma \, u_n(x)$ is quasi-uniformly convergent in the neighborhood of a point ζ in an interval $[a,b]$ if there exists a $\delta(\zeta) > 0$ such that $|r_n(x)| < \varepsilon$ for every arbitrarily small positive ε, every N and $n_0(\zeta,\delta,\varepsilon,N)$ greater than N and $\zeta - \delta(\zeta) \leqslant x \leqslant \zeta + \delta(\zeta)$.

In a manner similar to what has already been said for uniform convergence, also for quasi-uniform convergence we can think of a "global" definition for an interval.

B_1) A series $\Sigma \, u_n(x)$ is quasi-uniformly convergent in an interval $[a,b]$ if, for every arbitrarily small positive ε and every N, there exists a $n_0(\varepsilon,N)$ greater than N such that $|r_n(x)| < \varepsilon$ for $n = n_0$ and $a \leqslant x \leqslant b$.

In relation to Stokes' research, it is also important to introduce the definition (B_3) of quasi-uniform convergence at a point $x = \zeta$. This is written like (B_2), with the essential difference that δ depends on the preceding choice of ζ, ε and N.

The Italian mathematician Dini (1845-1918) in fact demonstrated that a necessary and sufficient condition for $s(x)$ to be continuous for $x = \zeta$ is that the series be quasi-uniformly convergent at the point ζ (Dini, 1878, pp. 107-8).

Thus, only after more than 20 years did the theorem set out by Cauchy in the *Cours d'analyse* begin to be clarified. Following Abel's counterexample and the silence of Dirichlet on the subject, three mathematicians, all unaware of each other's work and acting with different motives and objectives, produced different replies. Weierstrass introduced the idea of uniform convergence in an interval; Seidel that of uniform convergence in the neighborhood of a point; Stokes that of quasi-uniform convergence in the neighborhood of a point. Common to all three is the desire to surmount the idea of simple convergence as defined by Cauchy. For his part, Cauchy returned to his own theorem only in a note that appeared in the *Comptes rendus* of the Academy in 1853, observing simply that it is verifiable for a power series, but that it required restrictions "for other series." The example that he used is the same one that Abel had earlier given (see §3.5).

But Cauchy did not mention either the Norwegian mathematician or the more recent work of Seidel and Stokes. Instead he assumed a somewhat haughty tone. Certainly, in the enunciation of his theorem there was something that didn't work, he said, but after all, "it is easy to see how one can modify the statement of the theorem so that it will no longer have any exception. This is what I am going to explain in a few words" (Cauchy, 1853, pp. 31-2).

After recalling the definition of a continuous function that he had given in the *Cours d'analyse*, which is "generally adopted today," Cauchy observes that, "If one calls n' a whole number greater than n, the remainder r_n will be nothing else than the limit towards which the difference

$$(5.4.6) \qquad s_{n'} - s_n = u_n + u_{n+1} + \cdots + u_{n'-1}$$

will converge for increasing values of n'. We now consider that by attributing a sufficiently large value to n, we can render the modulus of the expression (5.4.6) (for any n') for all the values of x included between the given limits and, consequently, the modulus of r_n smaller than a number ε as small as one wishes" (1853, p. 32). This latter hypothesis on the convergence of a series in the interval, Cauchy says, is sufficient to guarantee the continuity of the sum-function $s(x)$. His old theorem must therefore be supplanted by the following:

Theorem (of uniform convergence). *If the different terms of the series $\Sigma\, u_n$ are functions of a real variable x, continuous with respect to this variable within the given limits; and if, in addition, the sum (5.4.6) always becomes infinitely small for infinitely large values of the whole numbers n and $n' > n$, then the series will be convergent and the sum s of the series will be, within the given limits, a continuous function of the* *variable x* (1853, p. 33).

But the language of infinites and infinitesimals that Cauchy used here seemed ever more inadequate to treat the sophisticated and complex questions then being posed by analysis. Furthermore, when Cauchy wrote this note in 1853, France was even less capable of preserving its leadership in research mathematics, as Lamé (1795-1870) pointed out that very year in an alarming report to the *Académie des sciences.* He exhorted mathematicians to do "pure" research, lamenting that the almost exclusive interest in applied questions impeded the development of mathematics. He thereby denounced the limitations inherent in the "polytechnic" tradition.

The problems posed by the study of nature, such as those Fourier had faced, now reappeared everywhere in the most delicate questions of "pure" analysis and necessarily led to the elaboration of techniques of inquiry considerably more refined than those that had served French mathematicians at the beginning of the century. Infinitesimals were to disappear from mathematical practice in the face of Weierstrass' ε and δ notation, and even the analysis of functions of complex variables, which Cauchy had initiated, was to be reformulated on the basis of the works of Riemann and the lectures of Weierstrass.

Notes to Chapter 5

[1]The Academy published this paper (Cauchy, 1827c) with as great a delay as it did (Cauchy, 1827a). In addition to the 13 notes in the original paper, Cauchy added "some new notes" which doubled the length of the paper that had actually won the prize.

[2]A detailed exposition of Poisson's and Cauchy's papers can be found in Burkhardt (1901-8, pp. 439-47).

[3]For the technical aspects of this intricate question see the accurate account given by Burkhardt (1901-8, pp. 454-63).

[4]Bucciarelli and Dworsky (1980) contains a scientific biography of Sophie Germain and a discussion of her contributions to the theory of elasticity.

[5]In an unpublished paper of 1820, Navier also studied the vibration of a plate. The following year, in an article presented to the Academy that was not published until 1827, he succeeded in determing the so-called 'Navier-Stokes equations,' the partial differential equations that express the conditions of equilibrium and vibration for isotropic elastic bodies. Allowing himself to be guided by the analogy with elasticity, Navier also found the equation of motion for fluids using the hypothesis of viscosity that Euler had neglected. In 1822 Cauchy (the examiner of the paper Navier

presented to the Academy) also applied himself to this issue and reproduced Navier's results. In 1828 he was also able to determine the equations of elasticity for the more difficult case of anisotropic bodies. In the same year Poisson also found the Navier-Stokes equations, beginning from a critique of Cauchy's molecular conceptual scheme.

These studies were naturally connected with the difficult and widely discussed problem of the propagation of light waves in the ether, which at that time was considered to be an elastic medium whose properties, however, were not well defined. The arguments and discussions that accompanied the emergence of the mathematical theory of elasticity were thus influenced by the scanty knowledge of the molecular structure of bodies available at the time. A treatment of this complex and fascinating story can be found in Burkhardt (1901-8) and Todhunter and Pearson (1960). See also Dahan (1985).

[6]Poisson's procedure has been revived by modern analysis for summing convergent series as well as summable divergent series. For a history of the latter theory see Tucciarone (1973).

[7]This appeared in the June, 1815 issue of the *Bulletin de la Société philomatique*.

[8]Schwarz gave a rigorous solution of 'Dirichlet's problem' for a circle. (See the Appendix.) In fact, if $u(x,y)$ is definite and continuous for all the points within *and* on the boundary of a surface T, and the partial derivatives

$$\frac{\partial u}{\partial x}, \frac{\partial u}{\partial y}, \frac{\partial^2 u}{\partial x^2}, \frac{\partial^2 u}{\partial y^2},$$

are finite and continuous and satisfy $\Delta u = 0$ *only* for the points within the surface T, Schwarz demonstrates, by using 'Poisson integral,' that, if T is a circle, the function u is uniquely determined by the values that it takes on the circumference.

Schwarz shows that these values can be given by any continuous function $f(\phi)$ by proving the theorem: "If along the boundary of a circular surface S, there is arbitrarily given a continuous and definite real function $f(\phi)$, finite for all values of the real argument ϕ, which with the increase of the argument periodically repeats about 2π, but with no other conditions imposed, then there always exists one (and according to the preceding only one) function u which satisfies the [above given] conditions for the surface S, and coincides with the given function $f(\phi)$ along the boundary of S.

This function will be represented by the integral

$$u(r,\phi) = \frac{1}{2\pi} \int_0^{2\pi} f(\psi) \frac{1 - r^2}{1 - 2r \cos(\psi - \phi) + r^2} d\psi$$

for all points $z = re^{i\phi}$ within S, $r < 1$" (Schwarz, 1872, p. 185).

On the basis of this result Schwarz was able to solve 'Dirichlet's problem' for a large class of domains, using the so-called "method of alternating procedures [*alternierendes Verfahren*]" (Monna, 1975, p. 48).

[9]Harnack first drew attention to it in 1888. Cauchy (1827d, pp. 398-400) gave the proof of the convergence of the Fourier series of a function $f(x)$ as an application of a more general theorem concerning residues. According to Harnack, "On the basis of Dirichlet's results, the older method of Cauchy acquired additional importance, in that it laid the essential foundations for the entire theory of the expansion of series of this kind in their most general form" (1888, p. 176).

[10]The second edition (1871) of his lectures on the theory of numbers, which was first published by Dedekind in 1863, contains numerous supplements by Dedekind himself.   Among these is the famous eleventh supplement, "On the theory of whole algebraic numbers," which first introduced many of the fundamental concepts of modern algebra.

[11]It is titled, *Sur la convergence des séries trigonométriques qui servent à représenter une fonction arbitraire entre des limites données.*
Dirichlet wrote "trigonometric" series, but really meant "Fourier" series.   The distinction between these and other trigonometric series was first clarified by Riemann (see §6.3).

[12]The other great mathematical journal at this time was the *Journal des mathématiques pures et appliquées,* which was founded by Liouville in 1836.   It succeeded Gergonne's *Annales.*

[13]Dirichlet adopted the attitude repeatedly expressed by Cauchy in his criticism of Lagrange's use of the Taylor series (see §3.6).

[14]Lebesgue (1905) has given a penetrating discussion of the various conditions Lipschitz, Jordan, and Dini used in order to make $f(x)$ the sum of the corresponding Fourier series.

[15]When the function becomes infinite at a point c, Dirichlet sets as a condition for the representability of the function in a Fourier series the convergence of the integral $\int \phi(\alpha)d\alpha$ in a small interval that includes the point c (Dirichlet, 1837b).

[16]The techniques that Du Bois Reymond used to arrive at this end are quite complicated and are founded on a theory of the increase and decrease of functions that he calls *Infinitärkalkül.*
Du Bois Reymond finds a function $f(x)$ such that:

(a) $f(0) = 0$
(b)   for $x \neq 0$, $f(x) = \rho(x)\sin \psi(x)$ where, as x tends to 0, $\psi(x)$ becomes infinite with infinite maxima and minima, and $\rho(x)$

tends to 0. "This function is continuous for an interval containing the point $x = 0$ and for a similar interval excluding the point $x = 0$, even with all their derivatives, and yet unrepresentable for $x = 0$.

Furthermore, one can obtain still other unrepresentable functions from the function $\rho(x)\sin \psi(x)$, whose expansion into Fourier series or expression by a Fourier integral (see 2.4.1) becomes infinite in every infinitesimal interval. We then form the function

$$f(\sin px) = \rho(\sin px)\cos \psi(\sin px).$$

With the condition that $f(0) = 0$, then this function is continuous for all x. The same is true of this:

$$F(x) = \sum_{p=1}^{\infty} \mu_p \rho(\sin px)\cos \psi(\sin px).$$

When we finally set

$$H(x) = \lim_{h\to\infty} \int_{-A}^{+B} d\alpha\, F(\alpha)\, \frac{\sin h(\alpha - x)}{\alpha - x};$$

then this function has, in every infinitesimal interval, a point in which it is infinite, or more precisely, the μ_p can always be determined in such a way that this is the case" (Du Bois Reymond, 1876, §§100-1).

In commenting on this example of a function in his *Lectures on the Theory of Simple and Multiple Integrals*, Kronecker writes, "It can nevertheless appear questionable whether such expressions, obtained from merely taking the limit [Grenzübergänge], can in any way still be considered functions" (1894, p. 94).

Simpler examples of continuous functions that are not representable by Fourier series were given by Schwarz (see Sachse, 1880, pp. 272-4) and Lebesgue (1906, p. 85 ff.).

The ideas and techniques elaborated by Du Bois Reymond were largely forgotten until they were taken up again by Hardy. He devoted a number of works to them, particularly (1910) and (1913). The latter can be considered a reworking in modern, rigorous terms of Du Bois Reymond's original paper. Nevertheless, as Hardy himself observes, "the method of proof is almost entirely different. ... The truth is that it is generally easier to find new proofs of Du Bois Reymond's assertions than to satisfy oneself that he has proved them" (Hardy, 1913, p. 282).

[17]The accounts of Casorati's conversations with Kronecker and Weierstrass, from which this citation is taken, are preserved in the Casorati papers in Pavia. They have been published in part in Neuenschwander (1978a) and Bottazzini (1977b).

[18]In his *Theorie der complexen Zahlensysteme*, Hankel states the principle of permanence as follows: "If two forms expressed in the general symbols of universal arithmetic are equal to each other, then they will also remain equal when the symbols cease to represent simple magnitudes, and the operations also consequently have another meaning of any kind" (1867, p. 11).

For a comment on the mathematical works of Hankel, and this principle in particular, see Monna (1973b).

[19]Weierstrass published it in his article, "Zur Funktionenlehre" (1880a). Today such a demonstration is obtained by applying the Heine-Borel theorem.

Chapter 6
RIEMANN'S THEORY OF FUNCTIONS

6.1. The Character of Bernhard Riemann's Works

There are times in history when a man flashes across the field of mathematics like a meteor, unexpectedly illuminating new domains before suddenly disappearing, although leaving behind him a durable and ineffaceable trail. It was so for Abel and for Galois; it was also so for Riemann (1826-1866). Although he was active in the mathematical world for little more than fifteen years, Riemann's influence on modern science has been enormous. Entire fields of mathematics were reformulated and placed on new foundations as a result of his work, if they were not founded completely *ex novo*.

In addition to fundamental work on complex analysis and the theories of trigonometric series and integration, Riemann initiated the field of algebraic topology and made substantial contributions to algebraic geometry (through the study of Abelian functions) and differential geometry, to mention only the more conspicuous examples.

The breadth of his research reflects the variety of interests and motives that underlay his conception of mathematics.

This is how Riemann himself described his activities in an undated note which was probably written around the middle of the 1850s:

The tasks which now particularly occupy me are

1. To introduce the imaginary (*Imaginäre*) into the theory of the other transcendental functions in the same way in which they have already been introduced into the algebraic functions, the exponential or trigonometric functions, the elliptic, and the Abelian functions, with such notable results. I have already set out the most important general preliminary work in my inaugural dissertation.

2. In connection with this there are new methods of integrating partial differential equations which I have already successfully

applied to many physical problems.

3. My major work involves a new interpretation of the known laws of nature -- this very expression requires the help of other fundamental concepts -- whereby the use of experimental data concerning the interaction between heat, light, magnetism, and electricity would make possible an investigation of their interrelationship. I was led to this primarily through the study of the works of Newton, Euler, and, on the other side, Herbart[1] (*Werke*, p. 507).

The central aspect of Riemann's work is in fact the unifying character that he continually brings to his mathematical research. Mathematics is the touchstone of his physical and philosophical ideas and, at the same time, the indispensable instrument of his "natural philosophy," taking this phrase in its classical sense. Thus, his "pure" mathematical results have deep roots elsewhere. He himself, for example, wrote that a large part of his initial research into Abelian functions began with the study of the conformal mapping of multiply connected surfaces, a study that was motivated by a subject quite different from that of Abelian functions (Riemann, 1857b, p. 102).

According to the historical reconstruction by Brill and Noether (1894), this subject involved the electrostatic equilibrium of cylindrical surfaces having circular cuts. They cited one of Riemann's posthumous notes that in fact contains almost all the theoretical tools that were used so extensively by Riemann in his published works (Riemann surfaces, crosscuts, etc.). Brill and Noether maintain that this can give us an idea of how Riemann arrived at his results (1894, p. 259).

When the tendency towards specialization was already widespread in the scientific world (and the mathematical world in particular), Riemann instead constantly aimed at a synthesis of differing theories, utilizing for this purpose devices that were often obtained in contexts seemingly far removed from his own field of research, such as mathematical physics.

The formative years he spent studying under Weber (1804-1891), a physicist who Gauss had wanted at Göttingen and with whom Riemann formed a lasting scientific association, seem to have been decisive for this. Equally important was the influence of Dirichlet, professor at Berlin and then at Göttingen after Gauss' death. From Dirichlet Riemann learned about potential theory, the essentials of trigonometric series, and the historical details of their most recent developments. He also completed and generalized Dirichlet's work on definite integrals and shared his (and Gauss') interest in the theory of numbers.[2]

In a talk delivered at Vienna in 1894, Felix Klein (1849-1925) emphasized the unified character of Riemann's works, finding in this the fecundity of Riemann's approach in contrast to the risks of specialization.

I believe that the longer our advancing knowledge continues under the influence of modern development, the greater danger it runs of becoming isolated. The close relationship between mathematics and theoretical natural science, which has benefited both fields since the emergence of modern analysis, is threatening to break up. In this lies a great, daily increasing danger (Klein, 1894, pp. 482-3).

In searching for the bases of Riemann's thought, a man "who, like no other, has determined the course of modern mathematics" (1894, p. 483), Klein recalled "the great tradition that is associated with the names of Gauss and Wilhelm Weber, and influenced by the philosophy of Herbart." Riemann had been raised in this tradition, "and he always sought to find in mathematical form a uniform formulation of the laws lying at the basis of all natural phenomena" (1894, p. 484). According to Klein, it is in this train of ideas that we must search for the origins of the developments in pure mathematics that are due to Riemann.

This unified conception of science and mathematics already emerges clearly in Riemann's first paper, his inaugural dissertation of 1851, on the foundations of a general theory of functions of a complex variable.

6.2. The Foundations of the Theory of Complex Functions

"Very few mathematical papers have exercised an influence on the later development of mathematics which is comparable to the stimulus received from Riemann's dissertation," Ahlfors asserted at the opening of a conference held in Princeton in 1951 to commemorate the centenary of Riemann's work. "It contains the germ to a major part of the modern theory of analytic functions, it initiated the systematic study of topology, it revolutionized algebraic geometry, and it paved the way for Riemann's own approach to differential geometry" (Ahlfors, 1953, p. 3).

The dissertation opens with the definition of a function of a real variable that had been established by the mathematical tradition from Euler to Dirichlet.

If we think of z as a variable quantity that can progressively assume all possible real values, and, if to every one of its values there corresponds a single value of the indeterminate quantity w, then w is called a function of z; and if, as z continuously passes through all the values lying between two fixed values, w also continuously changes, then this function is said to be continuous within this interval (Riemann, 1851, p. 3).

The following note, which was found among Riemann's papers, clarifies his ideas of continuity.

By the expression, "the quantity w varies continuously with z between the limits $z = a$ and $z = b$," we understand: in this interval every infinitely small change in z entails an infinitely small change in w, or, more clearly stated: for any given quantity one can always find a quantity α such that, within an interval for z, which is smaller than α, the difference between two values of w is never greater than ε. The continuity of the function thus implies that it is always finite, even if this is not specifically emphasized (1851, p. 46).

But according to Dirichlet, this was the content of a *theorem* on continuous functions that he had given in his lectures on integration at Berlin in 1854. At that time he had stated it as follows:

Let $y = f(x)$ be a function of x continuous in the *finite* interval from a to b. By *subinterval* I mean the difference between any two values of x, that is to say, every part of the abscissa between a and b. Then it is always possible to make correspond to a number ρ, as small as one wishes, a number σ ... in such a way that in every subinterval which is $\leqslant \sigma$, the function y varies at most by ρ" (In: Dugac, 1976a, p. 7).

This proposition, which, according to Dirichlet, had to be *demonstrated*, became the *definition* of continuity for Riemann, as it also did for Weierstrass. The latter had probably also been inspired by Dirichlet's course (Dugac, 1976a). It was in his lectures at Berlin in 1861 that Weierstrass gave his definition of continuous function in terms of $\varepsilon - \delta$ for the first time.

Now, if it is possible to determine a limit δ for h such that for *all* values of h which are smaller in absolute value than δ, $f(x + h) - f(x)$ will be smaller than any other arbitrarily small quantity ε, then we say that infinitely small changes of the function correspond to infinitely small changes of the argument (In: Dugac, 1973, p. 119).

According to Dugac (1976a, p. 7), it is from this moment that we can date the substitution of inequality, which implies the topological notion of neighborhood, for the intuitive idea of "tend towards."

We see here a process that is not infrequent in the history of mathematics. The central (and therefore the most profound) property of a class of objects emerges from practice, and it is subsequently natural to make precisely this property their *definition*.

After having so defined a function, Riemann comments, "This definition obviously does not set up any fixed law between the individual values of the function, since, once this function has been defined for a particular interval, the manner of its extension outside this interval remains completely arbitrary" (1851, p. 3).

These words recall those used by Dirichlet in (1837a). The dependence of *w* on *z*, Riemann says, can be arbitrary or given by mathematical laws. It is completely indifferent which is used, since "modern research has shown that there are analytical expressions by which any continuous function can be represented in a given interval" (*ibid.*). The researches he refers to are precisely those of Dirichlet on Fourier series.[3]

But if for Dirichlet, as had also been the case for Euler, the generality of the definition did not go along with a consistent practice in the study of equally "general" functions, the opposite is true for Riemann. In his thesis (*Habilitationsschrift*) of 1854 (see §6.3 below), he revealed to the mathematical world a universe extraordinarily rich in "pathological" functions.

If we look at the actual history of mathematics, then it seems to me undeniable that the modern concept of a function of a real variable, in its full generality, began to emerge in mathematical practice only towards the end of the 1860s as Riemann's writings became more widely known. At the same time the word spread that the Berlin mathematicians, primarily Weierstrass and Kronecker, were devoting special attention to the study of such functions in an effort to rigorize the fundamental concepts of the infinitesimal calculus, like those of continuity and derivative.

In his dissertation Riemann limited himself to the preceding few observations on real variability. The situation changes considerably, he observed, when we let the variable *z* assume complex values. In this case, if the dependence of $w = u + iv$ on $z = x + iy$ is arbitrary, the ratio

$$\frac{du + i \, dv}{dx + i \, dy}$$

will in general vary as dx and dy vary.

But, no matter how *w* may be determined as a function of *z* through a combination of simple algebraic operations [*einfachen Grössenoperationen*], the value of the derivative dw/dz will always be independent of the specific value of the differential dz.[4] Clearly not every arbitrary dependence between the complex quantity *w* and the complex quantity *z* can be expressed in this way (Riemann, 1851, p. 4).

Now, since we wish to consider functions independently of their analytic expressibility, this is a good property which we can use as a starting point for the definition of functional dependence, "without now proving its general validity and its adaptability [*Zugänglichkeit*] to the idea of a dependence expressible by algebraic operations [*Grössenoperationen*]" (*ibid.*).

For Riemann, then, "A complex variable quantity *w* is called a function of another complex variable quantity *z* when the one changes with the other in such a way that the value of the

derivative dw/dz is independent of the value of the differential dz"
(1851, p. 5).

This condition is satisfied when

$$(6.2.1) \quad \frac{\partial u}{\partial x} = \frac{\partial v}{\partial y} \quad \text{and} \quad \frac{\partial u}{\partial y} = -\frac{\partial v}{\partial x}$$

hence,

$$(6.2.2) \quad \frac{\partial^2 u}{\partial x^2} + \frac{\partial^2 u}{\partial y^2} = \Delta u = 0, \quad \frac{\partial^2 v}{\partial x^2} + \frac{\partial^2 v}{\partial y^2} = \Delta v = 0,$$

which derives immediately from being the value of

$$\frac{\left[\dfrac{\partial u}{\partial x} + i\,\dfrac{\partial v}{\partial x}\right]dx + i\left[\dfrac{\partial v}{\partial y} - \dfrac{\partial u}{\partial y}i\right]dy}{dx + i\,dy}$$

independent of $dz = dx + i\,dy$.[5]

Riemann thus places at the center of the concept of a function of
a complex variable the existence of (6.2.1), which for Cauchy was a
particular property of the function in question, that of being
monogenic.

Casorati emphasized the same thing by noting "the greater
simplicity" that the entire theory acquires from "the *immediate* [my
emphasis] establishment of the fundamental properties of functions
of a complex variable by the equation

$$(6.2.3) \quad \frac{\partial w}{\partial y} = i\,\frac{\partial w}{\partial x},$$

instead of first establishing the fundamental properties of the
component functions that satisfy the equation $\Delta u = 0$" (1868, pp.
139-40 n. 2). He continues,

> We believe that it is truly admirable with what assimilating
> power [Riemann] knew how to gather and establish in one
> compact, simple, and general theory, together with his own, all
> of the other studies that had an important relation to them. Of
> particular importance were Cauchy's many studies, spread over
> numerous publications, which were conducted with differing
> purposes and often wrapped in a heterogenous variety of terms
> and special notations. ... In particular, it is especially worth
> observing how [Riemann] always sets up his own conventions
> and definitions in such a way that every theorem can be stated
> as true without exception, or how many formulas or theorems
> ordinarily thought to be different from each other can be united
> into a single formula or theorem (Casorati, 1868, p. 140 n. 2).

In addition, Riemann is the first to have fully understood the
importance of equations (6.2.2), with which he became familiar

through his studies in physics. In his hands, these equations became the foundation of the entire theory.[6]

At this point Riemann has no difficulty showing how the dependence of w on z defined in this way allows one to interpret it geometrically as the conformal mapping of the plane z over the plane w, something that had been noted elsewhere.[7]

a) Riemann Surfaces

The initial paragraphs of Riemann's dissertation are followed by others dedicated to illustrating an entirely new concept of mathematics--the *Riemann surface*. Riemann introduces this as the geometrical realm best suited to presenting his thoughts about functions.

> We choose this covering [*Einkleidung*] so that it will be easy to speak of superimposed surfaces, thus leaving open the possibility that the place of point O extends many times over the same part of the plane. However, in such a case we assume that the superimposed parts of the surface are not connected along a line, so that a folding of the surface or a division into superimposed parts cannot occur (1851, p. 7).

This is the way, contrary to geometric intuition, in which Riemann introduced the idea of multiply covered surfaces, which turned out to be one of the most brilliant and profound achievements of his dissertation.

> The reader is led to believe that this is a commonplace convention, but there is no record of anyone having used a similar device before. As used by Riemann it is a skillful fusion of two distinct and equally important ideas: 1) A purely topological notion of covering surface, necessary to clarify the concept of mapping in the sense of multiple correspondence; 2) An abstract conception of the space of the variable; with a local structure defined by a uniformizing parameter. The latter aspect comes to the foreground in the treatment of branch points.
>
> From a modern point of view the introduction of Riemann surfaces foreshadows the use of arbitrary topological spaces, spaces with a structure, and covering spaces (Ahlfors, 1953, p. 4).

One of the intuitive reasons for introducing such surfaces lies in the idea that, when w is a multi-valued function, then there will be several points of the plane B of w that correspond to the same point of the plane A of z.

Riemann introduces these surfaces in such a way that, if w is multi-valued, every point of the surface represents a single value of

w corresponding to a single value of z.
On this Casorati comments,

> A geometrical realm constructed in such a manner is designed
> to be the intuitive field of the author's speculations. It is in his
> entirely original creation that one finds the reason for his
> language and for certain of his unusual procedures. At first this
> was not adequately explained outside of the Göttingen school, a
> fact which created a severe obstacle to the wider circulation of
> these admirable speculations (Casorati, 1868, p. 120).

Riemann himself recognized that his ideas were encountering
difficulties, even among German readers. He wrote of this to Betti
(1823-1892) in June of 1863, during his first stay in Italy (see note
18 below). We can see Riemann's constructions more clearly by
taking the example of the function $w^2 = z$.

To every point O of the plane A there correspond two points in the
plane B that vary continuously as z varies in A, and which coincides
with the origin for $z = 0$. We now imagine another plane A_1,
extended above A and cut along a line (for simplicity, a straight
line) that begins from the point $z = 0$ and extends to infinity
without repeatedly passing through one and the same point.

Now let A_1 be the locus of point O. We associate to any O one of
the two corresponding values of w in such a way that, by letting O
vary continuously, in every single successive position of O one can
always associate the value of w that we obtain by preserving the
continuity.

Thus to every point of A_1 there will be associated a value of w.
We now consider a second plane A_2 (in Riemann's terms, a sheet),
that is cut like A_1 from the origin to infinity.

If we now associate to every one of its points (representing a value
of z as in A and A_1) the value of w that is not associated with the
point underneath the plane A_1 and we further consider O to be
mobile in A_2, as we have seen above in the case of A_1, in such a way
as to associate to every point of A_2 a value of w that varies
continuously with z.

> In this way to every two points of A_1, A_2 superimposed   above
> one and the same A it is possible to place the two values of w
> corresponding to one and the same z. And, by considering the 2
> + 2 = 4 edges of the cuts of A_1 and A_2, we recognize that the
> values of w placed along the right edge of a plane are
> correspondingly equal to those placed along the left edge of the
> other. Therefore, by imagining that the right edge of A_1 is
> connected along its entire length to the left edge of A_2, and the
> right edge of A_2 with the left edge of A_1, we obtain a unique
> surface of two planes or sheets endowed with the property that
> *every distinct pair* of values of w and z, or *every distinct solution*
> of the equation $w^2 - z = 0$ has one *and only one* representative

point, and that any continuous and simultaneous variation of these two variables can be represented by a continuous movement of a point on the surface, and vice versa (Casorati, 1868, pp. 121-2).

Casorati was one of the most convincing advocates of Riemann's ideas in Italy. Together with Betti, he made an essential contribution to their diffusion and comprehension. We have followed his discussion of this example literally, because it seems admirably clear.

Casorati himself, in order to help his students understand the mathematical objects introduced by Riemann, used to give a visually effective representation of them in his lectures, that is, as far as this could be done within the limits of material models. He suggested imagining these surfaces "as nets, whose knots represent the points while the threads are variable in length, flexible, and overlapping," or in other words, essentially "to imagine them as systems of points insensibly separated from each other" (1868, p. 123 n.)

To this point, then, Riemann surfaces appear as instruments well suited to the study of algebraic functions, which we have already seen put forward by Cauchy and Puiseux (see §4.6).

It is this research, which matured within the ambience of Cauchy's "school," to which Markusevic (1908-79) was led in his search for the origins of Riemann's conceptions.[8] After observing that Riemann made hardly any explicit reference to other mathematicians in his thesis (apart from Gauss and Dirichlet), and that "Riemann behaved on the continent of analytic functions like Robinson Crusoe on his uninhabited island," Markusevic asserted that, notwithstanding his many innovations and the force of his genius, Riemann "was in many respects a mathematician of the old school" (1978, p. 1). He went on to claim that Riemann had found in Cauchy's research in general and in that of Puiseux in particular the "scaffolding" for his construction of many-sheeted surfaces.

As far as Cauchy is concerned, Markusevic recalled the discussion that Eisenstein (1823-1852) and Riemann had had concerning the fundamental concepts of complex analysis in 1847, while the latter was a student at Berlin attending the classes of Dirichlet, Jacobi, and Steiner (1796-1863). In that year Eisenstein was giving a course on the theory of elliptic functions. Riemann attended the lectures and formed a close friendship with the young professor.

In his biography of Riemann, Dedekind wrote that

Riemann later said that they had discussed with each other the introduction of complex magnitudes into the theory of functions, but that they had been of completely differing opinions as to what the fundamental principles should be. Eisenstein stood by the formal calculus, while [Riemann] himself saw the essential definition of a function of a complex variable in the partial differential equation (In: Riemann, 1876, p. 544).

It is very likely that the starting point for the reflections of the two young mathematicians was Cauchy's approach to the theory of complex functions, which emphasized to the formal aspect of the subject. Both men were familiar with Cauchy's *Cours d'analyse* as well as with his *Exercices de mathématiques*. But as far as Cauchy's "school" is concerned, Riemann's point of view had already been shaped in a completely original manner. In order to discover the deepest motivations for his work we must not look to Paris, but rather to Göttingen and Berlin. In fact, the differential equations in which Riemann saw the essence of the definition of functions of a complex variable are the conditions of monogeneity (6.2.1) and Laplace's equation (6.2.2). This latter played a fundamental role in potential theory. Not more than eight years before, in 1839, Gauss had written a fundamental work on this theory, and Weber had also worked on it in connection with his studies on magnetism (see the Appendix).

As Ahlfors had said, Riemann "virtually puts equality signs between two-dimensional potential theory and complex function theory" (1953, p. 4).

This is an approach that leads one to define a function from its singularities, as Riemann himself repeatedly and explicitly stressed. "This approach calls for existence and uniqueness theorems, in contrast to the classical conception of a function as a closed analytic expression," Ahlfors wrote (*ibid.*). It is a conception that Riemann attributed more to Euler than to Cauchy, as he wrote in his thesis and affirmed in his lectures. If we believe to the words of Prym (1841-1915), one of his students at Göttingen, Riemann said that he had been led to attribute decisive importance to equations (6.2.1) and (6.2.2) after he had asked under what conditions it was possible to "extend" the expansion in series of a function from one domain to another, and why, in so many cases, one obtains correct results when operating with divergent series, as Euler repeatedly did.

Dirichlet, when he was asked about this by Riemann, showed himself to be of the same opinion regarding the role of these equations.

It was just at this time, consistent with this point of view, that Riemann began to study the integral $\int X\,dx + Y\,dy$. He realized that the multiplicity of values of the integral depends on the fact that there are closed paths that do not bound a region. From this followed the introduction of crosscuts (to make the domain simply connected) and of surfaces of many sheets to represent multi-valued functions. Their determination was achieved by means of boundary conditions and discontinuities with the tools provided by potential theory and by Dirichlet's principle.

As far as the conjecture about Puiseux's influence on Riemann is concerned, Markusevic recognizes that there are no facts to support it,[9] other than the observation that many of Puiseux's results are found in a generalized and unified form in Riemann's dissertation, and that the very appearance of the concept of a many-sheeted surface at the beginning of Riemann's work, with the definition of the branch points of a surface and the description of the connections of the sheets with each other, distributed in cycles in the neighborhood of each branch point, must have seemed completely unexpected and extremely artificial to anyone who has not studied Puiseux's memoir.

But a shrewd reader of that memoir could discover in the new and unaccustomed concept an excellent geometrical explanation and, in its own terms, a summary of the fundamental content of the memoir (Markusevic, 1978, p. 8).

Certainly, by appealing to Riemann's surfaces we can better "see" the cycles and branch points of Puiseux, but the fundamental theoretical problem for Riemann was not the investigation of the properties of algebraic functions, as it had been for Puiseux. As Riemann himself repeatedly stated, he was searching for a general method capable of embracing larger classes of functions and treating them in a unified manner. It is precisely this that led him to his approach.

In his dissertation, for the many-sheeted surfaces that he introduced, Riemann defined *branch points* of the $(m - 1)$th order as points where the m sheets of the surface are connected in such a way as to return to the point of departure when the variable has completed m circuits, passing continuously from one sheet to the next in the neighborhood of the point in question.

After explaining what is meant by crosscuts ("lines which cut through the interior simply -- at no point multiply -- from one boundary point of a domain through the interior to another boundary point" (1851, p. 9)) and the connectivity of a surface ("A connected surface is called simply connected when it is divided into parts by every crosscut; otherwise it is multiply connected" (Riemann, 1851, p. 9)), he states and demonstrates the fundamental theorem:

"If a surface T is disconnected by n_1 crosscuts q_1 into a system T_1 of m_1 simply connected regions and by n_2 crosscuts q_2 into a system T_2 of m_2 regions, then $n_2 - m_2$ cannot be greater than $n_1 - m_1$" (*ibid.*, p. 10).

From this theorem Riemann obtained the definition of the connectivity number of a surface as the number $n - m$, where n is an indeterminate number of crosscuts and m is the number of simply connected regions into which the surface is disconnected by them.

In a later paper (1857b), Riemann prefaced the treatment of Abelian functions with a few observations of a topological nature (relative to *analysis situs*, as it was then called). On this occasion he gave a second definition of the connectivity number of a surface.

If n closed curves $a_1, a_2, ..., a_n$ can be drawn in a surface F, which [curves] neither by themselves nor with each other completely bound a region of this surface F, but with their

assistance every other closed curve can build a complete bound
of a region of the surface F, then this surface is said to be
$(n+1)$-ply connected (1857b, p. 92-3).

Here the decisive thing is to show that "this character of the surface
is independent of the choice of curves $a_1, a_2, ..., a_n$"[10] (*ibid.*).

Riemann was certainly not the first to concern himself with
topological questions. (They are already found in Euler, for example,
in Gauss, and above all in Listing (1808-1872), although in an
embryonic form). Nor are Riemann's definitions of the connectivity
number completely satisfactory from the modern point of view,
although the second contains the germ of the ideas that led Poincaré
(1854-1912) to the concept of homology in 1895. Nevertheless, it is with
Riemann that topology reveals its true fruitfulness for the study of
analysis and shows its deep links with geometry.

In comparing Riemann's approach with that of Cauchy, Hadamard
(1865-1963) has written that,

Cauchy's theory of functions does not throw the light on the nature
of algebraic functions that one would have expected and that, in
consequence, it has effectively shown itself capable of providing.
An essential element seems to have escaped him, which it is
necessary to remember in all important problems, namely that we
deal with algebraic functions and curves. When one studies the
geometry of one of these curves, or the expression of the
coordinates in functions of auxiliary variables, everywhere the same
whole number introduces itself, the *genus* of the curve. The theory
to which we have just alluded does not allow us to foresee its
introduction (Hadamard, 1909, pp. 813-4).

Thus, the "glory of laying the definitive foundations" of the theory
of algebraic functions belongs properly to Riemann (*ibid.*, p. 817).

It is in this geometrical context that Riemann places his study of
functions. In fact, he said that a variable magnitude which in general
(that is, except at isolated points and lines) assumes a well determined
value for every point O of the surface, and that varies continuously in
dependence on its position, can clearly be considered as a function.
This is how we will think about functions of x and y, Riemann wrote.
He then considers two functions X and Y of x, y, continuous in every
point of the surface T, and the integral

$$(6.2.4) \quad \int \left[\frac{\partial X}{\partial x} + \frac{\partial Y}{\partial y} \right] dT$$

extended to the surface T. He obtains the theorem already known to
Cauchy and Gauss, which is that the integral (6.2.4) is equal to the
integral $-\int (X \cos \xi + Y \cos \eta) ds$ taken around the boundary of T,
where ξ and η are the angles formed respectively with the x and y
axis by the normal at each point of the boundary, with the positive
orientation towards the interior, and that,

$$\int \left[\frac{\partial X}{\partial x} + \frac{\partial Y}{\partial y} \right] dT = -\int \left[X \frac{\partial x}{\partial p} + Y \frac{\partial y}{\partial p} \right] ds$$

(6.2.5)

$$= \int \left[X \frac{\partial y}{\partial s} - Y \frac{\partial x}{\partial s} \right] ds,$$

where s indicates the length of a part of the boundary from any fixed initial point to an arbitrary point O_0, and p indicates the distance of an indeterminate point O calculated along the normal to the boundary at O_0, oriented as stated. This is sufficient to express the coordinates x, y of the generic point O as functions of s and p.

With the additional hypothesis that the integrand of (6.2.4) be equal to zero, every integral of the second member of (6.2.5) will also be equal to zero when the integral is extended along the entire boundary of the surface T.

Also under the same hypotheses, the value of the integrals (6.2.5) will not be changed by increasing or decreasing the surface T in any manner whatsoever, provided that every time this operation does not add or substract any part of the surface in which the given hypothesis ceases to be valid (1851, p. 15). It is at this point that the importance of the connectivity of the surfaces emerges. If the surface is simply connected, the integral

(6.2.6) $$\int \left[X \frac{\partial x}{\partial p} + Y \frac{\partial y}{\partial p} \right] ds = \int \left[Y \frac{\partial x}{\partial s} - X \frac{\partial y}{\partial s} \right] ds$$

is zero when it is taken along the entire boundary of any part of the surface, and it retains the same value if it is taken along two different paths that run from any fixed point of the surface O_0 to any other point O, without leaving the surface T. If in T, where in general

(6.2.7) $$\frac{\partial X}{\partial x} + \frac{\partial Y}{\partial y} = 0.$$

is true, we omit the points of discontinuity by drawing closed paths that encircle them, then it will also be true that the integrals of the second member of (6.2.5) are zero in every remaining part of the surface T^*, which is obtained by making suitable cuts in such a way as to render it simply connected. Then, the integral

$$Z = \int_{O_0}^{O} \left[Y \frac{\partial x}{\partial s} - X \frac{\partial y}{\partial s} \right] ds \quad (O_0 \text{ fixed})$$

can be considered as a function of x, y (the coordinates of O), which assumes a well determined value whichever way one moves from O_0 to O in T^*. Moreover, it is a finite and continuous function, whose derivatives are respectively

$$\frac{\partial Z}{\partial x} = Y, \qquad \frac{\partial Z}{\partial y} = -X.$$

On the edges of every cut made in T the difference of the values taken by Z is given by a constant, whose number is equal to that of the crosscuts ($n - 1$ if T is n-ply connected). The constants or "modules of periodicity," as Riemann calls them in (1857b), are in addition completely independent of each other.

At this point Riemann studies the properties of the functions u, u' of x, y, introduced by setting

$$u \frac{\partial u'}{\partial x} - u' \frac{\partial u}{\partial x} = X \quad \text{and} \quad u \frac{\partial u'}{\partial y} - u' \frac{\partial u}{\partial y} = Y.$$

If u and u' are harmonic, one can apply the conclusions just obtained to the integral

$$(6.2.8) \quad \int \left[X \frac{\partial x}{\partial p} + Y \frac{\partial y}{\partial p} \right] ds = \int \left[u \frac{\partial u'}{\partial p} - u' \frac{\partial u}{\partial p} \right] ds,$$

since (6.2.7) is valid.

By now requiring that the function u and its first derivatives do not have discontinuities along a line and that, for every point where they are discontinuous, the quantities $\rho(\partial u / \partial x)$ and $\rho(\partial u / \partial y)$ become infinitely small together with ρ, ρ being the distance of O from the point of discontinuity, and also supposing that the surface T is composed of a single sheet, then Riemann succeeds in expressing the value of u at any point O_0 of T by means of the integral

$$u_0 = \frac{1}{2\pi} \int \left[\log r \frac{\partial u}{\partial p} - u \frac{\partial \log r}{\partial p} \right] ds,$$

the integral being taken along a closed path that enclosed the point $O_0 \equiv (x_0, y_0)$ in which u is otherwise supposed to be continuous and

$$\frac{1}{2} \log[(x - x_0)^2 + (y - y_0)^2] = \log r.$$

From this expression he obtains the theorem:

When a function u within a surface T that everywhere simply covers the plane A satisfies, in general, the differential equation

$$\frac{\partial^2 u}{\partial x^2} + \frac{\partial^2 u}{\partial y^2} = 0$$

in such a way that
1.) the points which do not satisfy this differential equation do not form a part of the surface,
2.) the points in which u, $\partial u / \partial x$, $\partial u / \partial y$ are discontinuous do not form a continuous line,
3.) for every point of discontinuity the magnitudes $\rho(\partial u / \partial x)$, $\rho(\partial u / \partial y)$ become infinitely small together with the distance ρ of the point O from these points, and
4.) it is impossible to remove the discontinuity of u by changing its value in isolated points,

then it is necessarily finite and continuous together with all of its derivatives for all points within this surface (1851, pp. 20-1).

Under the same conditions for u and T, Riemann proves the following propositions:

(I)   If u and $\partial u/\partial p$ are zero in all points of a line, u is zero everywhere in T.
(II)   If the value of u and of $\partial u/\partial p$ is given along a line, u remains determinate in all parts of T.
(III) The points within T where u has a constant value necessarily form, when u is not everywhere constant, lines that separate the parts of the surface where u is larger than the constant in question from those where u is smaller.

This proposition is a combination of the two following, which constitute "Riemann's maximum principle":

(a) u cannot have either a maximum or a minimum at a point within T,
(b) u cannot be constant *only in a part* of the surface.

At this point Riemann resumes the consideration of $w = u + iv$ as a function of z which he had earlier interrupted. w is continuous and satisfies the conditions (6.2.1) in general, except in lines or isolated points. First we consider the hypothesis that T is simply connected and is formed by a single sheet. Then, if the function w of z is never discontinuous along a line and, in addition, for every point O' of the surface where $z = z'$, $w(z - z')$ becomes infinitely small as O comes indefinitely close to O', this function is finite and continuous with all of its derivatives within the surface T.

In fact, (6.2.1) and hypotheses 1 to 4 of the preceding theorem are valid for u, v (components of w). In addition, we have

$$\int \left[u \frac{\partial x}{\partial s} - v \frac{\partial y}{\partial s} \right] ds = 0,$$

where the integral is taken along the complete boundary of any region of T. Therefore the function U of x, y (as O varies) defined by

$$U = \int_{O_0}^{O} \left[u \frac{\partial x}{\partial s} - v \frac{\partial y}{\partial s} \right] ds \quad \text{(for } O_0 \text{ fixed),}$$

whose derivatives are

$$\frac{\partial U}{\partial x} = u \quad \text{and} \quad \frac{\partial U}{\partial y} = -v,$$

is continuous and finite at every point of T, and the same is true for its derivatives.

Then the complex function $w = \partial U/\partial x - i(\partial U/\partial y)$ is also continuous and finite together with its derivatives.

Riemann next shows that if the product $w(z - z')$ does not tend to zero as $|z - z'|$ tends to zero, and if one can find an integer n for which $w(z - z')^n$ tends to zero as $|z - z'|$ tends to zero, then the function

$$w - \sum_{i=1}^{n-1} \frac{a_i}{(z - z')^i} \quad \text{(for } a_i \text{ constant),}$$

will be continuous and finite at the point O'.

The restrictions imposed on T to be a surface of a single sheet is not essential for these two theorems, which are also valid when one considers a many-sheeted surface and when the point O' is a branch point of the $(n - 1)$th order.

In the latter case Riemann in fact asserts that the region of the surface T which includes the point O' can be mapped on an auxiliary plane Λ by means of the function $\zeta = (z - z')^{1/n}$.

"In this way we obtain, as an image of this part of the surface T, a connected surface extended over Λ which has no branch point at point Θ', the image of point O'" (Riemann, 1851, p. 26). Riemann thus established Puiseux's uniformization theorem (see §4.6) by considering the uniformization parameter ζ, which realizes a map of a neighborhood of the point O' on the surface in a connected neighborhood of a point Θ' of the complex plane Λ, where $\Theta' \equiv (\varepsilon, \eta)$ and $\zeta = \varepsilon + i\eta$.

After giving a demonstration of this theorem in geometrical terms, Riemann adds, "If a function w of z becomes infinite as O approaches infinitely close to a branch point of the $(n - 1)$th order, then this infinite quantity is necessarily of the same order as the power of the distance whose exponent is a multiple of $1/n$. If this exponent is $= -m/n$, it can be changed into a [function] continuous at the point O' through the addition of an expression of the form

$$\sum_{k=1}^{m} \frac{a_k}{(z - z')^{k/n}}$$

where $a_1, a_2, ..., a_m$ are any complex quantities" (1851, p. 27). It follows as a corollary of this, Riemann adds, that w is continuous at O' if $(z - z')^{1/n}$ becomes infinitely small when O approaches indefinitely close to O'.

Having thus analyzed the behavior of the function at the poles and branch points, Riemann summarizes his results in the following manner. If the dependence of w on z (a variable in a surface T) is represented by means of a surface S in such a way that to every point O of T there corresponds a point Q of S with coordinates u, v, then the set of points Q will form the surface S in question as z varies.

Since a function $w = u + iv$ cannot be constant along a line without being so everywhere, (as Riemann easily proves) Q will generally move in S in a continuous manner as O varies continuously in T, and the boundary of S corresponds to the boundary of T and to the points of discontinuity. At every point Q in the surface S, z has a single determinate value that varies continuously with Q in such a way that dz/dw is independent of dw and consequently z is a function of w, continuous in the domain S. If O' and Q' are corresponding points respectively in T and S, the limit of the ratio

$$\frac{w - w'}{z - z'}$$

is finite as O tends to O' if O' and Q' are not branch points. The mapping is therefore conformal. If Q' is a branch point of order $n - 1$ and O' of order $m - 1$, then the limit

$$\frac{(w - w')^{1/n}}{(z - z')^{1/m}}$$

will be finite as O tends to O', and the mapping of T in S is given according to the "uniformization theorem" stated above.

b) Dirichlet's Principle

In the remaining part of his dissertation Riemann explains Dirichlet's principle (see the Appendix) and uses it to prove the existence of a complex function subject to opportune conditions.
  He begins by stating the theorem:

> Let α and β be any two functions of x and y for which the integral
>
> (6.2.9) $\quad \int\left[\left[\frac{\partial\alpha}{\partial x} - \frac{\partial\beta}{\partial y}\right]^2 + \left[\frac{\partial\alpha}{\partial y} + \frac{\partial\beta}{\partial x}\right]^2\right] dT$
>
> has a finite value for all parts of a surface T which covers A in any manner. If α is changed by continuous functions or by those discontinuous only in isolated points which equal zero at the boundary, then the integral always has a minimum value for one of these functions and, if we exclude discontinuities which can be removed by changes in isolated points, only for one (1851, p. 30).

This is the way in which Riemann utilizes Dirichlet's principle.

> This is precisely the instrument with which Riemann makes possible a more general determination of functions, by means of suitable systems of strictly necessary and sufficient conditions. Independently of the statement of an analytic expression, these permit ... the treatment of questions more with pure reasoning than with calculation. The use of Dirichlet's principle as an analytic instrument as well as the surface T as a geometrical support, is characteristic of the theory of functions taught in Göttingen (Casorati, 1868, pp. 132-3).

In regard to this principle Riemann observed some years later:

As a foundation for the study of a transcendent [function] it is above all necessary to set up a system of independent conditions sufficient to determine it. In many cases, particularly for integrals of algebraic functions and their inverse functions, we can use a principle which Dirichlet utilized to solve this problem for the case of a function of three variables satisfying Laplace's partial differential equation -- certainly inspired by a similar idea in Gauss. This was some years ago in his lectures on a force proportional to the inverse square of the distance[11] (Riemann, 1857b, p. 97).

The common attribution of this principle to Dirichlet begins here.

Riemann's demonstration of the stated theorem follows a path that had already been anticipated by Gauss. Riemann considered a function λ, continuous or at most discontinuous in isolated points, zero at the boundary, for which the integral

$$L = \int \left[\left(\frac{\partial \lambda}{\partial x} \right)^2 + \left(\frac{\partial \lambda}{\partial y} \right)^2 \right] dT$$

is finite relative to the surface T. He then indicates with $\omega + \lambda$ any of the functions $\alpha + \lambda$ and with Ω the integral

$$\Omega = \int \left[\left(\frac{\partial \omega}{\partial x} - \frac{\partial \beta}{\partial y} \right)^2 + \left(\frac{\partial \omega}{\partial y} + \frac{\partial \beta}{\partial x} \right)^2 \right] dT.$$

The set of the λ functions, Riemann says, is such that one can pass continuously from one to the other. None of them can tend indefinitely towards a discontinuous function along a line without L becoming infinite at the same time. (He reserves the demonstration of this for later.)

"Given $\omega = \alpha + \lambda$, Ω now has a finite value for every λ, which becomes infinite with L and which varies continuously with the form of λ but can never fall below zero. *Consequently*, Ω has *a minimum* at least for one form of the function ω" [my emphasis] (Riemann, 1851, p. 30).

The objections that can be raised against this conclusion are numerous and decisive (see the Appendix). However, having established the existence of a minimum in this manner, Riemann now has no difficulty showing that it is unique.

Moreover, having indicated with u the function that gives the minimum for Ω, we also find that $u + h\lambda$ (with h constant) satisfies the conditions imposed on the ω functions.

The integral Ω, calculated for the function $u + h\lambda$, gives

$$\Omega(u + h\lambda) = \int \left[\left(\frac{\partial u}{\partial x} - \frac{\partial \beta}{\partial y} \right)^2 + \left(\frac{\partial u}{\partial y} + \frac{\partial \beta}{\partial x} \right)^2 \right] dT$$

$$+ 2h \int \left[\left(\frac{\partial u}{\partial x} - \frac{\partial \beta}{\partial y} \right) \frac{\partial \lambda}{\partial x} + \left(\frac{\partial u}{\partial y} + \frac{\partial \beta}{\partial x} \right) \frac{\partial \lambda}{\partial y} \right] dT$$

$$+ h^2 \int \left[\left(\frac{\partial \lambda}{\partial x} \right)^2 + \left(\frac{\partial \lambda}{\partial y} \right)^2 \right] dT = M + 2hN + h^2 L.$$

By the definition of the minimum, N must $= 0$ for every λ, since in the contrary case

$$2hN + h^2 L = Lh^2 \left[1 + \frac{2N}{Lh} \right]$$

would be negative, provided that h is taken with a sign opposite to N, and $|h| < 2N/L$.

After having then demonstrated the stated property for λ, that is, that it cannot tend towards a discontinuous function γ along a line unless the integral L simultaneously becomes infinite, Riemann considers the integral $N = 0$. He separates the region T' containing the discontinuities of u, β, and λ from the surface T and lets

$$X = \left(\frac{\partial u}{\partial x} - \frac{\partial \beta}{\partial y} \right) \lambda, \qquad Y = \left(\frac{\partial u}{\partial y} + \frac{\partial \beta}{\partial x} \right) \lambda.$$

He finds that the part of N proper to the region of the surface $T'' = T - T'$ is given by

$$(6.2.10) \qquad -\int \lambda \left[\frac{\partial^2 u}{\partial x^2} + \frac{\partial^2 u}{\partial y^2} \right] dT - \int \left[\frac{\partial u}{\partial p} + \frac{\partial \beta}{\partial s} \right] \lambda \, ds$$

for (6.2.5). Since the second integral is zero for the conditions imposed on λ in the part of the boundary common to T'' and T, it follows that N can be considered to be given by the first integral of (6.2.10) relative to T'' and by the integral

$$(6.2.11) \qquad \int \left[\left[\frac{\partial u}{\partial x} - \frac{\partial \beta}{\partial y} \right] \frac{\partial \lambda}{\partial x} + \left[\frac{\partial u}{\partial y} + \frac{\partial \beta}{\partial x} \right] \frac{\partial \lambda}{\partial y} \right] dT + \int \left[\frac{\partial u}{\partial p} + \frac{\partial \beta}{\partial s} \right] \lambda \, ds$$

relative to T'.

Now if $\Delta u = 0$ everywhere in T, from the condition that $N = 0$ it follows that (6.2.11) is also equal to zero. Thus (6.2.7) and (6.2.6) are valid with respect to X and Y "in so far as this [second] expression really has a definite value" (1851, p. 34).

Let us now reduce the surface T, if it is multiply connected, to a simply connected surface T^* by means of crosscuts. The integral

$$-\int_{O_0}^{O} \left[\frac{\partial u}{\partial p} + \frac{\partial \beta}{\partial s} \right] ds \qquad \text{(for } O_0 \text{ fixed),}$$

taken along an arbitrary path from O_0 to O in T^*, can be considered as a continuous function ν in T, which is the same along the two edges of the crosscut.

We now consider the function $v = \beta + \nu$, whose partial derivatives are given by (6.2.1). Riemann states and demonstrates the fundamental theorem:

If a complex function $\alpha + \beta i$ of x, y is given in a connected surface T which can be reduced to a simply connected surface T^* by crosscuts, and if (6.2.9) has a finite value throughout the

entire surface, then it can always and only in one way be
changed to a function of z by adding a function $\mu + iv$ of x, y
which satisfies the following conditions:

1) $\mu = 0$ on the boundary or is different from it only in
isolated points, v given arbitrarily in a point,

2) the variations of μ in T and of v in T^* are discontinuous
only in isolated points and only in such a way that

$$\int \left[\left(\frac{\partial \mu}{\partial x} \right)^2 + \left(\frac{\partial \mu}{\partial y} \right)^2 \right] dT \text{ and } \int \left[\left(\frac{\partial v}{\partial x} \right)^2 + \left(\frac{\partial v}{\partial y} \right)^2 \right] dT$$

remains finite across the entire surface and the variations of v
are equal along both edges of the crosscut (1851, pp. 34-5).

These conditions are sufficient to determine the function $\mu + iv$.
The theorem just stated, Riemann says, "opens the way to an
investigation of determinate functions of complex variables
(*independent of an expression for them*)" [my emphasis] (1851, p. 35).

To give an idea of his method for a specific case, Riemann treats
the following: If T is a simply connected surface, the function $w = u$
$+ iv$ of z can be suitably determined from the following conditions:

1) a value is given for u at every boundary point, which value,
moreover, for an infinitely small change of position changes
arbitrarily by an infinitely small quantity of the same order;*

(*In themselves the changes of this value are only subject to the
restriction that they not be discontinuous along a part of the
boundary; an additional restriction is only made here in order to
avoid unnecessary prolixity.)

2) the value of v is arbitrarily given in any point;

3) the function must be finite and continuous in every point
(1851, p. 35).

Under these conditions w is completely determined. In fact, from
the preceding theorem we can determine $\alpha + i\beta$ in such a way that α
assumes the given values on the boundaries and $\alpha + i\beta$ is continuous
everywhere on the surface X. At the boundaries u can be given as a
completely arbitrary function of s, which fact also determines v (and
reciprocally) up to a determinable constant, when v is known in a
point.

Finally, indicating the generality and fruitfulness of his approach
to the theory of functions of complex variables, Riemann writes:

Until the present, the methods for handling these functions
have always used as a definition an expression of the function
in which its value is given for every value of its argument. Our
research has shown that, as a consequence of the general
character of a function of a complex variable, in a definition of

this kind a part of the determining elements [*Bestimmungsstücke*] is a result of the rest, namely, the range of the determining elements is traced back to those necessary for determining [the function] (1851, p. 38).

Moreover, "a theory of these functions on the foundations here provided would make the form of the function (i.e. its value for every value of its argument) independent of its determination by algebraic operations" (*ibid.*). (This was Cauchy's approach, for example.)

The unifying character of a class of functions which in the same way can be expressed through algebraic operations, presents itself then in the form of the boundary and discontinuity conditions imposed on them. For example, if the range of variability of the quantity z extends simply or multiply over the entire infinite plane A, and within it the function is discontinuous only in isolated points, namely it permits only infinities [*Unendlichwerden*] whose order is finite [that is, the function has only poles and no essential singularities] ..., then the function is necessarily algebraic. Reciprocally, every algebraic function fulfills these conditions (1851, p. 39).

We see here the profundity and generality of Riemann's approach to the theory of functions. Among other things, he is able to project a new light onto the entire theory of algebraic functions.

As far as this argument is concerned, in his dissertation Riemann limits himself to observing that, in order to be able to apply his results to the foundations of a general theory of algebraic functions, it will be necessary to demonstrate "that this fundamental definition of a function of a complex variable completely agrees with one of a dependence on expressible algebraic operations."[12] (1851, p. 39).

Nevertheless, subsequent research has shown that Riemann's conjecture is not sustainable. Seidel gave the first counterexamples (1871, p. 279) and Weierstrass later showed that "the concept of a monogenic function of a complex variable does not completely coincide with the concept of a dependency expressible by (arithmetic) operations" (1880a, p. 210). In fact,

If the domain of convergence of a series whose terms are rational functions of a variable x can be cut up into many parts in such a way that the series uniformly converges in the neighborhood of every place situated within one of these parts, then it represents a unique branch of a monogenic function of x in every individual part, but not necessarily branches of one and the same function in different parts (1880a, p. 221).

Finally, in the concluding paragraphs of his dissertation Riemann tackles the problem of the conformal mapping between two given

Riemann surfaces. "We limit ourselves to the solution of this problem for the case where every point of a surface corresponds to only one point of the other and the surfaces are simply connected" (1851, p. 40).

The solution, Riemann continues, is given by the following mapping theorem:

Two given simply connected plane surfaces can always be related to each other in such a way that, to every point of one there corresponds a point of the other moving continuously with it, and their smallest corresponding parts are alike; namely, the point corresponding to an interior point and to a boundary point can be arbitrarily given. Consequently, the relationship is determinate for all points (*ibid.*).

Riemann then observes that when two surfaces T and R are conformally mapped onto a surface S, T and R can also be conformally mapped on each other. Consequently, "the problem of relating two given surfaces to each other in such a way that the similarity occurs in the smallest parts is reduced to the problem of mapping any arbitrary surface onto a determinate [surface] similar to it in the smallest parts" (*ibid.*).

Now, taking a circle K in a plane B with center at the origin and radius 1, in order to demonstrate the theorem it is only necessary to show that, "An arbitrary simply connected surface T which covers A can always be mapped onto the circle K connected and similar in the smallest parts, but, only in such a way that an arbitrarily given interior point O_0 of the surface T corresponds to the center and an arbitrarily given boundary point O' to an arbitrarily given point of the circumference."[13] (*ibid.*).

Riemann demonstrates his claim in his peculiar geometrical language by using Dirichlet's principle. But Riemann states the "Riemann mapping theorem" in a form that, as Ahlfors has said, "would defy any attempt at proof, even with modern methods" (1953, p. 4).

This is, together with numerous other passages in Riemann's writings, an example of what Ahlfors called his "almost cryptic messages to the future"[14] (*ibid.*, p. 3).

c) A Glance Ahead

The "new method" set out in the inaugural dissertation "is basically applicable to every function which satisfies a linear differential equation with algebraic coefficients," Riemann wrote in a second paper devoted to the theory of functions representable by Gauss' hypergeometric series (1857a, p. 67) (see §3.2).

Both Gauss and Kummer (1836) had based their research on the differential equations (3.2.1). With his method, Riemann continued, "the results that had previously been found in part through fairly tiresome calculations can be derived almost directly from the definition" (*ibid.*). Riemann began by introducing the complex plane $\mathbb{C}$ in the usual manner and then outlined the method of analytic continuation. "If we consider the function w to be given in a part of this plane," he wrote, "then by an easily proved proposition there is only one way in which it can be continuously extended from there by the equation $\partial w/\partial y = i(\partial w/\partial x)$. Of course this extension must not occur in bare lines where a partial differential equation cannot be applied, but in ribbons [*Flächenstreifen*] of finite width" (1857a, p. 68).

He then defined the branch points ("round which the function continues in another") and the branches of a function.

In order to globally study the function $F(\alpha,\beta,\gamma,x)$ defined by (3.2.1), Riemann introduces certain P-functions that have three branch points with a prescribed behavior on the complex sphere and where any three satisfy a linear relationship with constant coefficients. He then succeeds in showing that a P-function verifies a hypergeometric differential equation and determines its coefficients.[15] This is a method that is basically still used in the "global" study of the solution of equation (3.2.1) (e.g. see Ahlfors, 1966, pp. 309-312).

In the same year Riemann presented another example of the power of his "global" method in his famous paper on Abelian functions (Riemann, 1857b). Here the Riemann surface was taken to cover the "Riemann sphere" obtained by adding the point ∞ to $\mathbb{C}$. A multiply connected surface was made simply connected with (in general) $2p$ crosscuts, where p, following Clebsch, is called the "genus" of the surface.

The fundamental theorem stated by Riemann in his dissertation could apply by simply assuming the behavior of a function at the poles (or at the logarithmic discontinuities) and at the branch points to be given, as well as the value of the real part of the $2p$ periods. Moreover, Riemann was able to classify Abelian functions into three fundamental types according to their singularities and further to determine the meromorphic functions on a surface, reobtaining Abel's theorem and laying the foundation for the modern theory of birational transformations.[16]

In the first three paragraphs of (1857b) Riemann summarized the main topological content of his dissertation (connectivity, crosscuts, ramified coverings, etc.) as well as Dirichlet's principle. He further clarified, in a more precise manner, the method of analytic continuation already discussed in his previous paper on the hypergeometric differential equation. From equation $\partial w/\partial y = i(\partial w/\partial x)$ it in fact follows "from a known proposition" that the function w "can be represented by a series of increasing whole powers of $z - a$ of the form $\Sigma_{n=0} a_n (z - a)^n$, provided that everywhere in the neighborhood of a it has *one* determinate value which varies continuously with z, and that this representability extends to a distance from a or a modulus of $z - a$ for which a discontinuity appears" (1857b, p. 88). With the method of indeterminate coefficients one can then completely determine the coefficients a_n.

"Combining the two considerations, one is easily persuaded of the correctness of the theorem:

"*A function of x + iy that is given in a region of the (x,y)-plane can be continuously extended from there in only one way*" (1857b, p. 89).

With the publication of these two works in 1857 Riemann's ideas were for the first time presented to mathematical circles outside of Göttingen. Following the completion of his doctoral work, Riemann treated the theory of complex, elliptic, and Abelian functions in various courses given at the University of Göttingen, beginning with the Winter semester of 1855/56. The part of his lectures devoted to Abelian functions in this year was used as a preliminary version of (1857b), while the part on elliptic functions was published by Stahl on the basis of his and Schering's lecture notes (Riemann, 1899).

In these two sets of lectures, Riemann presented complex function theory in a somewhat eclectic manner, combining the principles set out in his dissertation with Cauchy's and Laurent's results on residues and series expansions, openly utilizing the method of analytic continuation. Particularly interesting from this point of view is the last course given by Riemann at Göttingen in the Summer semester of 1861.[17]

Here, after introducing complex numbers and their geometric interpretation in a Gaussian plane, defining the derivative of a function, and reobtaining equations (6.2.1), Riemann abandons the path followed in his dissertation to adopt Cauchy's point of view, as it has been presented in a unified manner in the recent treatise by Briot and Bouquet (1859).

He thus introduces the notion of a curvilinear integral $\int_C f(z)dz$, demonstrates Cauchy's integral theorem, and then Cauchy's formula for residues

$$\int_C f(z)dz = 2\pi i \sum_k \text{Res}(f, z_k).$$

Cauchy's formula,

$$f(t) = \frac{1}{2\pi i} \int \frac{f(z)}{z - t}\, dt,$$

thus plays an important role in Riemann's lectures. It allows him to obtain the successive derivatives,

$$f'(t),\ f''(t),\ ...,\ f^{(n)}(t),\ ...,$$

and then to expand $f(z)$ in a Taylor series. In a similar manner he then obtains the expansion in a Laurent series by considering the expansion with negative powers for the case of a pole. He then determines the circle (or the circular corona) of convergence, obtains formula (4.6.8) on the number of poles and zeros of a meromorphic function, and finally confronts the problem of extending the domain of convergence by means of the method of "analytic continuation." "If $f(z)$ is given for an arbitrary small part of the surface and remains finite and continuous within it," Riemann says in his lectures, "then $f(z)$ can be extended beyond the limit of this part in only one way, which is, that the value of $f(z)$ is determined within a whole region of the surface when it is given for an arbitrarily small part of it, for example, a line" (Riemann, 1861, p. 28). These words recall what Riemann had written in (1857b).

In this latter work Riemann explicitly excluded thinking of a function defined by an analytic expression or by an equation and limited himself to asserting that however the value of the function was given in a limited part of the plane, one could continue it "in conformance with the differential equations (6.2.1)." In his lectures, however, he began with the expansion in power series defined above and presented the method made famous by Weierstrass, that of considering a point within the circle of convergence with the expansion centered on that point, its relative circle of convergence, and so on.

Only at the moment of studying "multiform" functions and their behavior in the neighborhood of the branch points does Riemann introduce in his lectures the idea of multi-sheeted surfaces, using for clarification the examples of $f(z) = \sqrt{z}$ and of a function defined implicitly by an algebraic equation of the 7th degree.

Having thus set the expansion of a function in a Taylor series as the foundation of his theory, unlike in his dissertation, Riemann no longer had any need either for the theorem of existence or for Dirichlet's principle, which was in fact never mentioned in his 1861 lectures.

The theory of functions of complex variables constituted the necessary premise for the theory of elliptic and Abelian functions which Riemann presented in the second part of his course. Here he returned to his peculiar geometrical treatment which, as Stahl has written, "visually displays the most important properties of elliptic functions" and constitutes "the best preparation for the study of Riemann's theory of Abelian functions, in that all the questions of the more generalized theory appear in it already in simplest form" (1899, p. iii).

After briefly introducing the properties of a doubly periodic function and the parallelogram P of the periods, Riemann considers the differential equation

$$(6.2.12) \quad C \, dz{:}dv = \sqrt{(z - a_1)(z - a_2)(z - a_3)(z - a_4)}$$

or, equivalently,

$$(6.2.13) \quad v = \int^z \frac{C \, dz}{\sqrt{(z - a_1)(z - a_2)(z - a_3)(z - a_4)}}$$

which he takes as the definition of a doubly periodic function $z = \phi(v)$. He then adds, "This definition can serve as the foundation for further analytic research on doubly periodic or elliptic functions. One acquires a new geometrical meaning of elliptic functions and integrals by representing the parallelogram P of the plane v on the plane z by means of equation (6.2.13)" (Riemann, 1899, p. 7).

In fact, if we consider v to be a function of z, $v = \psi(z)$, and its derivative $\psi'(z)$ to be given by (6.2.12), we immediately see that to every value of z there correspond two values of $\psi'(z)$ in the parallelogram P; therefore, "we must extend over the plane z a surface of two sheets" in which the points $z = a_i$ ($i = 1, 2, 3, 4$) represent the branch points of the function $\psi'(z)$ where the representation ceases to be conformal (Riemann, 1899, p. 7).

Riemann then resorts to the usual technique of "cutting" the sheets of the surface along the lines $a_1 a_2$ and $a_3 a_4$ and joining the sides of the cuts in a "crossing" manner in such a way as to construct a surface T able to geometrically represent the branching of the function $\psi'(z) = dv/dz$.

The function $v = \psi(z)$ further allows the representation of the boundary of the parallelogram P on the surface T. Riemann notes that to the curve v_0, $v_0 + k_1$ there in fact corresponds a curve b of the surface T, a loop rising from a point z_0 which includes two branch points. Similarly, to the line $v_0 + k_1$, $v_0 + k_1 + k_2$ there corresponds on the surface a curve a, a loop that runs partly on the upper sheet and partly on the lower (see Figure 8).

With the exception of the two closed curves a and b, the function $v = \psi(z)$ is defined everywhere and continuously on the surface T. Moreover, "the discontinuities of the function defined by the integral (6.2.13) are determined by the periods k_1 and k_2 of the doubly periodic function $z = \psi(v)$, which can be defined as the inverse function of the integral (6.2.13)" (Riemann, 1899, p. 8).

This "visual" example of the theoretical concepts already anticipated in Riemann's dissertation constitutes the natural geometrical setting in which to interpret his successive studies on

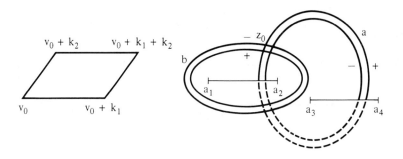

Figure 8

Abelian functions, from the analytic representation by means of trigonometric series, quotients of infinite products, and theta functions, to the problem of adding, multiplying, and transforming elliptic functions.

The path followed by Riemann in his lectures is probably a response to the need for didactic clarity, as he himself later wrote to Betti,[18] but certainly the attempt to "familiarize" his students with the ideas of "surfaces of many sheets" led him to make explicit the fertile interaction of diverse ideas -- the *global* view dominant in the geometrical theory presented in the dissertation and the *local* view based on expansions in series developed in his lectures. This global view would be taken up with necessary rigor by Weyl (1913) in his classic introduction to the modern theory of Riemann surfaces.

The importance of a "local" inquiry is a point that stands close to Riemann's heart and constitutes a central aspect of his conception of physics and mathematics.

In the introduction to his first course at the University of Göttingen, which was given in the Winter semester 1854-55 on the topic of partial differential equations and their application to physical questions, Riemann wrote:

Truly elementary laws for points of space and time can only be based on the infinitely small [*Unendlichkleinen*]. But such laws will in general be partial differential equations, and the derivation of laws for extended bodies and periods of time requires their integration. It is therefore necessary to have methods by which one can deduce these laws for the finite from the laws for the infinitely small, and I mean deduce with all rigor without permitting any sloppiness. Only then can one test them empirically[19] (Riemann, 1869, p. 4).

This is essentially the same point of view expressed in his inaugural lecture, delivered a short time before in 1854. Here he had explained his local conception of an *n*-dimensional geometry.[20] Turning to a discussion of the relationships between *n*-dimensional manifolds and real physical space he wrote that, "questions about the immeasurably great are idle questions for the understanding of nature. It is otherwise, however, for questions about the immeasurably small" (1867b, p. 285).

After emphasizing that the advances of the past into the understanding of nature were dependent on the study of the infinitely small combined with the powerful tools provided by infinitesimal analysis, Riemann concluded that "questions about the measurability (*Massverhältnisse*) of space in the immeasurably small are thus not idle ones" (*ibid.*).

Riemann's lectures had an enormous importance both for the development of modern differential geometry and for mathematical physics. According to Weyl, "*The principle of understanding the external world from its behavior in the infinitely small* is the driving epistemological motive of infinitesimal physics as of Riemann's geometry. It is also the driving motive in the rest of Riemann's magnificent lifework, and above all, in that directed towards the theory of complex functions" (1919, p. 82).

Riemann's ideas on manifolds (*mehrfach ausgedehnte Mannigfaltigkeiten*) would lead, with the emergence of the topology of sets of points, to the construction of the concept of differentiable manifolds of dimension n and class C^q, defined as a topological space T provided with a structure (atlas) of local coordinates $(V_i, \phi_i)_{i \in I}$, where V_i is an open set of T and ϕ_i is a homeomorphism of V_i on an open set of $\mathbb{R}^n$, with the condition that the change of local charts (local coordinates) be given by a differentiable $\phi_j \circ \phi_i^{-1}$ map of class C^q.

If the ϕ_i take values in a complex space $\mathbb{C}^n$ and the changes of coordinates are holomorphic, the manifold is holomorphic of complex dimension n. It is precisely this latter concept that, according to Weyl (1913), stands at the basis of the theory of Riemann surfaces, defined as complex connected manifolds of dimension 1. "It can be taken for certain that the ideas Riemann expressed in that lecture are closely connected to his research on function theory, even though he did not explicitly emphasize this connection" (Weyl, 1913, p. 36).

This conviction guided Weyl in his presentation of the concept of Riemann surfaces. They were thus freed from the original idea of a surface of many sheets extended over the complex plane or sphere, as well as from the necessity of being thought of as embedded in a three-dimensional Euclidean space. They were instead defined abstractly as an analytic realm in which to define functions.

Weyl states his own point of view with considerable vigor.

> We still encounter from time to time the view that the *Riemann surface* is nothing other than a "picture," a means (people say, "*very* valuable, *very* suggestive") for representing and illustrating the multi-valuedness of functions. This view is fundamentally backwards. The Riemann surface is an indispensable, *material* component of the theory; it is, frankly, its foundation. Nor is it something that has been extracted *a posteriori* and more or less artificially from analytical functions, rather it must throughout be regarded as the *prius*, as the mother earth in which functions can first of all grow and flourish[21] (Weyl, 1913, pp. vi-vii).

6.3. Research on Integration and Trigonometric Series

At the end of December, 1853, Riemann wrote to his brother, Wilhelm,

> My work at the moment stands about so: at the beginning of December I presented my *Habilitationsschrift*[22] and with it I had

to recommend three topics for the inaugural lecture, from which the faculty chooses one. I had completed the first two and hoped that they would pick one of these, but Gauss chose the third, and so I am now again in somewhat of a bind since I must still work on this[23] (*Werke*, p. 547).

The theme which Gauss chose for the inaugural lecture, which Riemann confessed to not having completely ready, is his now celebrated paper "On the Hypotheses that Lie at the Fundations of Geometry" (1867b). It was delivered in the spring of the following year, in Gauss' presence, but remained in manuscript until after Riemann's death. It was published for the first time by Dedekind in 1867.

Another of Riemann's manuscript works, which was published by Dedekind in the same volume as the inaugural lecture, was the paper "On the Representability of a Function by Means of Trigonometric Series" (1867a). According to Darboux (1842-1917), this is a "master work," one of whose pearls "is the definition of the definite integral. It is from this that I have drawn a mass of functions that have no derivatives."[24]

The first part of Riemann's paper provides a history of the topic, based primarily on Dirichlet's testimony about his Parisian period and what he had learned from Fourier himself at that time.

In Dedekind's biography of Riemann we read in fact that during the autumn holidays of 1852 Dirichlet spent some time in Göttingen and that he met with Riemann almost daily. The latter wrote to his father at this time, "The other morning Dirichlet was with me for nearly two hours. He gave me the notes that I need for my *Habilitationsschrift* and they are so complete that my work has been made considerably easier. Otherwise I could have spent a long time searching for many things in the library".[25]

The first important step towards the solution of the controversial problem of the representability of a function by means of trigonometric series, Riemann writes, was the recognition by Dirichlet that infinite series divide into two essentially distinct classes: those that are absolutely and those that are conditionally convergent. Only to the first can one apply the ordinary rules valid for finite sums of elements, something that completely escaped the mathematicians of the eighteenth century, primarily because the series of integral powers with which they worked generally (that is, excluding isolated values of the variable) belong to the first class.

This was certainly not the case for Fourier series, which also explains the failure of Poisson's and Cauchy's attempts to demonstrate their convergence. The idea that assured the success of Dirichlet's attempt was that of conducting an analysis of the demonstration given by Fourier for particular cases in the light of the standards of rigor adopted by Cauchy. This in fact provides an explanation of Dirichlet's results, which Riemann summarizes as follows:

A trigonometric series can represent every periodic function with the period 2π which

1) is integrable throughout,

2) does not have infinitely many maxima and minima, and

3) takes on the mean value between its two limiting values wherever its value changes abruptly (1867a, p. 237).

A function that does not satisfy condition (3) cannot, Riemann says, be represented by a trigonometric series. In fact, "the trigonometric series that represents it outside of the discontinuities would differ from it precisely at the points of discontinuity" (*ibid.*). But, when one or the other of the first two conditions is not satisfied, is it possible, and under which hypotheses, to represent the function by means of a trigonometric series? Riemann's studies grapple with this issue, even though, as he himself admitted, the question still remained open after his work. In fact, if (2) is not valid the conclusions on Dirichlet's integral are not valid; if instead (1) is dropped, where the integral is understood in Cauchy's sense, it is not possible to determine the coefficients of the Fourier series.

Riemann proposes to investigate the question without making any particular hypothesis about the nature of the function. His investigation will therefore be constrained by these conditions. "A path as direct as Dirichlet's," he writes, "is by the nature of the thing impossible" (*ibid.*, p. 239).

It is consequently very interesting to read of the motivations that Riemann himself assigned to his own work.

In fact, [the problem] was completely solved for all cases which present themselves in nature alone, because however great may be our ignorance about how the forces and states of matter vary in space and time in the infinitely small, we can certainly assume that the functions to which Dirichlet's research did not extend do not occur in nature.

Nevertheless, those cases that were unresolved by Dirichlet seem worthy of attention for two reasons.

The first is that, as Dirichlet himself remarked at the end of his paper, this subject stands in the closest relationship to the principles of the infinitesimal calculus and can serve to bring these principles to greater clarity and certainty. In this connection its treatment has an immediate interest.

But the second reason is that the applicability of Fourier series is not restricted to physical researches; it is now also being applied successfully to one area of pure mathematics, number theory. And just those functions whose representability by a trigonometric series Dirichlet did not explore seem here to be important (1867a, pp. 237-8).

A crucial point that follows from condition (1) above is the rigorous specification of the concept of the integral. Thus, after noting that much uncertainty still reigns over certain fundamental points of the infinitesimal calculus, Riemann asks the question, "What do we understand by $\int_b^a f(x)dx$?" (1867a, p. 239).

In fact, he says, since the work of Cauchy and Fourier the integral has no longer been understood intuitively as the inverse operation of the derivative, but as an area. Nevertheless, he continues, Cauchy's definition (which Dirichlet also adopted in his work of 1829) is applicable when the integrand $f(x)$ is continuous in the interval of integration or presents at most a finite number of points of discontinuity. However, it ceases to be valid if these points are infinite in number. But it is precisely this case that interests Riemann. Even Dirichlet had exhausted himself in vain over this problem.

The response that Riemann gives to the question "What do we understand by $\int_b^a f(x)dx$?" consists of the definition of the integral that carries his name.

Dividing the interval of integration $[a,b]$ into n parts $[x_{i-1},x_i]$ with

$$a < x_1 < x_2 < \cdots < x_{n-1} < b,$$

and letting $\delta = x_i - x_{i-1}$, Riemann considers the sum

$$S = \sum_1^n \delta_i f(x_{i-1} + \varepsilon_i \delta_i)$$

(where $x_0 = a$ and the ε_i are positive fractions). S depends on the choice of the partial intervals δ_i and of the numbers ε_i.

If, no matter how the δ_i and the ε_i are taken, S tends to a fixed limit A as δ_i tends to zero, this limit is what Riemann calls the integral $\int_a^b f(x)dx$. An extension of this definition "accepted by all mathematicians" occurs when, as $f(x)$ becomes infinitely large as x tends to a value c, there nevertheless exists the limit of

$$\int_a^{c-\alpha_1} f(x)dx + \int_{c+\alpha_2}^b f(x)dx$$

as α_1 and α_2 tend to zero.

But, Riemann asks, "in what cases is a function integrable and in what cases not?" (1867a, p. 240). We need to find a necessary and sufficient condition for the integrability of a function, a condition that Riemann sets out in the following manner:

In order that the sum S converges, if all the δ become infinitely small, besides the finiteness of the function $f(x)$, it is still necessary that the total size [*Gesamtgröße*] of the intervals in which the variations are $> \sigma$, whatever σ may be, can be made arbitrarily small through an appropriate choice of d [$= \sup \delta_i$].

And vice versa:

If the function $f(x)$ is always finite, and if, with the infinite decrease of all the quantities δ, the total size s of the intervals in which the variations of the function $f(x)$ are greater than a given quantity σ always becomes infinitely small in the end, then the sum S converges when all the δ become infinitely small[26] (1867a, p. 241).

It is not difficult to convince ourselves that this constitutes an effective extension of Cauchy's definition, in the sense that functions for which Cauchy's definition is not applicable are Riemann integrable.

As an example of a similar function, discontinuous in every rational point and nonetheless integrable, Riemann gives the following:

(α) $$f(x) = \sum_1^\infty \frac{[nx]}{n^2} \, ,$$

where $[nx] = x$ minus the nearest integer, or $[nx] = 0$ if x is equidistant from two integers.

The series is convergent, and if $x = p/2n$ (with p and n relatively prime) then,

$$f(x + 0) = f(x) - \frac{1}{2n^2}\left[1 + \frac{1}{9} + \frac{1}{25} + \cdots\right] = f(x) - \frac{\pi^2}{16n^2}$$

$$f(x - 0) = f(x) + \frac{1}{2n^2}\left[1 + \frac{1}{9} + \frac{1}{25} + \cdots\right] = f(x) + \frac{\pi^2}{16n^2} \, ,$$

while it is continuous in every other point.

The integrability of $f(x)$ results from the fact that the variation of the function is equal to $\pi^2/8n^2$ at the points $x = p/2n$ and consequently the points for which the variation is greater than any σ > 0 are finite in number.

Having defined the concept of the integral, Riemann next confronts the central argument of his investigation, in a manner that was entirely characteristic. Preceding works, he says, have followed this scheme: if a function has such and such properties, then it can be expanded in a Fourier series. "We must proceed from the inverse question: If a function is representable by a trigonometric series, what consequences does this have for its behavior, for the variation of its value with the continuous variation of the argument?" (1867a, p. 244).

In order to answer this he takes as given the series

(6.3.1) $A_0 + A_1 + A_2 + \cdots \, ,$

where for simplicity $A_0 = \frac{1}{2}b_0$ and $A_i = a_i \sin ix + b_i \cos ix$ ($i = 1,2,$ ...).

Let $f(x)$ be the value of the series in such a way that the function is determinate only for the values of x for which the series is convergent. A necessary condition for the convergence of a series is that its terms tend to zero as the index n increases. Now, we have two cases:

(A) the coefficients a_n and b_n tend to zero as n increases, and consequently the terms of the series become infinitely small for *any value whatever* of x (i.e., $A_n(x)$ converges to 0 uniformly as n increases),

(B) in the opposite case, they can become infinitely small only for particular values of x.

Riemann begins by examining the first of these. Under this hypothesis he considers the series

(6.3.2) $\qquad C + C'x + A_0 \dfrac{x^2}{2} - A_1 - \dfrac{A_2}{4} - \dfrac{A_3}{9} - \cdots = F(x)$

obtained by integrating (6.3.1) term-by-term twice. He demonstrates that $F(x)$ is continuous and integrable, and then gives several properties:

(1) When the series (6.3.1) is convergent, the expression

(6.3.3) $\qquad \dfrac{F(x + \alpha + \beta) - F(x + \alpha - \beta) - F(x - \alpha + \beta) + F(x - \alpha - \beta)}{4\alpha\beta}$

where α and β are infinitesimals whose ratio is finite, converges towards the same value to which the series (6.3.1) converges.

In the first place Riemann supposes $\alpha = \beta$, which allows him to write (6.3.3) as $A_0 + \Sigma_{i=1}^{\infty} A_i(\sin i\alpha/i\alpha)^2$, an expression that he demonstrates is convergent to $f(x)$ by showing that the series (6.3.1) implies

$$A_0 + A_1 + A_2 + \cdots + A_{n-1} = f(x) + \varepsilon_n.$$

By setting $A_n = \varepsilon_{n+1} - \varepsilon_n$, he obtains

(6.3.4) $\qquad \displaystyle\sum_0^{\infty} A_n \left(\dfrac{\sin n\alpha}{n\alpha}\right)^2 = f(x) + \sum_1^{\infty} \varepsilon_n \left\{ \left[\dfrac{\sin(n-1)\alpha}{(n-1)\alpha}\right]^2 - \left(\dfrac{\sin n\alpha}{n\alpha}\right)^2 \right\}.$

The series in the second member converges to zero, and this is sufficient to establish the proof. The demonstration is also valid when $\alpha \neq \beta$, seeing that

$$F(x + \alpha + \beta) - 2F(x) + F(x - \alpha - \beta) = (\alpha + \beta)^2 [f(x) + \delta_1]$$

$$F(x + \alpha - \beta) - 2F(x) + F(x - \alpha + \beta) = (\alpha - \beta)^2 [f(x) + \delta_2],$$

which is sufficient to apply the results obtained to this point.

In a similar manner Riemann proves the proposition

(2) $\lim\limits_{\alpha \to 0} \dfrac{F(x + 2\alpha) + F(x - 2\alpha) - 2F(x)}{2\alpha} = 0,$ for all x

He finally states the following proposition for $F(x)$:
(3) We designate with b and c two arbitrary constants where $b < c$, and with $\lambda(x)$ a function continuous between b and c that is null at b and c, whose first derivative has the same properties, and whose second derivative does not have an infinite number of maxima and minima. The integral

(6.3.5) $\mu^2 \displaystyle\int_b^c F(x)\cos \mu(x - a)\lambda(x)dx$

becomes smaller than any given magnitude as μ increases without limit.[27]

At this point Riemann can state the following propositions:

   I. If a periodic function $f(x)$ of period 2π can be represented by a trigonometric series whose terms ultimately become infinitely small for every value of x, then there must be a continuous function $F(x)$, on which $f(x)$ depends, such that (6.3.3) converges to $f(x)$ when α and β become infinitely small and thereby their ratio remains finite.
   In addition (6.3.5) must become infinitely small for increasing μ [under the same hypothesis for $\lambda(x)$, $\lambda'(x)$, and $\lambda''(x)$].
   II. If conversely these two requirements are fulfilled, then there exists a trigonometric series in which the coefficients ultimately become infinitely small and which represents the function wherever the series converges (1867a, p. 251).

Propositions I and II express the necessary and sufficient condition for which a trigonometric series exists for the given $f(x)$ that has $f(x)$ as its generalized Riemann sum. This is in fact the generalized second derivative of $F(x)$ given by expression (6.3.3).
Riemann develops the proof in the following manner:
We determine two quantities C' and A_0 in such a way that $F(x) - C'x - A_0 x^2/2$ is a periodic function of period 2π which we expand into the Fourier series $C - A_1/1 - A_2/4 - \cdots$ by setting

$$\frac{1}{2\pi} \int_{-\pi}^{\pi} [F(t) - C't - A_0 t^2/2]dt = C$$

$$\frac{1}{\pi} \int_{-\pi}^{\pi} [F(t) - C't - A_0 t^2/2]\cos n(x - t)dt = -A_n/n^2.$$

Then

(6.3.6) $A_n = -\dfrac{n^2}{\pi} \displaystyle\int_{-\pi}^{\pi} [F(t) - C't - A_0 t^2/2]\cos n(x - t)dt$

tends to zero as n increases, and consequently for theorem (1) the series $\Sigma\, A_i$ will have $f(x)$ as a sum when it is convergent.

Riemann's next theorem offers a method for studying the behavior of this series:

III. Let $b < x < c$, and let $\rho(t)$ be a function such that $\rho(t)$ and $\rho'(t)$ have the value 0 when $t = b$ and $t = c$ and varies continuously between these values, $\rho''(t)$ does not have infinitely many maxima and minima, and further that $\rho(t) = 1$, $\rho'(t) = 0$, and $\rho''(t) = 0$ for $t = x$, while $\rho'''(t)$ and $\rho''''(t)$ are finite and continuous. Then the difference between the series $A_0 + A_1 + \cdots + A_n$ and the integral

(6.3.7) $\qquad \dfrac{1}{2\pi} \displaystyle\int_b^c F(t)\, \dfrac{d^2\dfrac{\sin(2n+1)(x-t)/2}{\sin(x-t)/2}}{dt^2}\, \rho(t)dt$

becomes infinitely small as n increases. The series $A_0 + A_1 + A_2 + \cdots$ will then converge or not converge according to whether (6.3.7) approaches a fixed limit or not as n increases (1867a, p. 252).

We consider the sum of the terms (6.3.6) extended from 1 to n. Since

$$2\sum_{i=1}^{n} -i^2\cos\, i(x-t) = 2\sum_{i=1}^{n} \dfrac{d^2\cos\, i(x-t)}{dt^2}$$

where the second member is exactly equal to the second derivative with respect to t, which appears in the integrand of (6.3.7), we can write,

(6.3.8) $\qquad \displaystyle\sum_i^n A_i = \dfrac{1}{2\pi}\int_{-\pi}^{\pi}\left[F(t) - C't - \dfrac{A_0 t^2}{2}\right]\dfrac{d^2\sin\dfrac{(2n+1)(x-t)/2}{\sin(x-t)/2}}{dt^2}\,dt.$

On the basis of theorem (3), indicating the integrand of (6.3.8) by $H(t,x)$, we find that

$$\dfrac{1}{2\pi}\int_{-\pi}^{\pi} H(t,x)\lambda(t)dt$$

becomes infinitely small as n increases under the same hypotheses for $\lambda(t)$ and its derivatives and, moreover, under the hypothesis that for $t = x$,

$$\lambda(t) = 0, \qquad \lambda'(t) = \lambda''(t) = 0,$$

while $\lambda'''(t)$ and $\lambda''''(t)$ remain finite and continuous.

Having established this, if we take $\lambda(t) = 1 - \rho(t)$ between b and c and $\lambda(t) = 1$ elsewhere, we find that the difference between ΣA_i and the integral (6.3.7) becomes infinitely small as n increases, as was required.

This is the essential content of Riemann's work. In the same way we can also treat the case where the trigonometric series (6.3.1) has terms that become infinitely small like $1/n$ for a value of x of the argument, without this happening for *every* value of the argument. Riemann shows this by considering the series obtained by substituting $x + t$ and $x - t$ in (6.3.1). Then summing term-by-term he in fact obtains the series

$$2A_0 + 2A_1\cos t + 2A_2\cos 2t + \cdots$$

whose terms tend to zero as n increases for every t and for which we can therefore apply the results obtained.

It is also necessary to emphasize at this point that in establishing his theorems on the representability of a function in a trigonometric series Riemann did not make any hypothesis about the form of the coefficients in the series (6.3.1). Consequently, this need not be a Fourier series with the coefficients given by (2.3.14). The distinction between Fourier and trigonometric series, which was first introduced by Riemann, was to become important only after Cantor's work (see §7.2b below).

Riemann did not at this time consider the problem of all those cases which are excluded from Dirichlet's conditions, but limited himself to considering a few important examples. In actual fact, these examples raised more problems than Riemann was apparently able to solve, and this could be the reason why Riemann left his *Habilitationsschrift* unpublished during his lifetime.

As far as functions with an infinite number of maxima and minima are concerned, he asserts that he can give examples of integrable functions that are not representable in Fourier series, such as,

$$f(x) = \frac{d(x^\nu\cos(1/x))}{dx} \qquad (0 < \nu < 1/2).$$

He further notes that,

when the Fourier series of a function does not converge, in spite of the general possibility of integration, and its term itself can eventually become infinitely large, -- then, in spite of the general impossibility of integrating $f(x)$, there can be infinitely many values of x between any two values, no matter how close, for which the series (6.3.1) does converge (1867a, pp. 262-3).

An example of this is the function given by the series

$$\sum_1^\infty \frac{(nx)}{n},$$

where (nx) has the same meaning as it was given at the beginning. Riemann says that this function exists for every rational value of x and is representable by the trigonometric series,

$$\sum_{n=1}^{\infty} \frac{\Sigma_{\theta}(-(-1)^{\theta})}{n\pi} \sin 2n\pi x,$$

where all the divisors of n are set for θ.[28] But it is not contained in any interval, no matter how small, between finite limits, and consequently never integrable (1867a, p. 263).

Similar examples are given by the series

$$\sum_{0}^{\infty} c_n \cos n^2 x \quad \text{and} \quad \sum_{0}^{\infty} c_n \sin n^2 x,$$

where the c_i are positive quantities decreasing to zero, but for which the sum $\Sigma_{i=1}^{\infty} c_i$ becomes infinitely large.

Finally, "the trigonometric series can also converge infinitely often between any two arbitrarily close arguments if its coefficients do not ultimately become infinitely small" (1867a, p. 264). An example (with which he concludes the paper) is given by the series $\Sigma_{1}^{\infty} \sin(n!x\pi)$, which not only converges for every rational value of x, but also for infinite irrational values such as, for example, the multiples of $\sin 1$, $\cos 1$, $2/e$.

That Riemann had other "pathological" functions in mind is evident from what Weierstrass affirmed several years later in a communcation to the Berlin Academy of Sciences (1872).

On this occasion Weierstrass said that he had learned from those who had attended Riemann's lectures that in 1861 or perhaps even earlier he had maintained that the function represented by the trigonometric series

$$\sum_{n=1}^{\infty} \frac{\sin(n^2 x)}{n^2}$$

does not verify a property apparently admitted by mathematicians like Gauss, Cauchy, and Dirichlet, that is, that a continuous function of a real variable always has a first derivative (except in isolated points).

Unfortunately [Weierstrass continues] Riemann's proof of this has not been published, nor does it appear to have been preserved in his papers or by oral communication. ... Those mathematicians who have concerned themselves with the problem after Riemann's conjecture became known in wider circles, seem to be of the opinion (at least the majority thereof) that it is enough to prove the existence of functions that have points in every interval of their argument, no matter how small, where they are not differentiable. That there are functions of this kind is unusually easy to prove, and I consequently believe that Riemann had in mind only those functions that have no determinate derivative for any value of their argument (1872, pp. 71-2).

That the given function has this property is somewhat difficult to demonstrate, Weierstass adds.[29] It was simple enough to find functions that were continuous but not differentiable in any point. The example that Weierstrass produces in this context is the function

$$f(x) = \sum_{n}^{\infty} b^n \cos(a^n x \pi),$$

with a an odd integer, b positive and smaller than 1, $ab > 1 + \frac{3}{2}\pi$ (see §7.1).

There also are traces of the perplexities and difficulties raised by Riemann's function in several letters and notes written by the Italian mathematicians Betti and Casorati in the early 1860s. They probably learned of these from Riemann himself during his repeated stays in Italy during the years 1863-65. (This was due to his precarious state of health, which led to his death the following year during another trip to Italy.)

Betti communicated Riemann's ideas to Tardy (1816-1914), who replied that Dirichlet had already "noted similar things" during his trip to Italy in 1843. It is doubtful, however, that he really understood the matter. Indeed, Tardy's letter continues by citing the example of $y = x \sin(1/x)$, "which cannot be said to have any determinate derivative for $x = 0$." Tardy concludes with the statement, "I believe I can generally demonstrate the existence of the derivative in a way somewhat different from the one normally used."[30]

Uncertainties and misunderstandings over Riemann's works also appear in Casorati's notes, as for example in the report in his scientific diary of a discussion that he had with Prym on this subject in 1865.[31] But this was only one example of the complex and delicate questions, which were nevertheless fundamental for real analysis, that Riemann had been led to face by his studies of trigonometric series. These questions began to be answered a short time later, in the course of the rigorous reconstruction of analysis undertaken by Weierstrass and the Berlin mathematicians.

Notes to Chapter 6

[1]For Herbart's influence on Riemann, see Scholz (1982). After discussing how Herbart's philosophy was of particular interest for Riemann on the basis of their published works and of excerpts and notes found among Riemann's papers, Scholz concludes, "The assumption that Herbart's philosophy of space had an important impact on Riemann's formulation of the manifold concept cannot be confirmed. On the other hand, Herbart's epistemology and his ideas on the relationship between philosophy and science do seem to have influenced Riemann and thus Riemann's perception of the task of mathematics" (1982, p. 427).

[2]On this subject, however, Riemann wrote only a brief but profound note on the number of prime numbers smaller than a given number (1859). (See Edwards (1974).)

[3]Like Dirichlet, Riemann also seemed to share the then prevalent opinion that any function that is continuous in a given interval can be represented as a Fourier series and that Dirichlet's conditions are in a certain sense too restrictive (see §5.2).

[4]At this point Riemann adds the following note: "This assertion is clearly justified in all cases where an expression of dw/dz in z can be found from an expression of w in z by means of the rules of differentiation. Its rigorously general validity remains for the moment there established" (1851, p. 4 n).

[5]Peano (1858-1932), in his notes to A. Genocchi's *Trattato di calcolo differenziale e principi di calcolo integrale* (1884), has observed that "the conditions $\partial u/\partial x = \partial v/\partial y$ and $\partial u/\partial y = -\partial v/\partial x$ are necessary and sufficient for the existence of the derivative. The conditions $\partial^2 u/\partial x^2 + \partial^2 u/\partial y^2 = 0$, like those for v, are also necessary, but not sufficient."

Nevertheless, in 1936 Menchoff observed that, while "it is well known that a function $f(z)$, monogenic in a point z, satisfies the Cauchy-Riemann conditions (6.2.1) there ... Moreover, the reciprocal proposition is not true" (1936, p. 5). He illustrates his claim with the counterexample of $f(z) = z_0 + re^{5i\phi}$ where $z = z_0 + re^{i\phi}$, $r,\phi \in \mathbb{R}$ and $r \geq 0$. Hence, "in order to have a sufficient condition of monogeneity in a point, it is necessary to add to conditions (6.2.1) some other supplementary condition" (*ibid.*, 6). To this end he defines the concept of the Stoltz-Fréchet total differential for a function $f(z)$ in a point z if

(1) there are finite $\partial f(z)/\partial x$, $\partial f(z)/\partial y$ in z;

(2) $\lim\limits_{\Delta z \to 0} \dfrac{f(z + \Delta z) - f(z) - (\partial f/\partial x)\Delta x - (\partial f/\partial y)\Delta y}{\Delta z} = 0, \quad \Delta z = \Delta x + i\Delta y,$

and then shows that a function $f(z)$ is monogenic in a point z if it there possesses a Stoltz-Fréchet total differential and satisfies (6.2.1). For a discussion of this point see also Hille (1963, I, pp. 78-81) and Remmert (1984, pp. 35-44).

[6]Picard observes that, "In Cauchy's writings and that of the majority of his disciples the equation (6.2.2) seldom appears, and one reasons on the complex function itself." He then says of Riemann's concept, "This point of view is assuredly more philosophical; it has the great advantage of leaving aside every symbol, and the theory of complex functions is, definitively, the study of the two *associated* functions u and v" (Picard, 1893, p. 7).

This view was taken up more recently by Dieudonné when he wrote that Riemann "is moreover the first mathematician who knew how to profit from the fact that U and V are then harmonic functions" (1974, Vol. 1, p. 50).

[7]Riemann himself referred in a note to Gauss' paper of 1825. Nevertheless, it is well to point out, with Ahlfors, that "Riemann was the first to recognize the fundamental connection between conformal mappings and complex function theory: to Gauss, conformal mapping had definitely been a problem in differential geometry" (Ahlfors, 1953, p. 3).

This provides yet another demonstration of the vigorous thrust towards the synthesis of different theories present in Riemann's works.

[8]An inquiry into the origins of Riemann's ideas in the field of complex analysis formed the basis of A. I. Markusevic's communcation to the International Congress of Mathematicians at Helsinki in 1978. The citations are taken from an (unpaginated) copy graciously lent to me by H. Mehrtens. The original Russian text is printed in the proceedings of the congress.

[9]E. Neuenschwander's recent research on Riemann's papers has brought new evidence on the relationships between Riemann and the French mathematicians, in particular Cauchy and Puiseux. According to Neuenschwander, in a draft copy of the defense for his dissertation Riemann wrote about the definition of a complex function: "This is the view of Cauchy, who is the first and foremost among the French to have occupied himself with the theory of complex magnitudes, a view which he expressed in the session of the Paris Academy of March 31 this year on the occasion of a report on a work by Puiseux and which he pursued further in several later addresses" (In: Neuenschwander, 1981a, p. 91). See also Neuenschwander (1981b).

In my view, it is not at all astonishing that Riemann was aware of Cauchy's papers on "geometrical magnitudes" (see §4.6). Cauchy's work had been published in the *Comptes rendus* of the Paris Academy and through Cauchy's (1851c) Riemann could have learned of Puiseux's work. However, this does not undercut the view that Riemann's ideas developed by following a path that was more original than a simple translation of Puiseux's results into geometrical terminology.

[10]A. Tonelli, who was a student of Betti in the 1870s and studied for a period in Göttingen, was the first to note that "no rigorous demonstration of this important proposition has been given that can serve as a basis for the definition of the order of connectivity of a simply connected space" (Tonelli, 1873-4, p. 594). A history of algebraic topology from its origins to the end of the last century can

be found in Pont (1974). For Tonelli's criticisms in particular see pp. 82-3.

[11]According to Ahlfors, Riemann's approach can be expressed in modern terms as follows: "Given a closed differential α with given periods, singularities and boundary values, he assumes the existence of a closed differential β such that $\alpha + \beta^*$ (β^* denotes the conjugate differential) has a finite Dirichlet norm. Then he determines an exact differential ω, with zero boundary values, whose norm distance from $\alpha + \beta^*$ is a minimum. But this is equivalent to an orthogonal decomposition

$$\alpha + \beta^* = \omega_1 + \omega_2^* \qquad (\omega_1 \text{ exact, } \omega_2 \text{ closed})$$

from which it follows that

$$\alpha - \omega_1 = \omega_2^* - \beta^*$$

is simultaneously closed and co-closed, that is to say harmonic. Hence the existence theorem: there exists a harmonic differential with given boundary values, periods, and singularities" (Ahlfors, 1953, p. 5).

In order to render Riemann's reasoning rigorous we can place it in a Hilbert space, where the closed and exact differentials can be defined by conditions of orthogonality.

[12]To this passage Riemann adds the following note: "Under this will be understood any dependence expressible by a finite or infinite number of the four simple arithmetical operations [*Rechnungsoperationen*], addition, subtraction, multiplication, and division. The expression *Grössenoperationen* (in contrast to *Zahlenoperationen*) will mean arithmetic operations of this type, for which the commensurability of the magnitudes does not come into consideration" (1851, p. 39 n.).

[13]In modern terms Riemann's theorem states that every simply connected domain D of the complex plane is biholomorphically equivalent to K.

[14]This "cryptic" and inspiring "message" has found its natural and rigorous formulation in the modern theory of uniformization, one of the richest and most stimulating fields of research in complex analysis. Many distinguished mathematicians have contributed to it in the last one hundred years, from Klein and Poincaré to Koebe, Caratheodory, Bierberbach, Montel, and others. The principal result, the uniformization theorem, states that every simply connected Riemann surface is biholomorphically equivalent to either C, $C \cup \infty$, or the unit disk K. The three possibilities are distinct. As a consequence of Liouville's theorem (see §4.6), it follows that there

exists no biholomorphic mapping of the unit disk onto the whole plane C.

[15]Riemann's (1857a) paper, together with Weierstrass' lectures, had a strong influence on Fuch's work on differential equations (see Gray, 1984).

[16]Riemann's approach to Abelian functions has been thoroughly discussed by C. Houzel in Dieudonné (1978, II, pp. 1-113). In particular, see pp. 92-101.

[17]For the following presentation of Riemann's lectures for the Summer semester of 1861 I have utilized the unpublished notes taken by E. Schultze, who was one of Riemann's students at the time. I am grateful to Dr. Kirsten of the Akademie der Wissenschaften der DDR (Berlin) for having put this manuscript at my disposition.

[18]In 1863 Riemann wrote to Betti that, "My presentation of the branching of functions through surfaces in the form in which I set it out in Borchardt [Riemann, 1857b] has created difficulties even for German readers. It has therefore been necessary in my lectures to make this presentation clear and familiar to my listeners through simple, appropriately chosen examples" (In: Bottazzini, 1982, p. 255).

[19]In his lecture on Riemann, Klein, after recalling his research in the fields of optics and electromagnetic phenomena and his belief in the existence of the aether, was led to assert that, "What in physics is the banishment of action at a distance and the explanation of appearances through the internal forces of a space-filling aether, is in mathematics the understanding of functions *from their behavior in the infinitely small*, and thus in particular *from the differential equations* that they satisfy" (1894, p. 484). In order to emphasize this aspect even more strongly, he sought to draw a strict parallel between Faraday's work in physics and that of Riemann in mathematics.

[20]An historical account of Riemann's achievements in this field can be found in Torretti (1978) and in Scholz (1980).

[21]In 1913, Weyl took up the concept of an analytic element of a function according to Weierstrass. He then introduced Weierstrass' *analytisches Gebilde*, the set of (x,y) satisfying a polynomial equation $f(x,y) = 0$, including poles and branch points. After abstractly defining the notion of a bidimensional manifold in Riemann's sense, Weyl discusses under what conditions an *analytisches Gebilde* can be considered as a surface, and as a Riemann surface in particular. He further shows the relatively "secondary" character of the *analytisches Gebilde* concept.

With respect to Riemann surfaces, Weyl asserted: "The fundamental idea that lies behind its introduction is by no means limited to the theory of complex functions. A function of two real variables x, y is a *function in the plane*; but it is certainly just as justified to examine functions on a sphere, on a torus, or on a surface in general as those in the plane. Of course, as long as one only worries about the behavior of the function 'in the small' -- and most of the considerations of analysis are based on this -- the concept of a function of two real variables is general enough to represent the neighborhood of every point of a two-dimensional manifold by x, y (or $x + iy$). But as soon as one advances to study the behavior of functions "*in the large*," functions in the plane present an important but *special* case *among infinitely many other equally so.* Riemann and Klein have taught us not to stop at these special cases. Applied to the theory of complex functions this means: *before one begins to study any class of functions, the surface that furnishes the domain of the independent variable must always be defined; then it must be clarified what an "analytical function" means on the surface, whereby the surface becomes a Riemann surface. Only then can one finally approach the functions themselves*" (Weyl, 1913, p. 42; 1955, p. 38).

[22]This is the thesis required for promotion to *Privatdozent*, the first step in an academic career in the German universities.

[23]The other two topics Riemann prepared for his inaugural lecture were:

(1) The history of the question of the representability of a function by means of trigonometric series.
(2) On the solution of two equations of the second degree in two unknowns.

[24]This is found in a letter to Houël of March, 1873, printed in Dugac (1973, p. 150).

[25]Riemann, *Werke*, p. 546. Riemann's historical reconstruction remained standard until the end of the century, before it was corrected in several points and integrated into the works of Burkhardt (1901-8) and (1914-5).

[26]In modern terms, following Lebesgue, we can say that, "in order for a bounded function $f(x)$ to be integrable, it is necessary and sufficient that the set of its points of discontinuity be of measure zero" (1904, p. 29).

[27]The demonstration is found in Riemann (1867a, pp. 249-51), as well as in Tonelli (1928, pp. 76-82). The latter uses the uniform convergence of the series $\Sigma\ A_i$, which follows from the hypothesis that, for every x, $a_n \rightarrow 0$ and $b_n \rightarrow 0$. See also Lebesgue (1906, pp. 113-20).

[28]The symbol $\Sigma_\theta(- (-1)^\theta)$ signifies a sum of units, positive or negative according to whether the divisor of n is odd or even respectively.

[29]Weierstrass' intuition was correct. Gerver (1970) has recently shown that Riemann's function has a finite derivative $= -1/2$ for values $\zeta\pi$, with ζ a rational number of the form $(2A+1)/(2B+1)$. In 1916 Hardy had proved that the function does not have a derivative if ζ is an irrational or rational number of the form $2A/(4B+1)$ or $(2A+1)/2B$ (for integer A and B).

[30]This letter is printed in Bottazzini (1982, pp. 257). Tardy was one of the most distinguished mathematicians active in Italy at that time, and was also highly regarded abroad.

[31]Casorati's text is given in Neuenschwander (1978b).

Chapter 7
THE ARITHMETIZATION OF ANALYSIS

7.1. Discussions with Kronecker and Weierstrass

In the autumn of 1864 F. Casorati, then a young professor of analysis at the University of Pavia, traveled to Berlin to meet the men who at that time constituted the point of reference for European mathematics. Primary among them were the "triumvirate" of Kummer, Kronecker, and Weierstrass.

Kummer had succeeded to the chair of Dirichlet after the latter had moved to Göttingen in 1855, to fill Gauss' chair. Kronecker also gave lectures at the university, even though he did not have an official position, on the basis of a privilege that was granted to members of the Berlin Academy of Sciences. Weierstrass had been named an "extraordinary" professor at Berlin when he was already nearly forty, after having taught for many years in provincial gymnasia. The publication of a major paper on Abelian functions in Crelle's journal had brought him to the attention of the mathematical world and, with Kummer's support, he had been called to Berlin in 1856. Kronecker arrived in the capital in the same year, after accumulating a private fortune as a banker.

The notes that Casorati took during his conversations with Kronecker and Weierstrass, in which Schering (1833-1897), Kummer, Roch (1839-1866), and others also participated, provide a concise and penetrating account of the questions that were then at the forefront of their thought. In many points they anticipate results that would not come to public attention until much later.

The topics that they discussed centered on different aspects of real and complex analysis, from continuity to Dirichlet's principle to elliptic functions, but there was one point of view shared by the Italian mathematician and his distinguished colleagues -- it was necessary to introduce greater rigor into analysis. This was a problem that Weierstrass had begun to treat in his lectures beginning in 1860, after having dedicated his first courses to a variety of other subjects, from mathematical physics (in the first course during the Winter of 1856/57) to geometry to the theory of representing analytic functions by means of convergent series. (In the latter he set out results he had already obtained in 1841 but never published.) The integral calculus and the theory of elliptic functions were also among the topics covered in these first years. Elliptic and Abelian functions at that time seemed to be the principal objective of Weierstrass' scientific work. Indeed, in his inaugural lecture to the Berlin Academy of Sciences he had said,

> In a few words, I will now clarify the course of my earlier studies, and indicate the direction in which I henceforth intend to proceed.
> A comparatively younger branch of mathematical analysis, the theory of elliptic functions, has, from the time in which I first became acquainted with it under the direction of my honored teacher Gudermann, who I will always hold in grateful memory, exercised a strong attraction on me, and has retained a definite influence on the entire course of my mathematical development (*Werke* 1, pp. 223-4).

It was an influence that endured throughout Weierstrass' life and profoundly influenced the course of his research on the foundations of the theory of analytic functions.

Weierstrass' inaugural address is interesting in many respects. For example, he here forcefully emphasizes the need for a more profound understanding of the relationship between mathematics and natural sciences, a theme that he said "lies very close to my heart," but he did not say a word about the program of rigor, for which his name has become famous!

It would not be rash to suppose that Weierstrass began to feel the need for greater rigor in analysis when he began to face the problem of presenting the theory of analytic functions in his lectures. This theory was, of course, an indispensable prerequisite for the theory of elliptic and Abelian functions. In his 1859/60 lectures on the "Introduction to analysis," Weierstrass treated the problem of foundations for the first time. This course was followed in the next semester by one on "The integral calculus."[1] In 1863/64 the topic of his lectures was "The general theory of analytic functions." According to Dugac (1973, p. 56), it was at this time that Weierstrass began to set out his theory of the real numbers.

It is important to note that at almost the same time, Dedekind was also displaying a similar dissatisfaction with the foundations of the calculus in his lectures on the infinitesimal calculus at the Zurich Polytechnic.

Thus, in the second half of the last century, in the face of an extraordinary wealth of new results, those mathematicians who were most attentive to methodological questions were beginning to realize that the fundamental theorems of analysis and even those ideas which appeared to be the most secure lacked a rigorous foundation.

Examples were Cauchy's criteria for the convergence of series, the criteria for the term-by-term differentiability and integrability of a series of functions, and the theorem that every magnitude which continually increases, though not beyond all limits, necessarily approaches a limiting value. In regards to this theorem, Dedekind said that it "can to a certain degree be seen as a sufficient foundation for infinitesimal analysis" (1872, p. 316). Even the concept of continuity seemed uncertain, while the study of trigonometric series was far from complete.

There was the steadily growing conviction (although it did not become dominant without argument and stubborn opposition) that the desired rigor could only be achieved by abandoning the intuitive realm of geometric evidence, which had always been the natural frame of reference for analysis, and taking the arithmetic of natural numbers as the foundation on which the structure of analysis should be built.

It is therefore noteworthy that Kronecker, one of the most convinced advocates of this point of view, told Casorati that he had learned "to be more exact by cultivating the theory of numbers and algebra. In the use of infinitesimal analysis he had not found the occasion to acquire this exactitude" (Neuenschwander, 1978a, p. 75).

Weierstrass made a similar "confession of faith" about ten years later when he wrote to Schwarz in October of 1875,

> The more I think about the principles of function theory -- and I do it incessantly -- the more I am convinced that *this must be built on the foundation of algebraic truths*, and that it is consequently not correct when the "transcendent," to express myself briefly, is taken as the basis of simple and fundamental algebraic propositions. This view seems so attractive at first sight, in that through it Riemann was able to discover so many of the important properties of algebraic functions. (It is self-evident that, as long as he is working, the researcher must be allowed to follow every path he wishes; *it is only a matter of systematic foundations*)" [my emphasis] (*Werke* 2, p. 235).

Weierstrass then asserted he had been "especially strengthened [in this belief] by his continuing study of the theory of analytic functions of many variables" (*ibid.*).

When it is a matter of foundations, we do not need "brilliant intuition," which leads to discovery, but the rigor of simple "algebraic truths."[2] It is difficult to say how much Weierstrass was initially influenced in this way of thinking by Kronecker, whose contribution to this matter has perhaps not been fully appreciated by historians.

It is a well known fact that the two men later disagreed over the proper approach to the infinite in mathematics, in particular over the theory of real numbers and more generally Cantor's theory of infinite sets. But, as becomes clear from his conversations with Casorati, at the beginning of the 1860s Kronecker and Weierstrass agreed on the need for rigor in analysis and shared the conviction that the foundation was given by algebra and the arithmetic of natural numbers.

What do we mean when we say that the concept of natural number and the arithmetic of the natural numbers constitutes the foundation of analysis? Dedekind (1831-1916) expressed it with his usual clarity many years later when he wrote that with this conception

> it appears as something self-evident and completely without novelty, that every theorem of algebra and higher analysis, not matter how remote, can be expressed as a theorem about the natural numbers, a claim that I also heard from Dirichlet's lips many times. But I in no way see anything useful -- and this was also far from Dirichlet's intention -- in actually undertaking this wearisome reformation and insisting on using and recognizing nothing other than the natural numbers (1888, p. 338).

The point is otherwise. Here we are only concerned to establish, at least as a principle, the possibility of such a reduction and consequently, according to Dedekind, the existence of a secure foundation for all of analysis. The introduction of new concepts has instead a fundamental importance for the development of the theory:

> The greatest and most fruitful advances in mathematics and the other sciences have primarily been achieved through the creation and introduction of new concepts, after the frequent recurrence of complex phenomena, which were only poorly explained by the old concepts, had forced us to them (Dedekind, 1888, p. 338).

Thus the changes in the standard of rigor, like the tendency towards generalization and the introduction of new concepts, were motivated by necessity and were measured by the concrete standards of working mathematicians. How, for example, can we explain functions that are continuous in every point yet differentiable in none if we hold to the definition of continuity that could still be read in the *Traité élémentaire* of Lacroix, "One must understand by the *law of continuity* that which is observed in the description of lines through movement, according to which the consecutive points of a single line follow each other without any interval"?[3] (Lacroix, 1858, p. 88)

Such a theoretical notion renders the study of series of functions, and of trigonometric series in particular, impracticable. This is a topic that bristles with difficulty, but for which Riemann had shown there was a great potential for the development of real analysis.

These are the topics that wind through Casorati's discussions with Kronecker and Weierstrass. The Italian mathematician recorded them as they were expressed, still in the provisional and hypothetical form of research in progress. The account of his first meeting with Kronecker opens precisely with the question of continuity.

Continuity is still a confused idea. In class [Kronecker] defines a real function $\phi(x)$ of a variable x to be continuous when, by fixing a quantity δ as small as one wishes, we can make

$$\phi(x) - \phi(x') < \delta,$$

and this inequality persists when we set in place of x' any other value that is nearer to x than it (Neuenschwander, 1978a, p. 74).

This is the definition that we have seen Riemann give in his inaugural dissertation and Weierstrass in his lectures[4] (see §6.2).
"But is it really necessary," Casorati asks, "in order for a function to be called continuous, that $\lim_{x'=x}[\phi(x) - \phi(x')]/(x - x')$ be finite?" (*ibid.*).
In other words, is a continuous function necessarily differentiable?

On this question Kronecker shares with me the opinion that in a certain sense continuity exists even when

$$\lim \frac{\phi(x + k) - \phi(x)}{k^{1/2}} \quad \text{or} \quad \lim \frac{\phi(x + k) - \phi(x)}{k^{1/3}} \quad \text{or etc.}$$

is finite, adding that a function could probably be called continuous whenever $\lim[\phi(x + k) - \phi(x)]/f(k)$ is finite, where $f(k)$ is a certain function that vanishes together with k. However, it still remains to be decided what function should be taken for $f(k)$ (*ibid.*).

Casorati returned to the difficult and delicate question of continuity many times, pressing his colleagues with observations and questions. He asked Kronecker and Weierstrass if a continuous function must therefore be finite; he thought so and observed that one could omit the property of being finite from the definition of a continuous function. Weierstrass said he agreed, asserting that "continuity requires this condition: that to a finite portion of the plane of the variable there must correspond a finite portion of the plane of the function" (Neuenschwander, 1978a, p. 80). For his part, Kronecker specified that in the difference $\phi(x + a) - \phi(x)$, which is used to define continuity, $\phi(x)$ is taken to be a value of ϕ that is fully determinate, which would not be the case if $\phi(x) = \infty$.[5] But he nevertheless added that it would be better "not to omit the finite [*endlich*] so as not to believe that we can hold the two concepts of continuity and the finite to be identical, and to indicate that one cannot usually carry any judgement about continuity to infinity"

(Neuenschwander, 1978a, p. 75).

From here the discussion with Kronecker turned to the continuity of functions defined by series of continuous functions. Casorati noted that Kronecker

> was led to say that Abel, in his paper on the binomial series (where he does not define continuity precisely enough), although correcting Cauchy's error, gives a demonstration that is not valid. This is because it [the demonstration] rests essentially on this, that when, by taking for x any value in a given interval (for example from -1 to $+1$), we can always assign an upper limit to the value of a function $\phi(x)$, the function $\phi(x)$ must have a maximum in this interval. Kronecker says that Abel did not consider that if this upper limit depends on x, we cannot assert the existence of the maximum. He sees the defect in Abel's demonstration, but says he cannot see the means of obtaining a rigorous proof (Neuenschwander, 1978a, p. 76).

In reality, the rigor of a demonstration rests on the rigorous definition of the concepts one uses in it. It is therefore not surprising when Kronecker says that

> mathematicians ... are a bit arrogant (*hochmütig*) in using the concept of function. Even Riemann, who is generally very precise, is not beyond censure in this regard. If a function increases and then decreases or vice versa, Riemann says there must be a minimum or a maximum (see the demonstration of the so-called "Dirichlet's Principle"), while we should restrict the conclusion to the realm of functions which we might call "reasonable" [*vernünftig*] (Neuenschwander, 1978a, p. 74).

The implicit assumption here is to free analysis from constant reference to geometry, and in particular from the intuitive concept of the function $f(x)$ as a plane curve. In fact, this becomes unnecessary when one thinks of a function purely as a way of associating the values of a variable x with the values of a variable y. It is for this reason that the "pathological" functions to which Riemann called attention in his *Habilitationsschrift* aroused particular interest. (Although the dissertation had not been published, it was well known in Berlin.)

Thus Casorati reported that, "Kronecker knows functions that do not admit differential coefficients, that cannot represent lines, etc." (*ibid.*). And later, "The conversation moved to discontinuous functions, like, for example, those that have the value 1 for rational values of x and the value 0 for irrational values, which are generally said not to be integrable" (*ibid.*, pp. 76-7).

In regard to this subject, it was Schering who "lingered with Kronecker over a function that is discontinuous in the said way but which nevertheless has an integral" (*ibid.*, p. 77). However, the example that Schering gave was precisely the one that Riemann had proposed as a discontinuous yet integrable function (see §6.3).

Riemann seems to have been continually mentioned in Casorati's discussions with Kronecker and Weierstrass. In the majority of cases these concern criticisms *from the point of view of rigor*.

From Casorati's conversations we further learn that, "Riemann's things are creating difficulties at Berlin" (*ibid.*, p. 78), and that, in Weierstrass' view, "the disciples of Riemann are making the mistake of attributing everything to their master, while many [discoveries] had already been made by and are due to Cauchy, etc.; Riemann did nothing more than to dress them in his manner for his convenience" (*ibid.*, p. 79). This reveals, among other things, a substantial undervaluation and misunderstanding of the geometrical aspect of Riemann's theory on Weierstrass' part, and of Riemann surfaces in particular. Yet Weierstrass did not hesitate to assert that "he understood Riemann, because he already possessed the results of his research" (*ibid.*, p. 78).

The difference between Riemann's point of view and that of the Berlin mathematicians emerges even more clearly in the approach to the theory of functions of a complex variable, where Weierstrass and Kronecker anticipate the essential elements of the methods of power series and analytic continuation.

This is a topic that had already been presented in the mathematical literature in different ways. Riemann had referred to it in various places (see §6.2c), but it had never been treated with the necessary rigor. According to Casorati, Kronecker first observed that

we always suppose that a function can always be continued, in any part of the plane where the variable must go (Briot and Bouquet, Cauchy, ...), that is, that we can always give it a path that avoids the critical points, as if such points could not quite interrupt the connection between the parts of the plane. Riemann is a little more scrupulous, but maintains too much silence on this, so that his disciples can be thrown into the said error.[6]

As an example, Kronecker takes the function

$$\theta_0(q) = 1 + 2q + 2q^4 + \cdots,$$

which is defined only for $|q| < 1$, that is, in a circle with its center at the origin and a radius 1. "In order to know what the function is beyond this circumference it is necessary to resort to other means and not to those of making q follow a path joining a with b.[7]

In a later conversation Weierstrass agreed with Kronecker and further observed that Riemann had also examined the possibility of prolonging a function to any point in the plane.

But this is not possible, [Casorati writes] and it was precisely while searching for the demonstration of the general possibility that he [Weierstrass] realized it was in general impossible. He found the theorem that every function (monodromic understood) which does not have points in which it ceases to be definite is necessarily rational. (The function $e^{1/x}$ is not defined at the point $x = 0$ because it can have any possible value.) He believed that such points could not form a continuum and consequently that there is at least one point P where one can always pass from one closed portion of the plane to any other point M of it. But he [later] realized this was not the case (Neuenschwander, 1978a, pp. 79-80).

As for the function proposed by Kronecker, Weierstrass said it could not be extended outside of the circle where it is defined, asserting that

there is another expression for it that also has meaning outside of the circle, that is, inside and outside but not on the circumference. And yet this function satisfies a differential equation. A circumference of this type is therefore entirely composed of points in which the function is not defined; it can assume any value whatever there[8] (Neuenschwander, 1978a, p. 80).

The study of functions with unusual or "pathological" behavior thus has as its object the clarification and delimitation of concepts like continuity, analytic function, or the singularities of a function. (There here emerges the radical difference between the polar singularities and the essential ones, where, in Weierstrass' words, the function $e^{1/x}$ is not defined because it can have any value whatever (*ibid.*).) It also permits clarifying the concepts of minimum and greatest lower bound, following the criticism that Kronecker had made of Riemann's use of Dirichlet's principle. All of this illustrates the constant effort that Weierstrass made to state his theorems with the maximum possible generality and to rigorously deduce all the consequences derivable from them. This is an approach and a style that gradually became dominant in Germany, and which grew along with the increasing prestige of the University of Berlin, sited in the capital of the new nation that had recently been united under the political and military leadership of Prussia. Before long this approach to mathematics overflowed its national boundaries to become, towards the end of the century, *the* way of understanding mathematics, while Weierstrass' rigor became *the* model of rigor at which to aim.

7.2. Continuity and the Theory of Real Numbers

Several years later, by a variety of different routes, the topics that we have seen Casorati discussing with his colleagues in Berlin became part of the public domain of the mathematical world. In papers read to the Berlin Academy of Sciences in 1870 and 1872, Weierstrass published a counterexample to "Dirichlet's principle" (see the Appendix) and to the 'theorem' according to which a continuous function $f(x)$ is always differentiable except at isolated points (see §7.3). In the 1872 issue of Crelle's *Journal* Heine published the first article on the elements of the theory of functions according to Weierstrass. In this same year, Dedekind and Cantor (1845-1918) published their theories of the real numbers, which, joined to that of Weierstrass, gave a rigorously arithmetical systematization of continuity.[9] This proved to be a problem preliminary to any arithmetical treatment of analysis. Nevertheless, the different motives that lay behind Dedekind's, Cantor's, and Weierstrass' works led to the formation of three *conceptually* different theories, even if the numerical domains obtained were easily shown to be mutually isomorphic. For Dedekind the most urgent problem was that of finding a rigorous arithmetical foundation for the differential calculus; for Cantor it was to arrive at a uniqueness theorem for the representation of a function by trigonometric series; while Weierstrass considered the theory of real numbers to be an indispensable step in the construction of his theory of analytic functions.

a) 'Man Arithmetizes'

In 1888 Dedekind summarized the convictions that had animated him since the first draft of his work on continuity and irrational numbers in 1872 by modifying Plato's famous aphorism, "God geometrizes," to "Man arithmetizes." But, as he pointed out in the preface to his (1872), the first elaborations of his theory actually revert to 1858, the year in which he began to give lectures on the infinitesimal calculus at the Zurich Polytechnic. Nor should we ignore the influence of his formative period as a student at Göttingen. Dedekind had there been a student of Gauss, under whom he finished his doctoral degree in 1854, a few weeks after his friend Riemann. Nevertheless, the thing that made Dedekind "a new man," as he himself wrote, was the arrival of Dirichlet in Göttingen after Gauss' death. Dedekind attended Dirichlet's "deep and penetrating" seminars as well as his lectures on integration and potential theory.

Also of great importance for the future development of Dedekind's mathematical research were Dirichlet's lectures on the theory of numbers, which Dedekind edited in 1863. To this and later editions he added several long and important appendices which signaled a turning point in the history of modern algebra.

During his student years Dedekind had attended a course on the differential and integral calculus given by Stern (1807-1894). In the notes taken during this course in 1850, we find that Stern held that the difficulties inherent in the infinitesimal calculus were not mathematical by nature but logical, and that "Mathematics does not analyse the concept of the continuum ... "

"The differential calculus is concerned with representing, by means of mathematical formulas, what happens in the moment of the change from one condition to another. The difficulty of the concept of continuity again intrudes on it [the differential calculus]" (Dugac, 1976b, p. 154).

Dedekind was perhaps recalling these words of his old teacher at Göttingen when he wrote in the preface of his work,

> We often say that the differential calculus is concerned with continuous magnitudes, and yet a clarification of this continuity is nowhere given. Even the most rigorous presentations of the differential calculus do not base their proofs on continuity, but instead either appeal with greater or lesser awareness to geometrical images or to those induced by geometry, or rely on theorems which themselves are never proved *using purely arithmetical methods* [my emphasis] (Dedekind, 1872, p. 316).

Certainly the use of geometrical arguments is didactically effective and "indeed indispensable if one does not wish to lose too much time," Dedekind says, but the point is otherwise. "No one would want to maintain that this way of introducing the differential calculus can make any claim to being scientific" (*ibid.*).

Refusing all recourse to geometrical evidence, Dedekind sees the necessity of finding the true origin of the theorems of the differential calculus in the elements of arithmetic, "and thereby at the same time acquiring a real definition of the essence of continuity" (*ibid.*).

The first step consists of studying the properties of the rational numbers R, of which the following are of particular importance.

I. If $a > b$ and $b > c$, then $a > c$. ...

II. If a and c are two distinct numbers, then there are always infinitely many distinct numbers *lying* between a and c.

III. If A is any definite number, then all numbers of the system R divide into two classes, A_1 and A_2, each of which contains infinitely many elements; the first class A_1 includes all numbers a_1 that are $< a$, the second class A_2 includes all numbers a_2 that are $> a$ (Dedekind, 1872, p. 319).

Now, there is no difficulty in associating to every rational number a point on a line with a fixed origin and a unit measure of segments, but one discovers just as easily that there are points to which no rational number corresponds, that is, "The straight line L is infinitely richer in distinct points than the domain R of the rational numbers is in distinct numbers" (*ibid.*, p. 321).

Therefore, the rational numbers are not sufficient to "arithmetically" account for the properties of the line, and it "therefore becomes absolutely necessary" to extend the numerical field "by the creation of new numbers," in such a way as to obtain a field having "the same *continuity* as the straight line," to which we will attribute the property of being complete and without gaps (*ibid.*).

The road Dedekind followed is based on the ordered property of the line, by means of which we are able to formulate "a precise characterization of continuity, which can be used as the basis for valid deductions" (*ibid.*, p. 322).

In fact, every point p of the line divides the line into two parts in such a way that, once the orientation of the line has been fixed, every point of one part stands to the left of every point of the other.

I now find the essence of continuity in the inverse, that is, in the following principle:

If all the points of the straight line fall into two classes in such a way that every point of the first class lies to the left of every point of the second class, then there exists one and only one point which produces this division of all points into two classes, thus cutting the straight line into two parts (*ibid.*).

This is such an evident principle, Dedekind says, that the great majority of people would take it to be banal, and "will be very disappointed to learn that the secret of continuity should be revealed by this triviality" (*ibid.*, pp. 322-3). But that this principle is so clearly evident to everybody is a greater element of satisfaction, since "it is nothing other than an axiom ... through which we find continuity in the line" (*ibid.*, p. 323).

By defining continuity in an axiomatic manner, Dedekind also opens the way to an arithmetic characterization of it. In fact, every rational number determines a cut (A_1, A_2) of the rational field (as Dedekind now calls it). This has the property, 1) that every number of the class A_1 is smaller than every number of the class A_2, and 2) that either the class A_1 has a maximum or the class A_2 a minimum and vice versa. It is this second property that characterizes the cuts produced by rational numbers.

"But one is easily convinced that there also exist infinitely many cuts which cannot be produced by rational numbers" (*ibid.*, p. 324). An immediate example is given by the cut (A_1, A_2). In A_1 there are all the negative rational numbers, zero, and all the positive rational numbers p/q such that p^2/q^2 is less than 2; in class A_2 there are the rest of the rational numbers.

"Now every time a cut (A_1, A_2) appears which is produced by no rational number, then we *create*, a new, *irrational* number α, which we regard as being completely defined by this cut (A_1, A_2). We will say that the number α corresponds to this cut, or that it produces this cut" (*ibid.*, p. 325). In this way Dedekind introduces a new numerical domain, that of the real numbers, defined by cuts to each of which there corresponds one and only one number, rational or irrational.

Dedekind then demonstrates without difficulty the ordered property of the real numbers and subsequently the decisive theorem that the field $\mathbb{R}$ is continuous. "If the system $\mathbb{R}$ of all real numbers divides into two classes A_1, A_2 such that every number α_1 of the class A_1 is smaller than every number α_2 of the class A_2, then there exists one and only one number α by which this division is produced" (*ibid.*, p. 329).

In other words, he shows that the field of real numbers is closed with respect to the cutting operation. For these numbers Dedekind introduces the standard arithmetical operations. He then shows how, by using the real numbers, we can give a rigorous demonstration of one of the principal theorems of infinitesimal analysis, that of the existence of a least upper bound for every upperly bounded set of real numbers.

The publication of Dedekind's work, which was associated with the contemporary and similar works of Heine and Cantor, generated considerable interest and many lively reactions, not so much for the nature of the argument, which had been known to mathematicians since the time of Pythagoras, as for its author's point of view. The need for a reduction of the principles of analysis to the arithmetic of the natural numbers was certainly a widespread conviction, but it was for from being dominant. Even less widespread was the idea that the essence of the continuity of the line resides in Dedekind's axiom. It seemed indisputable that the basis of continuity lay in a primordial geometrical intuition.

These were precisely the issues that were developed during a brief but lively debate between Lipschitz (1832-1903) and Dedekind, which offered the latter an opportunity to state his opinions on irrationals and the problem of continuity with much greater clarity.

In June of 1876 Lipschitz wrote to Dedekind,[10]

I must confess that I do not deny the justification of your definition, but I am of the opinion that it differs from what the ancients established only in the form of the expression and not in the content. I can only say that I hold the definition given in Euclid V, 5, which I quote in Latin:

rationem habere inter se magnitudines dicuntur, quae possunt multiplicatae sese mutuo superare [magnitudes are said to have a ratio to one another which are capable, when multiplied, of *exceeding* one another] and what follows, to be entirely as satisfying as yours (In: Dugac, 1976b, p. 217).

Dedekind was resolutely opposed to this opinion. He responded that Euclid's starting point was completely insufficient for this purpose, "if we cannot add to Euclid's principles the crucial point of my work, which nowhere appears there, the essence of continuity" (Dedekind, *Werke* 3, p. 472).

Certainly the Euclidean definition allows one to treat the domain of incommensurable magnitudes, as is indeed done in book X of the *Elements*, "but nowhere in Euclid or in any later writer do we find the *closure* of such an extension, the concept of a *continuum*, that is an imaginably complete domain of magnitudes [*Grössen-Gebiete*]" whose essence is precisely given by the axiom of continuity (*ibid.*, p. 473). This is the decisive fact that separates the theory of Dedekind from that of Eudoxus and Euclid. It is here that Lipschitz' and Dedekind's conceptions of rigor diverge.

In a subsequent letter Lipschitz objected,

You have the intention *to set out from the beginning only rational numbers and magnitudes that can be measured by rational numbers.* Euclid proceeds otherwise in this point and this is the core of your difference with Euclid (In: Dugac, 1976b, p. 219).

Euclid thought that any given magnitude could be defined by the measure of a segment. From this point of view it was not difficult for him to show that there exist segments whose ratio is not given by whole numbers.

Lipschitz continues,

I know very well that you will object that it is not enough for you to derive the existence of a ratio from a geometrical construction. I answer this as follows. The human spirit has in large part drawn the strength that it now has from its occupation with geometry. The *rigor geometricus* has served the highest requirements for thousands of years. If we now set up other requirements, then we owe this in large part to the occupation with geometry, and these requirements are also not yet substantially different (*ibid.*).

In fact, Lipschitz concluded, "what you call the completeness of the domain that can be derived from your principles in fact coincides with the fundamental property of a line, without which no man can conceive of a line" (In: Dugac, 1976b, pp. 119-20).

Dedekind's reply, which concluded the debate, decisively clarified both the nature of the difference between the two, and the point of view that Dedekind firmly maintained. He observed first of all that the Euclidean definition of the ratio between homogeneous magnitudes is not sufficient if we "want to construct arithmetic on the concept of ratios between magnitudes (which was not Euclid's intention)" (Dedekind, *Werke* 3, p. 477).

This was also Weierstrass' opinion. In his lectures he had observed the necessity of freeing the concept of number from every geometrical reference to the Euclidean theory of ratios between magnitudes.

On the other hand, Dedekind was well aware that he had not illuminated any new phenomenon, seeing that, as he wrote to Lipschitz, "the appearance of cuts is cited in almost every textbook of arithmetic when it is necessary to approximate irrational numbers arbitrarily closely by means of rational numbers" (which, Dedekind emphasizes, commits a serious logical error, among other things) (*ibid.*, p. 475).

On the other hand, to approximate the real numbers by a succession of rational numbers corresponds to the concrete process of measuring magnitudes, and Dedekind's theory presents itself here as an "abstract" given of a concrete procedure that is at the base of modern science. But while the practical process of approximation presupposes the existence of a number to approximate, for Dedekind, on the contrary, the cut determines the real number.

We are faced with a fundamental reversal from the logical point of view, which implies a more substantial one. For the ancients, following Pythagoras, there was no problem of existence, since it went without saying that every magnitude could be associated with a measure, rational or not. It is precisely the question of measure that is posed and which justifies the axiomatic character of continuity, on which then the arguments for the existence and completeness of the real field are based.

In substance, in the processes of approximation (from Archimedes' method of exhaustion to Cauchy's demonstration of the convergence of series) the point was always to prove what we today call the unicity of the limit, not its existence. In his notes to the Italian translation of Dedekind's work, Zariski, in illustrating this aspect, observes that here resides

> the inadequacy of the methods of Greek mathematics in existential questions that are today resolved with the postulate of continuity. In all these examples we are faced with infinite sequences *in which the limit is already given as a concrete geometrical form* (length, area, volume), so that its existence lies before our eyes, so to speak (1926, pp. 252-3).

In other words, for Zariski, the Greeks knew that, given a magnitude, one could construct a sequence that admitted it as a limit, "but I do not see that they dared to consider the inverse thesis, that an infinite sequence, given *a priori* and satisfying some elementary convergence criteria, for example that of being bounded, admits a limit" (Zariski, 1926, p. 253). Expressing this fact in the language of set theory, Zariski concluded that "the Greeks understood that the set of points of the line was *dense*, but they did not understand its *integrity*, which is for us its true *continuity*"[11] (*ibid.*).

For Dedekind, the intuition of the continuity of physical space was subordinate in a natural manner to this concept of continuity. "If space *has any real existence at all*, [my emphasis] then it does not necessarily need to be continuous; innumerable of its properties would remain the same if it were also discontinuous," he had written in 1872 (p. 323). "I can represent to myself all of space and every line in it as discontinuous throughout," Dedekind now wrote to Lipschitz, adding that Prof. Cantor of Halle is certainly able to do so "and I hold that every man can do the same" (Dedekind, *Werke* 3, p. 478). The fact is that the continuity of space is not "a postulate that is inseparably tied to Euclid's geometry," Dedekind adds. "His entire system remains even without continuity--a result that is certainly surprising for many and to me therefore appears well worthy of mention" (*ibid.*, p. 479). Dedekind concludes his letter to Lipschitz with the assertion that he felt no need to add anything more.[12]

For Dedekind, therefore, the arithmetization of analysis succeeded in transporting our knowledge of physical space into the theoretical realm of our "mental constructions," while its properties of continuity, which had always been held to be intuitively evident, were attributed to the "creative power" of our thought, as he expressed it on numerous occasions. In fact, all of the arithmetic was reduced to this. Answering the question, *"Was sind und was sollen die Zahlen?"* [literally, "What are numbers and what should they be?" although usually translated as "The nature and meaning of numbers"], which formed the title of the 1888 study in which he gave an axiomatic presentation of arithmetic beginning from the concept of the set, he wrote that numbers are "creations of our thought."

b) Trigonometric Series and the Theory of Sets

From the moment of his arrival in Halle as a *Privatdozent* in 1869, Cantor set aside his interest in the theory of numbers. In Berlin, under the direction of Kronecker and Kummer, he had devoted both his dissertation (1867) and his *Habilitationsschrift* (1869) to this subject, but in Halle he instead turned his attention to the study of trigonometric series, stimulated by the work of his elder colleague, H. E. Heine (1821-1881). The latter had for some time been working on problems involving these series, which had never stopped attracting the attention of mathematicians.

In 1864, long before Riemann's *Habilitationsschrift* had been printed and apparently without knowledge of its existence, Lipschitz had taken up the same argument, seeking to extend the research in the directions set out by Dirichlet at the end of his 1829 paper (see §5.2).

As for Riemann, so also for Lipschitz the problem was to study those cases where the given $f(x)$ fails to satisfy Dirichlet's conditions. In the first two cases Lipschitz discussed, $f(x)$ might become infinite at isolated points in $(-\pi,\pi)$ or have an infinite number of discontinuities in this range. Following up the idea that Dirichlet had expressed in (1837b), Lipschitz enclosed each point c_i ($i = 1, \ldots, n$), where the f become infinite, in arbitrarily small intervals $(c_i - \delta_i, c_i + \delta_i)$, while in the rest of the original interval $f(x)$ satisfied Dirichlet's conditions. Lipschitz claimed that Dirichlet's proof could be applied provided that the integrals

$$\int_{c_i - \delta_i}^{c_i + \delta_i} f(x)dx \text{ converge.}$$

For the second case of a bounded and piecewise monotonic function $f(x)$ with infinitely many points of discontinuity, Lipschitz concluded that Dirichlet's proof would apply if the integral concept could be extended to include such an $f(x)$. According to Lipschitz, this could be done without difficulty if the set of points of discontinuity D was nowhere dense. He believed that by "a suitable argument" one could show that D' (the derived set of D) is finite. This was essentially the same argument Dirichlet had used.

Much more difficult was the case where, of all of Dirichlet's conditions, only the one regarding infinitely many maxima and minima was not satisfied. In studying the behavior of such infinitely oscillating functions Lipschitz emphasized that one was forced to be extremely careful because Dirichlet's techniques could not longer be successfully applied.

According to Lipschitz, $f(x)$ could have infinitely many oscillations within a fixed interval $[a,b]$ in two ways: 1) $a < r < b$ is given in such a way that, in the intervals $[a, r - \delta]$ and $[r + \delta, b]$ (with δ arbitrarily small), the function presents a finite number of oscillations, and an infinite number in $[r - \delta, r + \delta]$; 2) given any two values r and s, such that $a < r < s < b$, a finite number of oscillations cannot exist in $[r,s]$, i.e. the oscillations are densely displayed over $[a,b]$.

A third possibility was to combine cases 1) and 2). Lipschitz succeeded in proving Dirichlet's result by replacing the piecewise monotonicity with a new condition for integrability. He demonstrated the theorem that the Dirichlet integral

$$\int_g^h f(\beta) \frac{\sin k\beta}{\sin \beta} d\beta \qquad \left[0 \leqslant g < h < \frac{\pi}{2} \right]$$

tends to $(\pi/2)f(0)$ as k tends to infinity, if $f(x)$ satisfies the condition

$$|f(\beta + \delta) - f(\beta)| < B\delta^\alpha \qquad (\alpha > 0),$$

where B is a positive constant and δ is any positive, arbitrarily small constant (Lipschitz, 1864, p. 301).

The flaws in Lipschitz's paper as regards set theory (and nowhere dense sets in particular) clearly show the difficulties the mathematicians of the time faced in understanding the topology of the real line. Thus, "the tendency, suggested by Dirichlet's remarks, to regard 'nowhere dense' as synonymous with 'negligible for integration theory' and the tendency, evident in Lipschitz' paper, to conceive of a set of being dense either in the neighborhood of separated limit points or in an entire interval, continued until the early 1880's" (Hawkins, 1970, p. 15).

An uncertainty similar to that we have just seen in Lipschitz regarding the behavior of infinite sets of points is also found in Hankel (1870), an article that was clearly inspired by Riemann's *Habilitationsschrift*. As we have seen, Hankel was primarily interested in clarifying the function concept (see §5.3) and Riemann's concept of the definite integral. To this end, Hankel devoted particular attention to the study of singularities of real functions. He introduced the method of the "condensation of singularities" by which he was able to generate functions with an infinite number of singularities starting from a function with an assigned singularity at one point.

He considered a function $\phi(y)$ continuous and bounded on $[-1,1]$ except at the point $y = 0$ where $\phi(0) = 0$. He then defined

$$f(x) = \sum_{n=1}^{\infty} \frac{\phi(\sin nx\pi)}{n^s} \qquad (s > 3).$$

After analyzing the behavior of $f(x)$ at the points of the interval, Hankel concluded that the singularity of $\phi(y)$ at $y = 0$ was "transported" to $f(x)$ for all rational values of x, while for the irrational values f had a "legitimate" behavior.[13] "I have therefore deemed it possible to call the principle of going from such functions ϕ to f as that of the *condensation of the singularities*" (1870, p. 81). He then illustrated his "principle" with several examples, such as $\phi(y) = y \sin 1/y$ and $\phi(y) = y^2\sin 1/y$.

He next defined a function to be discontinuous at a point $x = a$ if for every $\varepsilon > 0$ there exists a δ such that $|\delta| < \varepsilon$ and $|f(a + \delta) - f(a)| > \sigma$, where σ is a certain positive number. According to Hankel, the function $f(x)$ "makes jumps at the point $x = a$ which are $> \sigma$" (1870, p. 84).

As far as discontinuities are concerned, Hankel says functions can be divided into two classes: 1) those that are pointwise discontinuous, and 2) those that are totally discontinuous. Functions of the first class are those "linearly discontinuous functions for which the points with jumps $> \sigma$ occur only *in a loose distribution* [*zerstreut*] and fill up *no interval*, however small the magnitudes $\sigma \neq 0$ may be" (1870, p. 89). Functions belonging to the second class are those "for which points with jumps which exceed a certain finite quantity *fill up entire intervals*" (*ibid.*, p. 91).

Turning to the integrability of discontinuous functions, Hankel thought (erroneously, as it were) that his distinction enabled him to conclude that a function would be Riemann integrable if and only if it were of the first class. He actually believed that he had proved "the necessary and sufficient condition for a linearly discontinuous function to be pointwise discontinuous is that the totality of all the intervals in which oscillations > σ occur, for every small $\sigma \neq 0$, can be made arbitrarily small" (1870, p. 89). But the sufficient condition is in fact incorrect.

As Hawkins has pointed out, "Hankel confounded 'topologically negligible' sets (i.e. nowhere dense sets) with sets that are negligible from a measure-theoretic point of view. ... And, as was the case with Dirichlet and Lipschitz, it was the inadequacy of his understanding of the possibilities of infinite sets -- in particular, nowhere dense sets -- that led him astray"[14] (1970, p. 32).

The clarification of these concepts and the development of the modern topology of infinite sets of points was undertaken in the 1870s by Cantor, arising from his study on trigonometric series.

Following the works of Dirichlet, Riemann, and Lipschitz, the situation had stood as follows: it had been demonstrated that a function, under hypotheses of a very general nature, can be represented as a Fourier series, but nothing was said about *how many methods of expansion there were*. It was this crucial problem that Heine, while working on it himself, called to the attention of his younger colleague Cantor. In an article published the year after Cantor arrived in Halle, Heine in fact wrote that Weierstrass' demonstration of the necessity of uniform convergence for the term-by-term integration of a series of functions had rendered untenable the proof of the theorem that a function $f(x)$ bounded on $[-\pi,\pi]$ can be represented in at most one way by a trigonometric series of the form

$$(7.2.1) \quad \frac{1}{2} a_0 + \sum_{i=1}^{\infty} (a_i \cos ix + b_i \sin ix).$$

The unfortunate fact, Heine lamented, is that the significance that had until then been attributed to the representation of a function by means of trigonometric series rested in large part on the uniqueness of the expansion, that is to say, on the certainty that one could find the same series no matter what method one used to transform the function into a series (Heine, 1870, p. 353).

Certainly a Fourier series that represents a *discontinuous* function, even a bounded one, cannot be uniformly convergent, but we do not yet know, Heine says, if a series that represents a continuous function must be uniformly convergent. Until now this had always been tacitly assumed.

"This matter will also not be clarified in what follows," Heine says, adding, "Furthermore, we still do not know today with certainty whether it is even possible to represent a given continuous function by a uniformly convergent trigonometric series" (*ibid.*, p. 354). A short time later Du Bois Reymond (1876) would give a negative response to this question (see §5.2).

In his paper on trigonometric series, Heine began with Dirichlet's results, which he reformulated in the following manner:

Proposition I. *The Fourier series for a finite function f(x) which has a finite number of maxima and minima ... converges uniformly when f(x) is continuous from −π to π (inclusive) and f(π) = f(−π). In all other cases it is only generally uniformly convergent* [where "generally" means "except in a finite number of points"] (1870, p. 354).

He then set out the following two propositions, derived from the first.

Proposition II. *A generally continuous but not necessarily finite function f(x) can be expanded as a trigonometric series of the form (7.2.1) in at most one way, if the series is subject to the condition that it is generally uniformly convergent. The series generally represents the function from −π to π.*

Proposition III. *If a trigonometric series (7.2.1) is generally uniformly convergent from −π to π and generally represents zero, ... then all coefficients a and b must vanish and the series consequently represents zero everywhere (ibid., p. 355).*

It is this last proposition which moved Cantor to prove his own uniqueness theorem, that is, that if a function can be represented by a trigonometric series, this representation is unique. Cantor's starting point, in March of 1870, was the so-called 'Cantor-Lebesgue theorem,' which generalized Heine's Proposition III.

If two infinite sequences: $a_1, a_2, ..., a_n, ...$ and $b_1, b_2, ..., b_n, ...$ are so constituted that the limit of

$$a_n \sin nx + b_n \cos nx,$$

for every value of x lying in a given interval $(a < x < b)$ of the real domain, is equal to zero as n increases, then both a_n and b_n converge to the limit zero as n increases[15] (Cantor, 1870a, p. 71).

The second instrument Cantor uses, adopting the conviction expressed by Heine (1870, p. 353), is Weierstrass' concept of the uniform convergence of a series.

Cantor observes that the problem of uniqueness cannot be resolved, as had previously been thought, by multiplying every term of the series (7.2.1) by cos $n(x − t)dt$, and then integrating term-by-term from −π to π (1870b, p. 80). Such a procedure would require not only that $f(x)$ be integrable, but also that the series converges uniformly.

In order to attain this objective we must take a different approach. Cantor begins his demonstration by supposing that there are *two* representations of the same function $f(x)$ in trigonometric series converging for every value of x. "There follows, by subtracting one from the other, a convergent representation of zero for every value of x:

$$0 = C_0 + C_1 + \cdots + C_n + \cdots ,$$

where $C_0 = \frac{1}{2}d_0$, $C_n = c_n\sin nx + d_n\cos nx$, and where the coefficients c_n, d_n become infinitely small as n increases" (1870b, p. 81).

At this point Cantor considers the Riemann function,

$$F(x) = C_0 \frac{x^2}{2} - C_1 - \cdots - \frac{C_n}{n^2} - \cdots ,$$

which, as Riemann had shown, is not only continuous and integrable, but also

$$\lim_{\alpha \to 0} \frac{F(x + \alpha) - 2F(x) + F(x - \alpha)}{\alpha^2} = 0 \quad \text{(see §6.3)}.$$

It follows from this that $F(x)$ is a linear function, $F(x) = cx + c'$. Cantor gives a demonstration that, he writes, was sent to him by Schwarz and is based on Weierstrass' theorem that if U is a compact set and f is continuous, $f(U)$ is compact.

Cantor could then conclude that

$$C_0 \frac{x^2}{2} - cx - c' = C_1 + \frac{C_2}{4} + \cdots + \frac{C_n}{n^2} + \cdots .$$

But the second member is periodic of 2π, and consequently $C_0 = c = 0$.

At this point Cantor was able to show that the series $\Sigma C_n/n^2$ is uniformly convergent. By multiplying by $\cos n(x - t)dx$ and integrating term-by-term from $-\pi$ to π, it follows from the property of the orthogonality of trigonometric functions that

$$c_n \sin nt + d_n \cos nt = 0, \quad t \in \mathbb{R},$$

and from the (1870a) theorem, $c_n = 0$, $d_n = 0$. Cantor could then conclude that "a representation of zero by a trigonometric series, convergent for every real value of x, is only possible if the coefficients d_0, c_n, d_n are all equal to zero" (Cantor, 1870b, p. 83).

From this there immediately follows the uniqueness theorem, Cantor's greatest result, which brought the admiration of all his contemporaries. "If a function of a real variable $f(x)$ is given by a trigonometric series convergent for every value of x, then there is no other series of the same form which likewise converges for every value of x and represents the function $f(x)$" (Cantor, 1870b, p. 83).

A short time later Cantor weakened the conditions of his uniqueness theorem by demonstrating that "we can now modify the assumption in the sense that, for certain values of x, fails either the representation of the zero by $[\Sigma_i C_i]$ or the convergence of the series" (1871, p. 85). It is these "exceptional" points that play a fundamental role in the extensions that Cantor would make fairly soon, both in the theory of real numbers and in the theory of sets. In his first work of 1871, which was published in the form of an addenda to the article of April 1870, Cantor limited himself to requiring that these exceptional points be finite in number for finite intervals. The great result that he obtained the following year (1872), was the discovery that his uniqueness theorem is also valid when there are an infinite number of exceptional points, provided these are distributed in a suitable manner.

The article in which Cantor announced this result contained, at the beginning, as its indispensable premiss, his theory of real numbers and of derived sets.

As had been the case for Dedekind, so also for Cantor the definition of real numbers was achieved by following an arithmetical route. He in fact based his considerations on infinite sequences of rational numbers $a_1, a_2, ..., a_n, ...$ with the property that, "for any positive, rational value ε there exists an integer n_1 such that $|a_{m+n} - a_n| < \varepsilon$ for $n \geqslant n_1$, for any positive integer m" (1872, p. 93). Every such sequence, which Cantor was to call "*fundamental*," and is today commonly called a Cauchy sequence, "has a determinate limit b," by which we mean that to the series $\{a_n\}$ there is associated a symbol b (*ibid.*).

This is the crucial point in the construction. Evidently Cantor meant to define b and was well aware of the logical error lying behind the practice of earlier mathematicians in introducing irrational numbers in terms of infinite series. Returning to this question in 1883, he wrote, "There is a logical error here, since the definition of the sum $\Sigma\, a_n$ is first found by equating it to the finished number b, necessarily defined beforehand. I believe that this logical error, first avoided by Weierstrass, was earlier committed quite generally, and was not noticed because it belongs to those rare cases in which real errors can cause no significant harm in calculations" (1883, p. 185).

For fundamental sequences Cantor then introduces an equivalence relation in the following manner: if to the sequence $\{a_n\}$ there is associated a limit b and to the sequence $\{a'_n\}$ the limit b', then if, for any $\varepsilon > 0$, there exists an n_1 such that $|a_n - a'_n| < \varepsilon$ for $n > n_1$, then the two sequences are equivalent in the sense that $b = b'$.

In a natural manner he then defines the order relations and the arithmetical operations for fundamental sequences. The set of symbols b so defined constitutes the new numerical domain B of the real numbers.

Cantor next considers sequences of elements $b_1, b_2, ..., b_n, ...$ of B and in a similar way defines a fundamental sequence $\{b_n\}$. He constructs the domain C of symbols c associated with every fundamental sequence $\{b_n\}$, and proves that it substantially coincides with B, in the sense that B and C have the same mathematical structure and can be put into one-to-one correspondence, even if (and this is a fundamental fact for Cantor), the elements of B and those of C have a different conceptual content. After demonstrating the closure of B with respect to the operation of the limit, Cantor, by reiterating this procedure λ times, arrives at the definition of the numerical domain L whose elements he calls "numerical quantities, values or limits of the λ-th kind" (1872, p. 95).

"The concept of number, in so far as it has been developed here, contains in itself the seeds of an infinite extension, necessary and absolute in itself" (*ibid.*), Cantor writes in commenting on this process of generating new elements. And it is precisely in this generative procedure that he recognizes the objectivity of those numbers, every one of which "is normally unobjective in itself, and only appears as a component of propositions that attain objectivity, for example, of the proposition that the said sequence has the numerical quantity as a limit" (*ibid.*).

We are faced with a kind of objectivity that is substantially different from that which we are used to recognizing in the natural or the rational numbers. It is one that is closely linked with Cantor's most general "formalistic" point of view on the question of the existence of mathematical entities.

"But what advantage will be gained by even a purely abstract definition of real numbers of a higher type, I am as yet unable to see, conceiving as I do the domain of real numbers as complete in itself," wrote Dedekind in regards to Cantor's construction (1872, p. 317). For Cantor such a "purely abstract" construction had a precise meaning, even if he could not at that time present it explicitly.

In fact, in Cantor's concept of real numbers of a higher-order there is already, in essence, the idea of generating "transfinite" sets that he later set out in many papers. When he turns to translate his thoughts about real numbers into the language of corresponding point sets on the line,[16] Cantor emphasizes the central importance of the concept of a "derived set" of a point set for the entire theory. He begins by defining the concept of the limit point of a point set (what is today usually called the accumulation point) in this way:

> By a "limit point of a point set P" I understand a point of the line so placed that in every neighborhood of it we can find *infinitely* many points of P where it is possible that it also belongs to the set. But by "the neighborhood of a point" should here be understood every interval which has the point *in its interior*. Thereafter it is easy to prove that a point set consisting of an infinite number of points always has at least *one* limit point (1872, p. 98).

This result is better known as the "Bolzano-Weierstrass theorem." Thus, for every point set P of the line, there is always its "derived" set P', which is the set of its limit points. (This will be empty if the point set P is finite.)

The repeated derivation of an infinite point set P presents itself conceptually to Cantor as the natural counterpart of the generation of the numerical domains B, C, ..., L. If, after n derivations, $P^{(n)}$ contains only a finite number of points (whereupon $P^{(n+1)}$ is empty), P is called a set of the n-th kind. An example of a set of this kind is "a single point, if its abscissa is given as a numerical quantity of the n-th kind" (*ibid.*). Indeed, Cantor wrote, by "solving" this number step-by-step in its constitutive terms, we arrive at an infinite sequence of rational numbers. By translating it into the corresponding language of point sets, we have a set of the n-th kind.[17]

In his 1872 paper Cantor limited himself to considering point sets of the first species, that is where $P^{(n)} = 0$ for $n < +\infty$. If the set of exceptional points of a trigonometric series is given by a set in this way, then his uniqueness theorem of representation is still valid. This was how he demonstrated it in the last part of the paper, but already at this time Cantor's primary interest was more concerned with the preliminaries of his theorem regarding derived sets. He had already considered the possibility of pushing the operation of derivation to the set $P^{(\infty)}$ and from here to go on to consider the sequence of derived sets of the second species, creating "a dialectical generation of concepts that always leads farther, and free from all arbitrariness, remains necessary and consequent in itself" (Cantor, 1880, p. 148). He wrote this a few years later when his attention was completely devoted to developing the theory of sets and of transfinite numbers, something that was only glimpsed here.

The possibility of an infinity of "exceptional" points admitted by his theorem of unicity had in fact posed for Cantor the problem of a more precise characterization of this infinity. The set of rational numbers was not continuous, as was that of the real, but how could one translate this into the language of infinite sets?

"The question is simply whether $\mathbb{N}$ and $\mathbb{R}$ may be set in correspondence in such a way that every member of one set corresponds to one and only one [member] of the other?" (Cantor/Dedekind, 1937, p. 12). So Cantor asked Dedekind in 1873, with whom he had begun to correspond some time after the publication of his 1872 paper.

The demonstration of the impossibility of doing this, which Cantor found in 1874, was the first of a series of important results on the topology of point sets which Cantor obtained in the following decade. The concept of a one-to-one correspondence and the power of a set provided him with the natural key for penetrating the universe of infinite sets and exploring their nature. He was thus able to clarify the weaknesses present in the fundamental fields of mathematics, from the theory of dimension to Riemann's theory of integration. He was likewise able to disentangle concepts like those of dense sets and continuous sets and the difference between nowhere dense sets and sets of zero content, which the mathematicians of the day had debated in vain.[18]

The fruitfulness of Cantor's ideas and findings for real analysis, together with those of Weierstrass and Dedekind, found their first and splendid confirmation in Dini's *Fondamenti* of 1878, a volume that gathered together the lectures Dini had given at the University of Pisa.[19] Dini's book made a strong impression on Cantor, who immediately suggested to his friend Dedekind that they jointly edit the German translation. (It appeared, however, independently in 1892.)

The *Fondamenti* was followed two years later by a second volume which set out new and profound results on Fourier series and trigonometric series which Dini had obtained with a penetrating and rigorous analysis (Dini, 1880).

The Italian mathematician belonged to that school of thought which had been nurtured by the "German mathematicians;" he sought to give "to the enunciations and to their demonstrations all the rigor appropriate to mathematics" (1878, p. iii). Dini's volumes confirmed that "the new methods which Weierstrass and the other German mathematicians had used in their demonstrations" (*ibid.*, p. iv) were becoming ever more widely diffused throughout Europe.

7.3. Weierstrass' Theory of Functions

One tends to assume that Riemann's approach to the theory of complex functions became dominant almost immediately in the second half of the last century, but this was not the case. On the contrary, Riemann's ideas were not widely accepted until much later. At first they had only a few isolated supporters among his students and a few others, to the extent that Betti, writing at the beginning of the 1863 academic year, could assert that

> [Riemann's] method has the advantage over the others of its immense generality and of completely satisfying the principal tendencies of modern analysis, since the mechanism of the calculus hardly enters at all and it is almost entirely a magnificent work of pure thought. But so great is the force of the mind, so great is the concision and obscurity of the style of this eminent geometer, that *at the moment it is as if his works did not exist in the scientific world* [my emphasis] (In: Bottazzini, 1982, p. 250).

For his course on analysis at the University of Pisa in 1860/61, Betti had chosen a different method for explaining the theory of elliptic functions.[20] He took up "another kind of function that is the simplest encountered in analysis, ... which is single-valued and does not become either infinite or discontinuous for finite values of the variable, which I call 'entire'" (*ibid.*). Such functions can be represented by power series in z, convergent for any real or complex value, which become infinite for $z = \infty$.

After proving that an entire function w always has a root and that, if this root is not finite the function is of the form e^w, Betti notes that if the infinite product

$$\prod_i \left(1 - \frac{z}{\alpha_i} \right),$$

where α_i assumes an infinite number of values which are the zeros of an entire function W, is convergent for every finite z, then it represents an entire function that has all and only the α_i for zeros and can be given in the form

$$W = e^w \prod_i \left(1 - \frac{z}{\alpha_i} \right),$$

where w is an entire function.[21]

Betti's next problem was to find an entire function when an infinite system of complex numbers α_i is given which are the zeros of the function. Under the supposition that these zeros are at a finite distance from each other, Betti concludes that, "the entire function can be split into an infinite number of factors of the first degree and of exponentials" (1860-61, p. 246).

For Betti, this study provided the necessary foundation for introducing Jacobi's theta functions in a more direct manner, that is, the entire functions whose zeros are of the form $m\omega + n\omega'$, with $m,n \in \mathbb{Z}$, and $\omega,\omega' \in \mathbb{C}$, Im $\omega/\omega' > 0$. In fact, with such functions one can readily define elliptic functions as the quotients of theta functions. "Their theory," Betti observes, "can be entirely derived from a single principle with simple reasoning, without ever needing to resort to the ingenious devices of the calculus" (ibid.).

The factorization theorem which Betti used was independently found in its full generality by Weierstrass a few years later.

Excerpts from Weierstrass' letters to Sonia Kowalewski (Mittag-Leffler, 1923) show that towards the end of 1874 he had discovered that for any sequence of complex numbers $\{a_n\}$, $a_n \to \infty$ as $n \to \infty$, there exists an entire function which has its zeros at the points a_n. The multiplicity of zeros is, moreover, given by the frequency of the a_n in the sequence.

Weierstrass presented this result in his lectures in 1875 and published it the following year. This paper (Weierstrass, 1876), one of his most influential, deals with the problem of infinite-product expansions of single-valued complex functions, a problem to which Weierstrass had been led by his studies on Gauss' and Euler's Γ-function (see §3.2). This is a meromorphic function, whose reciprocal is entire.

"According to the definition given by Gauss," Weierstrass noted, "its expression is the everywhere convergent infinite product

$$\prod_{n=1}^{\infty} \left\{ \left[1 + \frac{z}{n} \right] \left[\frac{n+1}{n} \right]^{-z} \right\} \quad \text{or} \quad \prod_{n=1}^{\infty} \left\{ \left[1 + \frac{z}{n} \right] e^{-z \, \log(n+1/n)} \right\},$$

that is, the function is representable as the product of infinitely many factors which, although not entire linear functions of z, nevertheless like these *are single-valued functions with only one singularity* (∞) *and only one zero*.

"Proceeding from this remark I asked myself the question whether it might not be true that every function $G(z)$ may be composed of

factors of the form $(kz + \ell)^{\overline{G}(z)}$, and, by pursuing this thought, I finally arrived at a result through which the theory of single-valued functions with a finite number of essential singularities becomes a satisfying completion (*Abschluss*)" (1876, p. 91).

Weierstrass' paper follows his characteristic style of systematic exposition. He begins by establishing the "characteristic properties" of rational functions. He asserts that a function is regular in the neighborhood of a point $z = a$ if it is there expandable in a power series $\Sigma_n a_n(z - a)^n$ and defines the domain of continuity of a function. He then notes that "for every function $f(z)$ ... there necessarily exist *singularities*, as I will call them, which are the boundary points of the domain of continuity of the function, without themselves belonging to it" (*ibid.*, p. 78).

He distinguishes between polar and essential singularities according to whether or not there exists an integer n such that $(z - a)^n f(z)$ is regular in a, and shows that one can construct a function in such a way that it has infinite essential singularities. Then the class of rational (meromorphic) functions can be defined as "the totality of those single-valued functions of x which have only poles [*ausserwesentliche singuläre Stellen*] in the domain of this quantity" (1876, p. 79).

It follows then that such a function $f(z)$ can be represented as a

quotient of the form $G(z)/\Pi_{k=1}^{r}(z - a_k)^{m_k}$ where $G(z)$ is a polynomial and the points a_k are the poles of $f(z)$ in the finite plane. Weierstrass remarked that his discussion "also gives a hint for the study and classification of *transcendental* single-valued functions of *one* variable" (1876, p. 81).

If one adds to the domain of continuity of a function $f(z)$ those points that are poles, then one finds a domain A' in which "$f(z)$ everywhere behaves like a rational function. This domain is unbounded or bounded according to whether $f(z)$ is a *rational* or a *transcendental* function" (*ibid.*). In the latter case the boundary consists of the essential singularities of the function.

All those functions which have the same domain A' are taken to belong to the same class. Then the class of rational functions consists of precisely *one* of these classes. But, Weierstrass adds, there are also "countless classes of transcendental functions" (*ibid.*). He limits himself here to analyzing those single-valued functions that have only a *finite number* of essential singularities, "which are most like rational functions" (*ibid.*, p. 82).

The path Weierstrass follows is based on an analogy with the case of rational functions. Thus, given that a rational function is expressible as a quotient of two polynomials, is it possible, he asks, that something like this is also true for single-valued functions having a finite number of essential singularities? In other words, "is it possible to form arithmetical expressions of the variable z and indeterminate constants which represent all functions of a certain class, and only those?" (*ibid.*, p. 83).

The simplest way of finding an answer to this question is to consider those functions having only one singular point. If such a point is at ∞, then a function $G(z)$ is representable in series $\sum_{n=0} A_n z^n$ convergent for every finite value of z. Weierstrass calls this an "entire" function. $G(z)$ is rational or transcendent according to whether the point ∞ is a polar or an essential singularity. The case of n singularities can be reduced to that of only one.

At this point Weierstrass states the following theorems, which provide the answer to his question.

A. The general expression of a single-valued function of z having only *one* (essential or inessential) singularity (c) is $G(1/z-c)$, where, if $c = \infty$, $1/(z-c)$ is to be replaced by z. The singularity is essential or not [a pole] according to whether G is a transcendental or a rational entire [polynomial] function of $(1/z-c)$.

B. The general expression of a single-valued function of z with n (essential or inessential) singularities $(c_1, ..., c_n)$ can be formed in various ways by n functions each having only *one* singularity. The simplest forms are, however, the following:

$$1)\ \sum_{v=1}^{n} G_v\left[\frac{1}{z - c_v}\right], \quad 2)\ \prod_{v=1}^{n} G_v\left[\frac{1}{z - c_v}\right] R^*(z),$$

where $R^*(z)$ is a rational function which becomes zero or infinite only at the essential singularities of the function to be represented.

C. Every single-valued function of z that has n *essential* singularities $(c_1, ..., c_n)$ and in addition arbitrarily many (even infinitely many) [poles] can be expressed in each of the following forms:

$$1)\ \frac{\sum_{v=1}^{n} G_v\left[\frac{1}{z - c_v}\right]}{\sum_{v=1}^{n} G_{n+v}\left[\frac{1}{z - c_v}\right]} \quad 2)\ \frac{\prod_{v=1}^{n} G_v\left[\frac{1}{z - c_v}\right]}{\prod_{v=1}^{n} G_{n+v}\left[\frac{1}{z - c_v}\right]} \cdot R^*(z),$$

in such a way that the numerator and denominator do not both vanish for any value of z.

Conversely, if the functions $G_1, ..., G_{2n}$ are given arbitrarily, each of these expressions represents a single-valued function of z, which in general has n, in special cases fewer than n, essential singularities, while the number of [poles] in which the function becomes infinite is unlimited" (1876, pp. 84-5).

The demonstration of these theorems constitutes the essential portion of Weierstrass' paper. Theorem A in fact was fairly well known, but in order to prove the others in a more general and rigorous manner, "I had first ... to fill in a gap which existed in the theory of transcendental entire functions. I was first able to do this in a satisfactory way not long ago following many unsuccessful attempts" (1876, p. 85).

The "gap" which Weierstrass refers to is that we have seen discussed by Betti. Until what point is a function $G(z)$ determinate from its zeros? And also the theorem of factorization stated above: given an infinite sequence of constants $\{a_i\}$, with lim $a_n = \infty$, does there always exist a function $G(z)$ whose zeros are given by $\{a_i\}$?

The infinitely many functions of the form $G(z)e^{\overline{G}(z)}$, where $\overline{G}(z)$ is an arbitrarily given entire function, answers the first question, while Weierstrass' reflections on Gauss' function lead him to answer the second. In order to do this he introduced certain "prime functions" [Primfunktionen] which are single-valued functions having one singularity and at most one zero. Their most general expression is

$$\left[\frac{k}{z - c} + \ell \right] e^{G(1/z-c)} ,$$

where k and ℓ are constants and c is the point of singularity.

Weierstrass then shows that every single-valued function $f(z)$ having only one singularity (a pole or an essential one) "is itself either a prime function or a product of prime functions with this same singularity"[22] (ibid., p. 92). The expressions (B, 2) and (C, 2) then show how such a function with the properties required by theorems B and C can be formed from the multiplication and division of prime functions.

In the final paragraph of his paper Weierstrass studied the behavior of a single-valued function in the neighborhood of an essential singularity. As a consequence of the theorems given above he asserts that, "It follows that the function $f(z)$ varies discontinuously in an infinitely small neighborhood of the point c and that it can come arbitrarily close to any given value; consequently it has no determinate value for $x = c$. In the expanded expressions of the function it is seen that these [expressions] cease to have a meaning when $x = c$" (1876, p. 124). This is the famous 'Casorati-Weierstrass theorem,' whose basic form had appeared in discussions between these two men in 1864 (see §7.1). Casorati published it in (1868b) and included it in his lectures on complex functions in 1868.[23]

Representation theorems for entire functions like those given by Weierstrass were immediately extended to the case of meromorphic functions by G. Mittag-Leffler (1846-1927), who had attended Weierstrass' lectures in 1875. Mittag-Leffler took as his starting point a variation of the demonstration of the existence of a function with prescribed zeros which Weierstrass had given in his lectures and included in his published paper (1876, p. 93).

In the paper, by letting $a_v \neq 0$, Weierstrass sets

$$F(z) = \sum_{\nu=1}^{\infty} \frac{1}{z - a_\nu} \left(\frac{z}{a_\nu}\right)^{m_\nu},$$

where $F(z)$ is analytic in the complex plane, except for the (simple) poles given by the points a_ν, where the residues of $F(z)$ equal the frequency of a_ν in $\{a_\nu\}$. He further says, referring to one of his earliest papers, that there exists an entire function $G(z)$ whose logarithmic derivative $G'(z)/G(z)$ equals $F(z)$. $G(z)$ therefore has the desired zeros.

But at the same time Weierstrass had also proved that there exists a meromorphic function $F(z)$ with given simple poles of arbitrary residue $m \in \mathbb{N}$. The question that Mittag-Leffler posed was whether it was possible to demonstrate the existence of a meromorphic function with arbitrarily prescribed principal parts,

$$G_n\left(\frac{1}{z - a_n}\right) = \sum_{k=1}^{\nu(n)} C_{\nu(n)}^n/(z - a_n)^k$$

at the arbitrary given poles $\{a_n\}$?

Mittag-Leffler published the first demonstrations of his theorem in several articles in Swedish after his return to Stockholm, extending it to the case in which $\{a_n\}$ has a finite number of accumulation points. He eventually assembled his findings into a paper (1884) in which he proposed a series of general topological notions on infinite point sets according to Cantor's new theory.

Mittag-Leffler stated the main theorem in the following way:[24]

Let Q be an isolated set in the extended complex z-plane whose points will be denoted by $a_1, a_2, ..., a_r, ...$; further let

$$G_1\left(\frac{1}{z - a_1}\right), G_2\left(\frac{1}{z - a_2}\right), ..., G_r\left(\frac{1}{z - a_r}\right), ...$$

be a sequence of analytic functions, where $G_r(1/(z-a_r))$ denotes an entire function of $1/(z-a_r)$ that vanishes when $1/(z-a_r) = 0$. Then one can always form an analytic expression which is regular everywhere except in the neighborhood of the points belonging to $Q + Q'$ [where Q' is the derived set of Q] and which, for all r, can be put in the form $G_r(1/(z-a_r)) + (z-a_r)$ in the neighborhood of $z = a_r$, [where $P(z-a_r)$ is a converging power series of $z - a_r$] (Mittag-Leffler, 1984, p. 8; trans. Birkhoff, 1973, pp. 88-9).

The study of the behavior of analytic functions in the neighborhood of an essential singularity, which was undertaken by Weierstrass at the end of his (1876), was extended three years later by an additional result found by the young French mathematician E. Picard (1856-1941). Utilizing methods appropriate for the theory of elliptic function, Picard showed, in a short note that appeared in the *Comptes rendus* of the Academy (1879), that "there is at most one value which an entire function does not take on anywhere in the finite [complex] plane" without reducing to a constant (*ibid.*, p. 1024).

In a later, extended version of his work, which appeared the following year in the *Annales de l'Ecole normale supérieure*, Picard also proved that if there are at least two values, each of which is taken on only a finite number of times, then the function is a polynomial. A meromorphic function (taking infinity as a possible value) can omit two values without being reduced to a constant.

Moreover, Picard demonstrated that in any neighborhood of an isolated essential singularity, a function takes on every value except at most one. Thus, even in France, where mathematicians had for a long time remained wedded to the tradition founded by Cauchy, people began to utilize Weierstrass' arithmetical style of analysis. The authority of the Berlin mathematician, in Hermite's words, "the great legislator of mathematics" (In: Dugac, 1973, p. 152), was openly acknowledged with the French translation of his 1876 paper, which appeared in 1879. This was followed two years later by the translation of Weierstrass' (1880a).

Riemann's achievements in complex function theory, on the contrary, were viewed with suspicion in France, and his peculiar idea of covering surfaces was judged at most to be a useful geometrical framework for analytic research. Already in 1875, in the preface to the second edition of Briot's and Bouquet's *Traité*, they wrote, "The idea of surfaces of many sheets presents difficulties. In spite of the fine results Riemann achieved with this method, it seems to us to present no advantage for the object we have in view" (Briot and Bouquet, 1875, p. iv).

Casorati (1887) protested vigorously against this opinion. As we have seen, he was a strong advocate of Riemann's approach. Yet he himself had nevertheless to forego publishing the second part of his *Teorica*, a book that Klein placed first among those inspired by Riemann's ideas (Klein, 1926, p. 274), primarily because of his inability to overcome the difficulties Weierstrass had raised against Dirichlet's principle.

At the beginning of the 1880s Klein took up Riemann's theory of algebraic functions and their integrals in his course at Leipzig (Klein, 1882). In order to retain the global, geometrical point of view of Riemann's dissertation, he had to avoid the use of Dirichlet's principle. Klein proposed using a physical model consisting of stationary currents on closed surfaces, made simply connected by suitable cuts.

Klein's work was received with much perplexity by mathematicians, given his repeated assertion that he had thereby found the deepest core of Riemann's ideas. Klein's physical model might have had some heuristic value, but it was certainly no substitute for a rigorous demonstration.[25]

Parallel with the growing tendency towards the arithmetization of analysis, Weierstrass' arithmetical view was becoming dominant. As a consequence, his Berlin lectures became the reference point for European mathematicians and a goal for many gifted young men from German and foreign universities. Indeed, it was primarily through the lecture notes that these young mathematicians took in Weierstrass' lectures that his views were spread throughout Europe. The *Saggio* of Pincherle (1880) is a work of this type. It was written after a study trip to Berlin, where Pincherle (1853-1936) had followed Weierstrass' course in 1877-78, and can be used to give us an idea of Weierstrass' introduction to the theory of functions.

The *Saggio* is divided into four parts. The first sets out the fundamental principles of arithmetic, the basis of the theory of functions. It includes the theory of integral, rational, and negative numbers, as well as Weierstrass' theory of the real numbers, which in his terminology are numbers composed of an infinity of elements. It also discusses the theory of numbers formed from two principal units, that is, the complex numbers, and finally presents the theory of infinite numeric series and products.

The second part sets out *a few theorems about magnitudes in general*, among which is found the so-called 'Bolzano-Weierstrass theorem.' This is stated in this way: if x is a letter to which we can assign any real value whatever (which is to say that the set of the values that x can take forms a simply infinite manifold), and we indicate by a *place* or *point* of the manifold a special value attributed to x (and we still define in the usual manner the concept of a neighborhood of a place, even if this is at infinity), then the following *fundamental proposition* is valid:

If, in a manifold of one dimension, there are infinitely many points satisfying one common definition, we will find in this manifold at least one point having the property that in any one of its neighborhoods, no matter how small it is, there exists an infinite number of points satisfying this definition (Pincherle, 1880, p. 237).

Only in the third part of Pincherle's account does the *concept of function* appear. The definition that we read there is the following:

If a variable quantity, real or complex, that we will call y, is associated with another variable quantity, either real or complex, x, in such a way that to a value of x there corresponds within certain limits one or more determinate values of y, we say that y *is a function of x in the most general sense of the word*, and we will write it $y = f(x)$ (*ibid.*, p. 246).

In commenting on the definitions of continuity and derivative that Weierstrass gave, which is done in a manner like that used today, Pincherle writes:

It has been thought until recently that being continuous is enough for a function to be differentiable; many treatises of the differential calculus even give a demonstration of the theorem: "every continuous function is differentiable." But all these demonstrations implicitly admit certain properties that are not contained in the general concept of a function (*ibid.*, p. 247).

As a counterexample, Pincherle gives the function

$$(7.3.1) \qquad f(x) = \sum_{1}^{\infty} \frac{\sin(n!x)}{n!}$$

which is continuous and finite for every $x \in \mathbb{R}$, but is not differentiable in any point.

Weierstrass' subsequent observations on the concept of function are also particularly interesting. Pincherle notes them as follows:

> If we imagine between x and y a relation [*relazione di dipendenza*] subject only to the condition that, given a value of x, there follows a corresponding value for y, we find that the links thus established between x and y are so vague and indeterminate that it is impossible to find any properties common to all the *functions* (*ibid.*, p. 253).

After claiming that he has a "*purely analytic* study of functions" in mind (*ibid.*), Weierstrass continues, "We cannot create a theory of functions if we do not in some way limit the class of functions for which we want to give common properties" (*ibid.*, p. 254). These will be *analytic functions*. Functions that instead lose their meaning for complex values of the variable, like (7.3.1), "will not be treated as truly *analytic functions*, but only as limiting cases of them" (*ibid.*).

In the light of this declared objective, which occupies the fourth part of the *Saggio* dealing with rational functions and power series, Weierstrass introduces the concepts of rational functions and of functions defined as infinite sums of rational functions, that is, series of rational functions. For these he establishes the concepts of the domain of convergence of a series and of uniform convergence, setting out the theorem: "A sum of infinite rational functions uniformly convergent within a certain domain represents a continuous function within this domain" (*ibid.*, p. 325).

In Weierstrass' approach a fundamental role is played by the theory of the power series of a variable for which he defines the circle of convergence and enunciates the "classic" theorem that if a series converges in a circle of radius R, then it converges uniformly in a circle of radius r, with $r < R$.

After setting out the conditions of identity for two power series and the term-by-term derivation of power series, under the hypothesis of uniform convergence, Weierstrass goes on to illuminate the foundations of the principle of analytic continuation. Using $p(x,a)$ to indicate a series of powers that converges in a circle of center a and radius R, he considers any point b of the circle. For all values of x such that $|b - a| + |x - b| < R$, we can find an expansion of the series given by the powers of $x - b$, and this series will converge within a circle of center b and radius $R - |b - a|$, but it can also converge in a circle of radius greater than this, and in points common to the two circles the two series $p(x,a)$, $p(x,a,b)$ (which Weierstrass uses to indicate the series derived from $p(x,a)$) give the same value.

This is the basic idea of *analytic continuation*. In fact,

> A series of powers $f(x,a)$, convergent within a circle a of radius r, determines for all values of x within the circle a function having an entirely rational character.
>
> Taking a point b within this circle we can deduce from the primitive series a series $f(x,a,b)$ with respect to point b. This will give a circle of convergence r' that can also be extended outside of the circle (a,r). We say that this series gives the *continuation* of the function defined by $f(x,a)$ for the points of the circle (b,r') outside of the circle (a,r) (Pincherle, 1880, p. 353).

The series introduced in this way constitute the *elements of an analytic function*, by means of which, through analytic continuation, we can obtain the *analytic function* in its totality. The *singular* points are those that do not belong to the "domain of validity" of the function. Weierstrass finally distinguished between *monodromic* functions and *polydromic* ones. The first are those for which, beginning from a first element $f(x,a)$ and following any path "traced by the points that are used for the intermediate expansion, we always arrive at a unique expansion $f(x,a, ..., x_0)$ with respect to every point x_0 of the domain of validity" (*ibid.*, p. 354). *Polydromic* functions are those where the contrary applies.

In explaining his own point of view and his difference from Riemann's approach, Weierstrass asserts that,

> The functions that we call *analytic* are those defined by Cauchy and Riemann by the following property: if $w = u + iv$ is a function of $z = x + iy$, we must have
> $$\frac{\partial u}{\partial y} = -\frac{\partial v}{\partial x}, \quad \frac{\partial u}{\partial x} = \frac{\partial v}{\partial y}.$$
>
> This general definition of a function appears to be based on a property of an arbitrary character, whose generality cannot be demonstrated *a priori*. It further requires that the functions u and v be chosen between those functions of two real variables that admit partial derivatives, whereas in the current state of knowledge those functions of real variables subject to differentiation do not constitute a class that can be precisely delimited. For these reasons this definition of an analytic function will not be adopted (*ibid.*, p. 317-8).

Thus by refuting the *global* point of view maintained by Riemann in (1851), Weierstrass comes to conceive of the study of analytic functions as a *local* theory. As Boutroux (1880-1922) has written,

> This involves representing a function and recognizing its properties in the immediate neighborhood of a given point. From this follows the privileged role attributed to the expansions converging in a circle or in a corona described around a point. For the Weierstrass school, to define a function is essentially to give a Taylor series, since from this series one can *theoretically*, by the method of analytic continuation, deduce the value of the function at every point where it is defined (Boutroux, 1908, p. 2).

At the end of the century the German word *Funktionenlehre* became virtually synonymous with the theory of functions of a complex variable *according to Weierstrass' principles*. However, there was no lack of opposition. Lie, for example, did not hesitate to write to Darboux that it was entirely because of Weierstrass and his school that there was no serious research in geometry in Germany (In: Dugac, 1973, pp. 147-8). He even spoke of the "great stupidities" being uttered on this topic, which, apart from the polemics, reveals a profound difference in his understanding of mathematics. This difference was also shared (although in another way), by men like Poincaré and Klein. The latter, in a talk delivered to the Göttingen Society of Science on the occasion of Weierstrass' eightieth birthday, recalled the "impulse" that had been given to contemporary mathematics by Weierstrass' rigor and Kronecker's extreme tendency "to ban irrational numbers and reduce mathematical knowledge to relations between *whole* numbers alone." He then continued,

> I would like to include all of these developments under one word: the *arithmetization of mathematics*. ... In this there lies, as you well know, both a complete understanding of the extraordinary importance of the developments connected with this, and a rejection of the view that the true contents of mathematics should already be completely contained in a sort of extract of arithmetic. I must accordingly divide my views into two parts, those positive and concurring and those negative and dissenting. Since I do not see the arithmetical form of the evolution of thought as the essence of the subject, but rather as a logical sharpening achieved by this means, there follows the challenge -- and this is the positive side of my lecture -- of subjecting the usual disciplines of mathematics to a *reworking* with reference to the arithmetical foundations of analysis. But on the other hand -- and this is the negative side -- I have to maintain and firmly stress that mathematics will never be completed by logical deduction, that in relation to this *intuition* also retains its full specific importance (1895, pp. 233-4).

Even though he maintained that " he did not have anything particularly new" to say, Klein's words in fact partook of a debate that already foresaw the developments of mathematics in our own century.

Notes to Chapter 7

[1]We can obtain an idea of the content of Weierstrass' lectures from the lecture notes taken by Schwarz. Extracts of Schwarz's manuscript have been published by Dugac (1973, pp. 118-125). In the same paper, Dugac gives a detailed analysis of Weierstrass' courses at Berlin, with particular attention to the foundations of the theory of real numbers.

[2]"Riemann is a man of brilliant intuition. When his interest is awakened he begins fresh, without letting himself be diverted by the tradition and without acknowledging the requirements of systematization. Weierstrass is first of all a logician; he proceeds slowly, systematically, step-by-step. When he works, he strives for the definitive form." Thus did Klein see the essential character of these two mathematicians' methods (1926, p. 70), both of which were to have such a major influence on the future development of analysis.

[3]Lacroix continues, "The way of conceiving the quantities of the calculus does not at first seem to permit this law, since we always assume an interval between two consecutive values of the same variable; but by letting them vanish in order to pass to the limit, we say that there is continuity here" (1858, p. 88). According to him, this "*métaphysique*" contains "the philosophical explanation of the properties of the differential calculus" (*ibid.*). Although it was 60 years old, Lacroix' textbook was still widely used at this time. The edition I have cited here includes among its addenda a long (more than 100 page!) note on elliptic functions by Hermite.
Other treatises were not much different. Thus, for example, Duhamel (1860, I, p. 222): "A variable is continuous when it cannot pass from one value to another without passing through all the intermediate values." And, "a function is said to be continuous when, on allowing the dependent quantities to vary in a continuous manner, it is always real and finite, and itself varies in a continuous manner, that is to say, that it cannot pass from one value to another without passing through all the intermediaries" (1860, I, p. 222-223).

[4]Of course, in Kronecker's definition we have to add the symbol for absolute value and the hypothesis that $\delta > 0$. It is also probable that this imprecision results from the private nature of Casorati's notes.

[5]"Of the infinite we do not have any determinate concept; we understand it only as a limit," is the meaningful note that Casorati adds in the margin.

[6]This citation is taken from Casorati's original manuscript in Pavia. See also Bottazzini (1977b, pp. 64-7).

[7]The function $\theta_0(q) = 1 + 2\Sigma_{\nu=1} q^{\nu^2}$ discussed here by Kronecker was introduced by Jacobi in his study of elliptic modular functions. See also Kronecker (1894, p. 182). For a detailed discussion of Kronecker's contribution to the theory of elliptic functions presented in terms appropriate for the modern reader, see Weil (1976).

[8]The circumference of a circle with center at the origin and radius 1 is composed of singular points and is the so-called "natural boundary" for $\theta_0(q)$. Weierstrass did not publish these results until 1880 (1880a, p. 227). In the same work he also showed that the circumference is the natural boundary for all series of the form

(1) $\sum_{\nu} b^{\nu} x^{a^{\nu}}$,

where a is an odd integer $\neq 1$, $0 < b < 1$ and $ab > 1 + \frac{3}{2}\pi$. In fact, a series of this type "defines a function that cannot be extended beyond the domain of convergence of the series and consequently exists only for those values of x whose absolute value does not exceed one" (1880a, p. 223). It is in this context that Weierstrass says he has found a proof that the real part of (1) given by $\Sigma_{\nu=0} b^{\nu} \cos a^{\nu} t$ is not differentiable for any value of t. Series like (1) are called "lacunary series." In 1892 Hadamard demonstrated that a lacunary series $\Sigma_{k=1} a_k z^{n_k}$ ($a_k \neq 0$, $\lim_{k \to \infty} k/n_k = 0$) has the circle of convergence as a natural boundary provided that there exists a fixed $\lambda > 1$ such that for all k

$$\frac{n_{k+1}}{n_k} \geq \lambda.$$

For a proof and further properties of such a series, see Hille (1962, II, pp. 87-92).

[9]In 1872 C. Méray, a student of Bouquet, published a book on the infinitesimal calculus in which he independently presented a theory of the real numbers analogous to that of Weierstrass. He too had been dissatisfied by the lack of rigor in earlier treatments of the irrationals. For a comparison of Weierstrass' and Méray's theories, see Dugac (1973).

[10]Lipschitz's letters to Dedekind were published by Dugac (1976b), while those of Dedekind appeared in his *Werke*, III, pp. 464-482.

[11]See Dugac (1976b, p. 58) for a discussion of and interpretation of this point in modern algebraic terms. See also Krull (1971, pp. 34-5).

[12]The relationship between Dedekind and Lipschitz nevertheless remained one of the highest mutual regard. Dugac (1976b, p. 53) tells us that at his death, Lipschitz expressed the desire to have Dedekind edit his works.

[13]Hankel's analysis of the behavior of $f(x)$ was nevertheless less than rigorous, as Dini later showed (1878, pp. 117-147). He gave a precise analysis of Hankel's arguments.

[14]For a discussion of Hankel's paper and its importance for the theory of integration, see Hawkins (1970, pp. 28-33).

[15]Cantor demonstrated the theorem in a fairly laborious manner. His proof is summarized in Dauben (1971, pp. 184-189).

[16]Naturally, in Cantor's theory of the real numbers, the one-to-one correspondence between the points of the line and the real numbers is assured by means of an axiom of continuity similar to that of Dedekind, but it is not equivalent to it, in the sense that Archimedes' postulate can be derived from Dedekind's axiom, but not from Cantor's.

[17]"It might perhaps also have been worthwhile to have deliberately emphasized that this is not always the case," Zermelo, the editor of Cantor's *Abhandlungen*, added in a footnote. "In general, the point set arising in any way from a numerical quantity of the n-th kind can not only be lower than but also higher than the n-th kind or it cannot be of any definite kind" (Cantor, 1872, p. 99, n. 1).

[18]Cantor's contributions to set theory and the theory of transfinite numbers have been amply studied by historians. For a detailed account and the related bibliography, see Dauben (1979).

[19]There is a survey of Dini's work on foundations in Bottazzini (1985).

[20]Betti's lessons were published in the issues of the *Annali di matematica pura ed applicata*. See Betti (1860-61).

[21]There are similar observations in Briot and Bouquet (1859, p. 135). In this regard Betti noted in a letter to Tardy in 1859 that Briot's and Bouquet's theorem that two functions which have the same zeros and infinites can differ only by a constant factor was true for rational functions which have a finite number of zeros and poles, but that it seemed to him not rigorous enough for the case of

an infinite number of zeros and infinites. In this case the two functions differ at most by a factor e^w, where w is an entire function.

[22]A modern presentation of Weierstrass' theorem of factorization by means of "prime functions" can be found in Hille (1963, I, pp. 225-229).

[23]This theorem was also published at about the same time by the Russian mathematician Y. V. Sokhotskii (1842-1927) in his Master's thesis on "The theory of integral residues with applications." For the history of the 'Weierstrass-Casorati-Sokhotskii theorem,' see Neuenschwander (1978c).

[24]Mittag-Leffler sent his theorem to Weierstrass in 1877, who was very interested in the result. It permitted him, among other things, to solve a question that had been left open in his 1876 paper, namely, that a single-valued function with infinite poles and finite essential singularities could be represented by a finite sum of functions, each with an essential singularity. Weierstrass published a simpler version of Mittag-Leffler's theorem, together with the above observations, in (1880b).

[25]A summary of Klein's basic work in (1882) is given in Springer (1957). As noted (§6.2c), the integration between Riemann's and Weierstrass' conceptions was not achieved until the first decade of this century, following the vigorous impulse given by Weyl's book of 1913.

Appendix
ON THE HISTORY OF "DIRICHLET'S PRINCIPLE"

Following the publication of Newton's *Principia mathematica philosophiae naturalis* in 1687, one of the problems on which the scientists of the eighteenth century worked was that of determining the gravitational attraction of celestial bodies, particularly that of the earth on the moon. This problem could, in a first approximation, be reduced to that of determining the force of attraction of a body on a point mass.

In classical Newtonian theory, we first determine the force of attraction of an infinitesimal mass of the body on a unit mass placed at a point P and then obtain the force of attraction exerted by the body as the sum of the forces exerted by the infinitesimal masses. This, however, leads immediately to a difficult problem of integration. In fact, by introducing a suitable coordinate system, the components of the force along the axes are given by the volume integrals

$$f_x = -k \iiint \rho \, \frac{(x - \xi)}{r^3} dv$$

$$(1) \qquad f_y = -k \iiint \rho \, \frac{(y - \eta)}{r^3} dv$$

$$f_z = -k \iiint \rho \, \frac{(z - \zeta)}{r^3} dv,$$

where (ξ, η, ζ) is any point of the body, ρ as its density, $r = \{(x - \xi)^2 + (y - \eta)^2 + (z - \zeta)^2\}^{1/2}$, and k is Newton's constant.

In order to calculate these integrals it is first necessary to know the shape of the body. This is the reason for the large number of studies both on the shape of the earth, the subject of Cairaut's (1713-1765) famous *Théorie de la figure de la terre* of 1743, and on the attraction of ellipsoids (or more generally, of solids of revolution), since it was clear that the form of the earth was not exactly spherical but ellipsoidal. Naturally, in the case of those studies on the attraction of the earth, the problem was complicated by not knowing the value of ρ.

But there was a more practical way of arriving at this goal. In his *Hydrodynamica* (1738) Daniel Bernoulli suggested taking the function $V = \Sigma(m_i\mu/r_i)$ for a point P of mass μ attracted by a finite number of point masses m_i, where r_i is the distance from M to m_i. By replacing the finite number of masses m_i with a solid M of density ρ and supposing P to lie outside of M, the function V becomes

$$(2) \qquad V(x,y,z) = \mu \iiint_M \frac{\rho}{r} d\zeta \, d\eta \, d\xi.$$

In (1773) Lagrange observed that from a knowledge of the function V one could without difficulty obtain the components of the attractive force exerted on P by means of a derivation under the integral sign. This idea was generally accepted and the determination of the properties of the function V became a major subject of research.

The first successes were reported by Legendre (1782), who determined the attractive force for solids of revolution by introducing infinite series of certain polynomials P_n that are today called "Legendre polynomials," "Laplace coefficients," or "local harmonics."[1] Later, Legendre showed how to simplify his calculations by "using a theorem kindly communicated to me by M. de la Place." This enabled him to obtain the attractive force as the gradient of the potential function V.

Laplace was also working on these problems at this time. In the 1770s he had published several papers on the force of attraction exerted by solids of revolution. Stimulated by Legendre's results, though without mentioning him by name, he succeeded in determining V for the case of an arbitrary spheroid by assuming that V satisfies the equation

$$(3) \qquad \Delta V = \frac{\partial^2 V}{\partial x^2} + \frac{\partial^2 V}{\partial y^2} + \frac{\partial^2 V}{\partial z^2} = 0$$

for the points within and outside the body[2] (Laplace, 1782). This had certainly been known to him earlier since it had already been known to Euler. The latter, in a work on the motion of fluids in 1752, had shown that if u, v, and w are the velocity components of any point in an incompressible fluid, then the expression

$$u \, dx + v \, dy + w \, dz$$

must be an exact differential dS, and therefore

$$u = \frac{\partial S}{\partial x}, \qquad v = \frac{\partial S}{\partial y}, \qquad w = \frac{\partial S}{\partial z}.$$

From the incompressibility of the fluid, it follows from the "law of continuity" that, during the movement,

$$\frac{\partial^2 S}{\partial x^2} + \frac{\partial^2 S}{\partial y^2} + \frac{\partial^2 S}{\partial z^2} = 0,$$

an equation that Euler was able to integrate for the particular case in which S was given by a polynomial in x, y, z.[3]

The problem was taken up by Poisson in an article that appeared in 1813. After establishing equation (3), "which Laplace has made the basis of his beautiful investigations on the attraction of spheroids of any form," Poisson observed that (3), "holds when the attracted point lies outside the solid under consideration, or even when, this body being hollow, the attracted point is situated in the interior cavity: actually, these two cases are the only ones to which equation (3) has been applied. It is nevertheless not superfluous to observe that it no longer holds if the attracted point is an interior point of the solid. This is especially strange since, according to the usual proof, it would seem that equation (3) must hold identically in the coordinates x, y, and z" (Poisson, 1813, p. 388; Eng. trans. Birkhoff, 1973, p. 343).

Poisson then showed by "direct integration" that if the point $P(x,y,z)$ lies *within* a homogeneous or heterogenous body V, then the equation

(4) $\Delta V = -4\pi\rho$

holds, where ρ, the density of the attracting body, is a continuous function of x, y, z. Although correct, Poisson's statement was not demonstrated in a rigorous manner, since the continuity of the function $\rho(x,y,z)$ is not enough, as Hölder later pointed out (1882).

In the same year Gauss published a paper (1813b) on the attraction of homogeneous ellipsoids in which he obtained Green's formula (6) (see below) for particular cases. This kind of problem, which is linked to the astronomical problem of determining the orbit of the moon, was one reason for Gauss' interest in potential theory.

Gauss was also concerned with two-dimensional potential theory and the solution of the equation

(5) $\dfrac{\partial^2 V}{\partial x^2} + \dfrac{\partial^2 V}{\partial y^2} = 0$

because of his work on the problem of conformal mapping, in particular, the conformal mapping of the interior of an ellipse onto that of a circle (*Werke*, 10(1), pp. 311-20). He recognized that the solution of equation (5) was found by replacing the function (2) by

$$V = \iint_D \rho \, \log \frac{1}{r} d\xi \, d\eta,$$

the so-called "logarithmic potential," for a bounded domain D in the plane.

The problem is related to complex function theory, because every solution $u(x,y)$ of (5) is such that in D,

$$-\frac{\partial u}{\partial y}dx + \frac{\partial u}{\partial x}dy$$

is an exact differential. Moreover, up to an additive constant there exists an harmonic function v conjugate to u given by

$$v = \int\left(-\frac{\partial u}{\partial y}dx + \frac{\partial u}{\partial x}dz\right),$$

such that $u + iv$ is an analytic function of $x + iy$ (see §6.2b).

But the principal reason for Gauss' interest in potential theory lay in his studies of terrestrial magnetism. The study of equations (3) and (5) became particularly important in the first decades of the nineteenth century when physicists discovered the role these equations played in electrostatics (Coulomb's law in particular), and in the theories of magnetism and heat[4] (see §2.2).

Gauss in particular studied terrestrial magnetism intensively during the 1830s, while working at Göttingen with Weber. Together they founded the German "Magnetische Verein" and published a scientific journal, *Resultate aus der Beobachtungen des Magnetischen Vereins*. It was here that Gauss published his (1839), which became a landmark in the history of modern potential theory.

In this work, Gauss gave a proof of (4), a theory of equation (3), and numerous results on the potential of a distribution of mass or electricity on a closed surface S.

Among these, Gauss was particularly concerned with the equilibrium problem -- to determine a distribution of electric charges (or masses) on the surface S such that the corresponding potential is constant on S. On the basis of physical arguments, Gauss asserted that the problem always has a solution.

In seeking to give a mathematical demonstration of this, Gauss introduces an idea that was to become central in potential theory. Taking a potential given by

$$V = \iint_S \frac{\rho}{r}ds \quad (\rho > 0)$$

and any continuous function U on S, Gauss considers the family of integrals

$$\Omega = \iint_S (V - 2U)\rho\,ds,$$

and shows that if ρ is chosen in such a way that Ω has a minimum, then $V - U$ is a constant on S. Gauss then adds that the existence of such a ρ is evident. This was the first case of what later came to be called "Dirichlet's problem." In terms of potentials, Gauss' method solved the problem of determining an harmonic function in the domain bounded by S continuous in the closure of the domain and equal on S to a given function U. However, as Monna has pointed out, "there is an essential difference between the problems of Gauss and Dirichlet: the latter had to prove that the minimizing function satisfies Laplace's equation. For Gauss this was no problem because he only considered potentials (harmonic functions), generated by the admissible functions (the density [ρ])" (Monna, 1975, p. 32), and it was quite obvious that they satisfy Laplace's equation outside the body.

Before Gauss, and unknown to him, a self-taught English mathematician, Green (1793-1841), had concerned himself with the study of electricity and magnetism. He had even published an essay on the subject at his own expense in 1828. This work remained unknown until it was republished, after the author's death, by W. Thomson (Lord Kelvin) in Crelle's *Journal* in 1850. Among its principal results is a demonstration of "Green's formula,"

$$(6) \qquad \iiint U \Delta V \, dv + \iint U \, \frac{\partial V}{\partial n} d\sigma = \iiint V \Delta U \, dv + \iint V \, \frac{\partial U}{\partial n} \, d\sigma,$$

where U and V are continuous functions with finite derivatives in the domain, and n is the normal to the surface directed towards the interior. This formula was obtained in the same year by the Russian mathematician Ostrogradski (1801-1861).

In a subsequent work of 1833, Green showed that if a continuous function V is given on the boundary of a body, then there exists a function that is harmonic in the interior of the body and assumes the given values on the boundary. In his demonstration Green made use of the fact that there always exists a function that minimizes the integral

$$\iiint \left[\left(\frac{\partial V}{\partial x} \right)^2 + \left(\frac{\partial V}{\partial y} \right)^2 + \left(\frac{\partial V}{\partial z} \right)^2 \right] dv.^5$$

This idea was explicitly formulated a few years later by Thomson (1847), as well as by Dirichlet in his lectures on potential theory, which were delivered in Göttingen in the Winter of 1856-57.

In these lectures, which were published by Grube in 1876, we read, "For an arbitrary bounded domain there is always one and only one function u of x, y, z which, together with its first-order derivatives, is continuous, satisfies the equation [$\Delta u = 0$] within this entire domain, and reduces to a given value at every point of the [boundary] surface" (Dirichlet, 1876, p. 127). "This proposition is known under the name of 'Dirichlet's principle'," Grube comments at this point (*ibid.*, p. 182, n. 28). However, the problem of determining a harmonic function in the interior of a domain, assuming values given continuously on the boundary, is today called "Dirichlet's problem." In modern terminology it can be stated as follows: "Let Ω be an open set in $\mathbb{R}^3$ and let there be given a continuous real function f on the boundary $\partial\Omega$ of Ω. The problem is whether there exists a function F, defined on the closure $\bar{\Omega}$ of Ω, harmonic in Ω and such that $f = F$ on $\partial\Omega$" (Monna, 1975, p. 29).

'Dirichlet's principle' is a *method* of resolving this problem, as is clear from Dirichlet's own words. He in fact continued by observing that "the problem of finding this function u cannot be solved: we can only speak of an existence proof for it. The latter presents no difficulty." At this point he stated 'Dirichlet's principle' in these words:

> For every bounded connected domain T there are clearly infinitely many functions u continuous together with their first-order derivatives, for x, y, z which reduce to a given value on this surface. Among these functions there will be at least one which reduces the following integral

(7) $$U = \int\left[\left[\frac{\partial u}{\partial x}\right]^2 + \left[\frac{\partial u}{\partial y}\right]^2 + \left[\frac{\partial u}{\partial z}\right]^2\right]dT,$$

> extended over the domain T, to a minimum; it is evident that this integral has a minimum since it cannot become negative. We can now show the following:
>   1. Every such function u which minimizes U, satisfies the differential equation

$$\frac{\partial^2 u}{\partial x^2} + \frac{\partial^2 u}{\partial y^2} + \frac{\partial^2 u}{\partial z^2} = 0.$$

> everywhere in the domain T. This already makes it clear that there always exists a function u having the desired property, namely that function for which U becomes a minimum.
>   2. Every function u which satisfies the differential equation [$\Delta u = 0$] within the domain T, minimizes the integral U.
>   3. The integral U can have only one minimum.
>   It follows from 2 and 3 that there is only one function u with the desired property (Dirichlet, 1876, pp. 127-8).

Dirichlet's argument is open to several criticisms. 1) It is not in fact evident, as Dirichlet assumes, that the class H of the functions u is not empty. 2) It is not stated that the integral (7) is finite for every function of the class H. Finally, and this is the crucial point, 3) it is not said that the greatest lower bound of the class of the functions u is a function that belongs to H, that is, that a minimum exists. Weierstrass clearly formulated this objection in 1870 by giving a counterexample. This is the integral

(8) $$J = \int_{-1}^{1} x^2\left[\frac{dy}{dx}\right]^2 dx,$$

in which the functions $y(x)$ are taken to be continuous together with their first derivatives between $x = -1$, $x = 1$, such that $y(-1) = a \neq b = y(1)$. He shows that, among these functions, there exist those that satisfy the inequality $J < \varepsilon$ for every $\varepsilon > 0$, however small, and therefore that the lower bound of the integral (as y varies) is zero. And yet there is no minimum, since in this case y would equal a constant, which is impossible given that $a \neq b$. Using, as Weierstrass does, ε to indicate a positive constant, we consider the function,

(9)
$$y = \frac{a + b}{2} + \frac{b - a}{2} \frac{\text{arctg } x/\varepsilon}{\text{arctg } 1/\varepsilon} \, ,$$

which satisfies the required conditions. By substituting (9) into (8) in place of y, we obtain,

$$J < \int_{-1}^{1} (x^2 + \varepsilon^2) \left[\frac{dy}{dx}\right]^2 dx = \varepsilon^2 \frac{(b - a)^2}{(2 \text{ arctg } 1/\varepsilon)^2} \int_{-1}^{1} \frac{dx}{x^2 + \varepsilon^2}$$

$$= \frac{\varepsilon}{2} \frac{(b - a)^2}{\text{arctg } 1/\varepsilon}.$$

Therefore, J will become as small as we like when ε is taken arbitrarily small, but there does not exist a function y that minimizes the integral and is continuous. In fact, as ε tends to zero, $y = a$ for $x < 0$ and $y = b$ for $x > 0$ (for $x = 0$, $y = (a+b)/2$). Therefore this *discontinuous* function is that for which J is zero. "Dirichlet's method of proof thus leads in this case to an obviously false result," Weierstrass concludes (1870, p. 54).

"With this a large part of Riemann's developments come to nought," Klein asserts (1894, p. 492), recalling that Riemann had based his existence proofs for the theory of functions of complex variables and for Abelian functions on Dirichlet's principle.

Nevertheless, the far-reaching results that Riemann based on the said principle are all correct, as Carl Neumann [1877] and Schwarz [1870b, 1872] later amply showed with rigorous methods. We must conclude that Riemann originally derived the theorems themselves from physical intuition, which here again proved its value as a heuristic principle, and only afterwards based it on the said method of reasoning in order to have an entirely mathematical train of thought. In doing so he clearly experienced certain difficulties, as long passages of his dissertation show. But since he saw this method of reasoning unhesitatingly accepted in similar cases in his surroundings, even by Gauss himself, he did not pursue it as far as would have been necessary (1894, p. 492).

And in a note to this passage Klein added,

I remember that Weierstrass once told me that Riemann had never laid any particular value on finding his existence proofs with "Dirichlet's principle." Because of this his (Weierstrass') critique of "Dirichlet's principle" would have not made any particular impression on him. In any case, the problem arose of demonstrating the existence proofs in another way (1894, p. 492 n. 8).

Weierstrass' counterexample revealed the necessity of treating the calculus of variations with Weierstrassian rigor. Weierstrass and many of his students, among them Du Bois Reymond (1879) and

Schwarz, dedicated themselves to this task.[6]

As for Dirichlet's problem in particular, the works of Neumann and Schwarz[7] mentioned by Klein should be joined by the contemporary works of Dini (1873, 1876), who had probably been inspired by Schwarz and Betti (Bottazzini, 1982, pp. 270-2; 1985).

In fact, Weierstrass' critique constrained mathematicians to do without 'Dirichlet's principle' as a demonstrative argument. As a testimony to the situation this produced we can cite the words of Betti. In the preface to his *Teoria delle forze newtoniane* (1879), which collected in one volume results that had appeared in articles in the *Nuovo Cimento* over many years, he wrote that in order to prove the existence of a harmonic function in a domain with values given on the boundary, he had at first used the "Dirichlet-Riemann theorem," but that he had subsequently renounced it in the face of criticism and used other methods.

But if Weierstrass' rigor, as convincingly demonstrated by his critique of 'Dirichlet's principle,' became the norm for mathematicians, it was not so for physicists. Helmholtz (1821-1894), in summarizing their point of view, once said to Klein that "for we physicists Dirichlet's principle continues to remain a demonstration."

In reporting what Helmholtz had said to him, Klein noted, "He thus clearly distinguished between proofs for mathematicians and proofs for physicists, as it is on the whole a general fact that physicists are less concerned with mathematical details; for them the 'evidence' is sufficient" (1926, p. 264). Klein further added that he had reported this episode precisely in order to illustrate the slowness with which new mathematical ideas were accepted.

However, towards the end of the century, there were numerous studies that had Dirichlet's principle as their objective, from Poincaré's works of 1887 and 1890, which were based on his method of *balayage* ("sweeping out"), to those conducted with methods appropriate to the new functional analysis.

Finally, with two papers published sucessively in 1900 and 1901, Hilbert (1862-1943) "resurrected" Dirichlet's principle (as he himself said in the first paper). Dirichlet's problem is only a special problem of the calculus of variations, he wrote, and we hence arrive at a statement of Dirichlet's principle in the following more general form: "Every regular problem of the calculus of variations has a solution, provided restrictive assumptions regarding the nature of the given boundary conditions are fulfilled and, when necessary, the concept of solution has been suitable extended" (Hilbert, 1900, p. 11).

In his famous lecture to the International Congress of Mathematicians in Paris in 1900, Hilbert restated Dirichlet's problem and included it among the more general problems involving the calculus of variations. With this, he inaugurated the period of modern research on this topic.[8]

Notes to the Appendix

[1]For a detailed analysis of Legendre's paper, see Kline (1972, pp. 525-8).

[2]Actually, Laplace gave equation (3) in rectangular coordinates only in a later paper (1787). In (1782) he used polar coordinates.

[3]For a careful and penetrating discussion of Euler's work on fluid mechanics, see Truesdell's Preface to Euler's *Opera omnia*, (2) 12.

[4]For a history of potential theory from its beginnings to 1880, see Bacharach (1883). For Gauss' contributions in particular see Geppert (1933).

[5]Actually, Green considers a function v on n variables $x_1, ..., x_n$.

[6]For a history of the calculus of variations, see Kneser (1900) and Zermelo and Hahn (1904). See also Goldstine (1980).

[7]The methods of Schwarz and Neumann and their relations with Riemann's theory of complex functions are fully treated in Forsyth (1918, II, Chaps. 17-18).

[8]Variational problems are becoming a major research area of contemporary mathematics and the literature on this field is consequently enormous. For the history and the modern developments of Dirichlet's principle in particular, see Monna (1975). For its connections with the history of functional analysis, see Dieudonné (1981).

BIBLIOGRAPHY

Collected editions of a particular author's works are listed first and cited under the shortened forms, *Opera*, *Oeuvres*, *Werke*, etc. Individual articles and monographs are listed thereafter in chronological order of appearance. The dates used are those printed in the book or journal. In many cases this is not the date of actual publication.

Abel, N. H. (1839). *Oeuvres complètes*, 2 Vols., B. Holmboe (ed.). Christiania [Oslo].
___ (1881). *Oeuvres complètes de Niels Henrik Abel*, 2 Vols., L. Sylow and S. Lie (eds.). Christiania [Oslo].
Note: References to Abel's individual works are to the second edition by Sylow and Lie (1881).
___ (1826). Recherches sur la série ... In: *Oeuvres* 1, pp. 219-250.
___ (1841). Mémoire sur une propriété générale d'une classe très-étendue de fonctions transcendantes. In: *Oeuvres* 1, pp. 145-211.
___ (1881). Sur les séries. In: *Oeuvres* 2, pp. 197-205.
___ (1902). Mémorial publié à l'occasion du centenaire de sa naissance. Christiania [Oslo].
Ahlfors, L. V. (1953). Development of the Theory of Conformal Mapping and Riemann Surfaces through a Century. *Ann. of Math. Stud.* 30, pp. 3-13.
___ (1966). *Complex Analysis*. New York.
Ahrens, W. (1906-7). Ein Beitrag zur Biographie C. G. J. Jacobis. *Bibl. math.* (3) 7, pp. 157-192.
Ampère, A. M. (1806a). Recherches sur quelques points de la théorie des fonctions dérivées ... *J. École Polytech.* 6 (Cahier 13), pp. 148-181.
___ (1806b). Démonstration générale du principe des vitesses virtuelles, dégagée de la consideration des infiniment petit. *ibid.*, pp. 247-269.

Arbogast, L. F. (1789). Essai sur des nouveaux principes du calcul différential et du calcul intégral, indépendents de la théorie des infiniment petits ... Paris (manuscript).

____ (1791). Mémoire sur la nature des fonctions arbitraires qui entrent dans les intégrales des équations aux différences partielles. St. Petersburg.

Argand, R. (1874). Essai sur une manière de représenter les quantités imaginaires dans les constructions géométriques, 2nd ed. Paris.

Arzelà, C. (1897). Sul principio di Dirichlet. *Rend. Accad. Sci. Bologna* **5**, pp. 225-244.

Bacharach, M. (1883). Abriss der Geschichte der Potentialtheorie. Göttingen.

Belhoste, B. (1984). *Cauchy 1789-1857. Un mathématicien légitimiste au XIXe siècle.* Paris.

Belhoste, B. and Lützen, J. (1984). Joseph Liouville et le Collège de France. Matematisk Institut Odense Universiteit, Preprint No. 3.

Bell, E. T. (1940). *The Development of Mathematics.* New York. [2nd ed. 1945.]

Bernoulli, D. (1755a). Réflexions et éclaircissements sur les nouvelles vibrations des cordes ... *Mem. Acad. Sci. Berlin* **9**, pp. 147-172.

____ (1755b). Sur le mélange de plusieurs espèces de vibrations simples isocrones. *Mém. Acad. Sci. Berlin* **9**, pp. 173-195.

____ (1758). Lettre de Mr. Daniel Bernoulli ... *J. Savants*, pp. 157-166.

Bernoulli, J. (1742). *Opera omnia*, 4 Vols. Lausanne et Geneve.

____ (1727). Theoremata selecta pro conservatione virium vivarum demonstrandi ... In: *Opera* 3, pp. 124-130.

____ (1728). Meditationes de chordis vibrantibus ... In: *Opera* 3, pp. 198-210.

Betti, E. (1903-13). *Opere matematiche*, 2 Vols. Milano.

____ (1860-61). La teorica delle funzioni ellittiche. In: *Opere mat.* 1, pp. 228-412.

____ (1879). Teorica delle forze newtoniane. Pisa.

Biot, J. B. (1805). Mémoire sur la propagation de la chaleur. *J. des Mines* **17**, pp. 203-224.

____ (1816). *Traité de physique experimentale et mathématique*, 4 Vols. Paris.

Birkhoff, G. (1973). *A Source Book in Classical Analysis.* Cambridge, Mass.

Bolzano, B. (1816). Der Binomische Lehrsatz ... In: (1981a), pp. 253-416.

____ (1817). Rein analytischer Beweis des Lehrsatzes ... *Abh. Ges. Wiss. Prague* (5)3, pp. 1-60. Repr. in: Ostwald's Klassiker **153**, pp. 3-43, (Leipzig 1905), and in: (1981a), pp. 417-476.

____ (1851). *Paradoxien des Unendlichen*, F. Prihonsky (ed.). Leipzig. Eng. trans. D. A. Steele, *Paradoxes of the Infinite.* New Haven 1950.

____ (1930). *Funktionenlehre*, K. Rychlik (ed.). Prague.

____ (1962). *Theorie der reenen Zahlen ill Bolzanos Handschriflichem Nachlass*, K. Rychlik (ed.) Prague.

____ (1981a). Early Mathematical Works. Lubos Nový (ed.) [*Acta Historiae Rerum Naturalium necnon Technicarum*, Special Issue 12] Prague.

____ (1981b). Impact of Bolzano's epoch on the development of science. [*Acta Historiae Rerum Naturalium necnon Technicarum*, Special Issue 13] Prague.

Borchart, C. W. (1880). Lecons sur les fonctions doublement périodiques faites en 1847 par M. J. Liouville. *J. reine angew. Math.* **88**, pp. 277-310.

Bos, H. J. M. (1974). Differentials, higher-order differentials and the derivative in the Leibnizian Calculus. *Arch. Hist. Exact Sci.* **14**, pp. 1-90.

Bottazzini, U. (1977a). Riemanns Einfluss auf E. Betti und F. Casorati. *Arch. Hist. Exact Sci.* **18**, pp. 27-37.

____ (1977b). Le funzioni a periodi multipli nella corrispondenza tra Hermite e Casorati. *Arch. Hist. Exact Sci.* **18**, pp. 39-88.

____ (1978). Ricerche di P. Tardy sui differenziali di indice qualunque, 1844-1868. *Historia Math.* **5**, pp. 411-418.

____ (1981). Aspects of Italian mathematics in the early 19th century: the heritage of Lagrange and Cauchy's "modern analysis." In: Bolzano, (1981b), pp. 271-282.

____ (1982). Enrico Betti e la formazione della scuola matematica pisana. In: *Atti del convegno su "La storia delle matematiche in Italia"* Cagliari, pp. 229-276.

____ (1985). Dinis Arbeiten auf dem Gebiet der Analysis: Schwerpunkte seiner Erforschung der Grundlagen. In: M. Folkerts and U. Lindgren (eds.). Mathemata: Festschrift für Helmuth Gericke, Stuttgart, pp. 591-605.

Bourbaki, N. (1939). *Élements de mathématique*. Première partie: Les structures fondamentales de l'analyse. Livre I: Théorie des ensembles. Fascicule des résultats. Paris. Eng. trans. *Elements of Mathematics. Theory of Sets*. Paris, (1968).

____ (1960). *Éléments d'histoire des mathématiques*. Paris.

Boutroux, P. (1908). *Lecons sur les fonctions définies par les équations différentielles du premier ordre*. Paris.

Boyer, C. B. (1959). *The History of the Calculus and its Conceptual Development*. New York.

____ (1968). *A History of Mathematics*. New York.

Brill, A. and Nöether, M. (1894). Die Entwicklung der Theorie der algebraischen Funktionen in älterer und neuerer Zeit. *Jahresber. Deutsch. Math.-Verein.* **3**, pp. 107-565.

Briot, C. and Bouquet, C. (1859). *Théorie des fonctions doublement périodiques et, en particulier, des fonctions elliptiques*. Paris.

____ (1875). Théorie des fonctions elliptiques. Paris.

Bucciarelli, L. L. and Dworsky, N. (1980). *Sophie Germain: An Essay in the History of the Theory of Elasticity*. Dordrecht.

Bucée, M. (1806). Mémoire sur les quantités imaginaires. *Phil. Trans. Royal Soc. London*, pp. 13-88.

Burkhardt, H. (1900). (with W. F. Meyer) Potentialtheorie. *Enzykl. math. Wiss.* [Art II A 7b], Vol. II, Teil 1, 1, pp. 464-503.

___ (1901-8). Entwicklungen nach oszillierenden Funktionen und Integration der Differentialgleichungen der mathematischen Physik. *Jahresber. Deutsch. Math.-Verein.* 10, pp. 1-1804.

___ (1914-5). Trigonometrische Reihen und Integrale. *Enzykl. math. Wiss.* [Art. II A 12], Vol. II, Teil 1, 2, pp. 819-1354.

Burzio, F. (1942). *Lagrange.* Torino.

Cantor, G. (1932). *Gesammelte Abhandlungen mathematischen und philosophischen Inhalts.* E. Zermelo (ed.). Berlin.

___ (1867). De aequationibus secundi gradus indeterminatis. In: *Gesam. Abh.*, pp. 1-30.

___ (1869). De transformatione formarum ternarium quadraticarum. In: *Gesam. Abh.*, pp. 51-62.

___ (1870a). Über einen die trigonometrischen Reihen betreffenden Lehrsatz. In: *Gesam. Abh.*, pp. 71-79.

___ (1870b). Beweis, daß eine für jeden reellen Wert von *x* durch eine trigonometrische Reihe gegebene Funktion *f(x)* sich nur auf eine einzige Weise in dieser Form darstellen läßt. In: *Gesam. Abh.*, pp. 80-83.

___ (1871). Notiz zu dem Aufsatz: Beweis, daß eine für jeden reellen Wert... In: *Gesam. Abh.*, pp. 84-86.

___ (1872). Über die Ausdehnung eines Satzes aus der Theorie der trigonometrischen Reihen. In: *Gesam. Abh.*, pp. 92-102.

___ (1880a). Fernere Bemerkung über trigonometrische Reihen. In: *Gesam. Abh.*, pp. 104-106.

___ (1880b). Uber unendliche lineare Punktmannigfaltigkeiten, Nr. 2. In: *Gesam. Abh.*, pp. 145-148.

Cantor/Dedekind (1937). *Briefwechsel.* E. Nöether and J. Cavaillès (eds.). Paris.

Carslaw, H. S. (1921). *Introduction to the Theory of Fourier's Series and Integrals* (2nd ed.). Cambridge. Repr. New York 1950.

Cartan, H. (1961). *Théorie élémentaire des fonctions analytiques d'une ou plusieurs variables complexes.* Paris.

Casorati, F. (1951-52). *Opere.* 2 Vols. Roma.

___ (1868a). *Teorica delle funzioni di variabili complesse.* Pavia.

___ (1868b). Un teorema fondamentale nella teorica delle discontinuità delle funzioni. In: *Opere* 1, pp. 279-281.

___ (1887). Sopra le 'coupures' del sig. Hermite, i 'Querschnitte' e le superficie di Riemann, ed i concetti d'integrazione si reale che complessa. In: *Opere* 1, pp. 385-418.

Cauchy, A.-L. (1882-1974). Oeuvres complètes d'Augustin Cauchy. Sér. 1, 12 vols.; Sér. 2, 15 vols. Paris.

___ (1821). Cours d'analyse algébrique ... In: *Oeuvres* (2) 3.

___ (1823a) Résumé des lecons ... sur le calcul infinitésimal ... In: *Oeuvres* (2) 4, pp. 5-261.

___ (1823b). Mémoire sur l'intégration des équations linéaires aux différentielles partielles et à coefficients constants. In: *Oeuvres* (2) 1, pp. 275-357.

___ (1825). Mémoire sur les intégrales définies, prises entre des limites imaginaires. In: *Oeuvres* (2) **15**, pp. 41-89.

___ (1826a). Sur un nouveau genre de calcul analogue au calcul infinitésimal. In: *Oeuvres* (2) **6**, pp. 23-37.

___ (1826b). Sur diverses relations qui existent entre les résidus des fonctions et les intégrales définies. In: *Oeuvres* (2) **6**, pp. 124-145.

___ (1826-30). Exercices de mathématique. In: *Oeuvres* (2) **6-9**.

___ (1827a). Mémoire sur les intégrales définies. In: *Oeuvres* (1) **1**, pp. 319-506.

___ (1827b). Mémoire sur les développements des fonctions en séries périodiques. In: *Oeuvres* (1) **2**, pp. 12-19.

___ (1827c). Théorie de la propagation des ondes à la surface d'un fluide ... In: *Oeuvres* (1) **1**, pp. 3-312.

___ (1827d). Sur les résidus des fonctions exprimées par des intégrales définies. In: *Oeuvres* (2) **7**, pp. 393-430.

___ (1829). Lecons sur le calcul différentiel. In: *Oeuvres* (2) **4**, pp. 263-609.

___ (1830). Sui metodi analitici. In: *Oeuvres* (2) **15**, pp. 149-181.

___ (1831). Extrait du mémoire présenté à l'Académie de Turin le 11 octobre 1831 par M. Augustin Cauchy, membre de l'Institut de France. In: *Oeuvres* (2) **15**, pp. 262-411.

___ (1833). Résumés analytiques. In: *Oeuvres* (2) **10**, pp. 5-184.

___ (1834). Sulla meccanica celeste e sopra un nuovo calcolo chiamato calcolo dei limiti. Trans. G. Piola and P. Frisiani. *Opuse. mat. fis.* 2, pp. 1-84, 133-202, 261-316.

___ (1835). Mémoire sur l'intégration des équations différentielles. In: *Oeuvres* (2) **11**, pp. 399-465.

___ (1837). Extrait d'une lettre à M. Coriolis. In: *Oeuvres* (1) **4**, pp. 38-42.

___ (1840-47). Exercices d'analyse et de physique mathématique. In: *Oeuvres* (2) **11-14**.

___ (1841a). Resumé d'une Mémoire sur la Mécanique céleste et sur un nouveau calcul appelé *calcul des limites*. In: *Oeuvres* (2) **12**, pp. 48-112.

___ (1841b). Note sur la nature des problèmes que présente le calcul intégral. In: *Oeuvres* (2) **12**, pp. 263-271.

___ (1843a). Rapport sur un Mémoire de M. Laurent, qui a pour titre: Extension du théorème de M. Cauchy ... In: *Oeuvres* (1) **8**, pp. 115-117.

___ (1843b). Note sur le développement des fonctions en series convergentes ordonnées suivant les puissances entières des variables. In: *Oeuvres* (1) **8**, pp. 117-120.

___ (1844a). Mémoire sur les fonctions continues. In: *Oeuvres* (1) **8**, pp. 145-160.

___ (1844b). Mémoire sur quelques propositions fondamentales du calcul des résidus et sur la théorie des intégrales singulières. In: *Oeuvres* (1) **8**, pp. 366-375.

___ (1846a). Sur les integrals qui s'étendent à tous les points d'une courbe fermée. In: *Oeuvres* (1) **10**, pp. 70-74.

___ (1846b). Mémoire sur les intégrales dans lesquelles la fonction sous le signe ∫ change brusquement de valeur. In: *Oeuvres* (1) **10**, pp. 133-143.

___ (1846c). Considérations nouvelles sur les intégrales definies qui s'étendent à tous les points d'une courbe fermée, et sur celles qui sont prises entre des limites imaginaires. In: *Oeuvres* (1) **10**, pp. 153-168.

___ (1847a). Mémoire sur une nouvelle théorie des imaginaires, et sur les racines symboliques des équations et des équivalences. In: *Oeuvres* (1) **10**, pp. 312-323.

___ (1847b). Mémoire sur la théorie des équivalences algébriques substituée à la théorie des imaginaires. In: *Oeuvres* (2) **14**, pp. 93-120.

___ (1847c). Mémoire sur les quantités géométriques. In: *Oeuvres* (2) **14**, pp. 175-202.

___ (1847d). Sur les fonctions des quantités géométriques. In: *Oeuvres* (2) **14**, pp. 359-365.

___ (1849a). Sur les quantités géométriques, et sur une méthode nouvelle pour la résolution des équations algébriques du degré quelconque. In: *Oeuvres* (1) **11**, pp. 152-160.

___ (1851a). Mémoire sur les fonctions irrationnelles. In: *Oeuvres* (1) **11**, pp. 292-300.

___ (1851b). Sur les fonctions de variables imaginaires. In: *Oeuvres* (1) **11**, pp. 301-304.

___ (1851c). Rapport sur un Mémoire présenté à l'Académie par M. Puiseux .. In: *Oeuvres* (1) **11**, pp. 325-335.

___ (1851d). Rapport sur un Mémoire présenté à l'Académie par M. Hermite, et relatif aux fonctions à double periode. In: *Oeuvres* (1) **11**, pp. 363-373.

___ (1851e). Note de M. Augustin Cauchy relative aux observations présentées à l'Académie par M. Liouville. In: *Oeuvres* (1) **11**, pp. 373-376.

___ (1853). Note sur les séries convergentes dont les divers termes sont des fonctions continues d'une variable réelle ou imaginaire, entre des limites données. In: *Oeuvres* (1) **12**, pp. 30-36.

___ (1855). Sur les compteurs logarithmiques. In: *Oeuvres* (1) **12**, pp. 285-292.

___ (1857). Théorie nouvelle des résidus. In: *Oeuvres* (1) **12**, pp. 433-444.

___ (1981). *Equations différentielles ordinaires*. C. Gilain (ed.). Paris.

Cavaillès, J. (1938). *Remarques sur la formation de la théorie abstraite des ensembles*. Paris.

Chasles, M. (1837). *Aperu historique sur l'origine et le développement des méthodes en géométrie*, Paris (2nd ed. 1875).

Clagett, M. (1959). *The Science of Mechanics in the Middle Ages*. Madison.

Condorcet, M.-A. N. (1765). *Traité du calcul intégral*. Paris.

Cooke, R. (1984). *The Mathematics of Sonya Kovalevskaya.* New York.

Cox, D. A. (1984). The arithmetic-geometric mean of Gauss. *L'enseign. math.* (2) **30**, (3-4), pp. 275-330.

Cramer, G. (1745). *Virorum celeberr. Got. Gul. Leibnitii et Joh. Bernoulli commercium philosophicum et mathematicum*, 2 Vols. Lausanne and Geneva.

Crombie, A. C. (1959-61). *Augustine to Galileo*, 2 Vols. London.

Crowe, M. (1967). *A History of Vector Analysis.* Notre Dame.

Dahan, A. (1985). La mathématisation des théories de l'élasticité par A. L. Cauchy et les débats dans la physique mathématique francaise (1800-1840). Prépublication scientifique. Paris.

D'Alembert, J.-B. le R. (1747a). *Traité de dynamique.* Paris.

___ (1747b). Réflexions sur la cause générale des vents. Pièce qui a remportée le prix proposé par l'Academie de Berlin pour l'année 1746. Berlin and Paris.

___ (1748). Recherches sur le calcul intégral. *Mém. Acad. Sci. Berlin* **2**, pp. 182-224.

___ (1749a). Recherches sur la courbe que forme une corde tendue mise en vibration. *Mém. Acad. Sci. Berlin* **3**, pp. 214-219.

___ (1749b). Suite des recherches sur la courbe ... *ibid.*, pp. 220-253.

___ (1752a). Essai d'une nouvelle théorie de la résistance des fluides. Paris.

___ (1752b). Addition au mémoire sur la courbe ... *ibid.* **6**, pp. 355-360.

___ (1761a). Sur les logarithms des quantités negatives. *Opusc. math.* **1**, pp. 180-230.

___ (1761b). Recherches sur les vibrations des cordes sonores. *Opusc. math.* **1**, pp. 1-73.

___ (1780). Sur les fonctions discontinues. *Opusc. math.* **8**, pp. 372-377.

Darboux, J.-G. (1875). Sur les fonctions discontinues. *Ann. Sci. École Norm. Super.* (2) **4**, pp. 57-112.

Dauben, J. W. (1971). The Trigonometric Background to Georg Cantor's Theory of Sets. *Arch. Hist. Exact Sci.* **7**, pp. 181-216.

___ (1979). *Georg Cantor* ... Cambridge, Mass.

___ (1981a). (ed.) *Mathematical Perspectives* ... New York.

___ (1981b). Progress of Mathematics in the Early 19th Century: Context, Contents and Consequences. In: Bolzano, (1981b), pp. 223-260.

Dedekind, J. W. R. (1930-32). *Gesammelte mathematische Werke*, 3 Vols. R. Fricke, E. Nöether and O. Ore (eds.). Braunschweig.

___ (1872). Stetigkeit und irrationale Zahlen (5th ed. 1927). In: *Werke* **3**, pp. 315-334.

___ (1888). Was sind und was sollen die Zahlen? (6th ed. 1930). In: *Werke* **3**, pp. 335-391.

___ (1876). Bernhard Riemann's Lebenslauf. In: Riemann, *Werke*, **pp.** 541-558.

Delambre, J.-B. (1810). Rapport historique sur le progrés des sciences ... Paris.

De Prony, R. (1798). Sur le principe des vitesses virtuelles et la décomposition des mouvements circulaires. *J. École Polytech.* **2** [Cahier 5], pp. 191-208.

Dhombres, N. (1981-2). Une communauté scientifique en 1800? *Sci. Tech. en Perspective*, Vol. I. Univ. de Nantes.

Dikstein, S. (1899). Zur Geschichte der Prinzipien der Infinitesimalrechnung. Kritiker der "Théorie des fonctions analytiques" von Lagrange. *Abh. Gesch. Mat.* **9**, pp. 65-79.

Dieudonné, J. (1974). *Cours de géometrie algébrique*, 2 Vols. Paris.

_____ (1978). *Abrégé d'histoire des mathématiques 1700-1900*, 2 Vols. Paris.

_____ (1981). *History of functional analysis.* Amsterdam.

Dini, U. (1953-59). *Opere.* 5 Vols. Roma.

_____ (1873). Sull'integrazione dell'equazione $\Delta u = 0$. *Ann. Mat. pura appl.* (2) **5**, pp. 305-345. In: *Opere* 2, pp. 264-310.

_____ (1876). Su una funzione analoga a quella di Green. *Atti Acad. Lincei* (2) **3**:2, pp. 129-137. In: *Opere* 2, pp. 311-322.

_____ (1878). *Fondamenti per la teorica delle funzioni di variabili reali.* Pisa.

_____ (1880). *Serie di Fourier e altre rappresentazioni analitiche* ... Pisa.

Dirichlet, P. G. L. (1889-97). *G. Lejeune Dirichlet's Werke*, 2 Vols. L. Kronecker and L. Fuchs (eds.). Repr. New York 1969.

_____ (1829). Sur la convergence des séries trigonométriques qui servent à représenter une fonction arbitraire entre les limites données. In: *Werke* 1, pp. 117-132.

_____ (1837a). Über die Darstellung ganz willkürlicher Funktionen durch Sinus-und Cosinusreihen. In: *Werke* 1, pp. 133-160.

_____ (1837b). Sur les séries dont le terme général dépende de deux angles ... In: *Werke* 1, pp. 283-306.

_____ (1863). Démonstration d'un théorème d'Abel. In: *Werke* 2, pp. 305-306.

_____ (1876). *Vorlesungen über die im umgekehrten Verhältniss des Quadrats der Entfernung wirkenden Kräfte,* F. Grube (ed.). Leipzig (2nd ed. 1887).

_____ (1897). Briefwechsel zwischen Lejeune Dirichlet und Gauss. In: *Werke* 2, pp. 373-387.

Du Bois Reymond P. (1870). Antrittsprogramm, enthaltend neue Lehrsätze über die Summen unendlicher Reihen ... Berlin. Repr. in: Ostwald's Klassiker, P. E. B. Jourdain (ed.), **185**, pp. 3-42. Leipzig 1912.

_____ (1874). Beweis, daß die Koeffizienten der trigonometrischen Reihe ... *Abh. bayer. Akad. Wiss.*, II Kl., Vol. 12, Pt. 1, pp. 117-66. Repr. in: Ostwald's Klassiker, P. E. B. Jourdain (ed.), **185**, pp. 43-91. Leipzig 1912.

___ (1876). Untersuchungen über die Konvergenz und Divergenz der Fourierschen Darstellungsformeln. *Abh. bayer. Akad. Wiss.*, II Kl., Vol. 12, Pt 2, pp. i-xxiv, 1-102. Repr. in: Ostwald's Klassiker, P. E. B. Jourdain (ed.), **186**, Leipzig 1913.

___ (1879). Erläuterungen zu den Anfangsgründen der Variationsrechnung. *Math. Annalen* **15**, pp. 283-314.

Dugac, P. (1973). Eléments d'analyse de Karl Weierstrass. *Arch. Hist. Exact Sci.* **10**, pp. 41-176.

___ (1976a). Problèmes d'histoire de l'analyse mathématique au XIX siècle. Cas de Weierstrass et Dedekind. *Historia Math.* **3**, pp. 5-19.

___ (1976b). *Richard Dedekind et les fondements des mathématiques.* Paris.

___ (1978). Fondements de l'analyse. In: Dieudonné (1978), Vol. 1, pp. 335-392.

Duhamel, M. (1860). *Eléments de calcul infinitesimal*, 2nd ed., 2 Vols. Paris.

Eccarius, W. (1976). A. L. Crelle als Herausgeber des Crellschen Journal. *J. reine angew. Math.* **186-7**, pp. 5-25.

Edwards, H. M. (1974). *Riemann's zeta function*, New York.

Enros, P. C. (1983). The Analytical Society (1812-1813): Precursor of the Renewal of Cambridge Mathematics. *Historia Math.* **10**, pp. 24-47.

Euler, L. (1911). *Opera omnia.* Ser. 1, 29 Vols.; Ser. 2, 31, Vols.; Ser. 3, 12 Vols.; Ser. 4, in progress. Leipzig/Berlin/Zurich/Basel.

___ (1734a). De infinitis curvis eiusdem generis, seu methodus inveniendi aequationes ... In: *Opera* (1) 22, pp. 36-56.

___ (1734b). Additamentum ad dissertationem de infinitis curvis eiusdem generis. In: *Opera* (1) 22, pp. 57-75.

___ (1748). Introductio in analysin infinitorum. In: *Opera* (1) 8-9.

___ (1749). De vibratione chordarum exercitatio. In: *Opera* (2) 10, pp. 50-62.

___ (1750). Sur la vibration des cordes. In: *Opera* (2) 10, pp. 63-77.

___ (1751). Recherches sur les racines imaginaires des equations. In: *Opera* (1) 6, pp. 78-147.

___ (1752). Principia motus fluidorum. In: *Opera* (2) 12, pp. 133-168.

___ (1755a). Remarques sur les mémoires précédens de M. Bernoulli. In: *Opera* (2) 10, pp. 233-254.

___ (1755b). Institutiones calculi differentialis. In: *Opera* (1) 10.

___ (1767a). Eclaircissements plus detaillés sur la generation et la propagation du son, et sur la formation de l'eco. In: *Opera* (3) 1, pp. 540-567.

___ (1767b). De usu functionum discontinuarum in analysi. In: *Opera* (1) 23, pp. 74-91.

___ (1767c). Sur le mouvement d'une corde, qui au commencement n'a été ébranlée que dans une partie. In: *Opera* (2) 10, pp. 426-450.

___ (1794). Specimen transformationis singularis serierum. In: *Opera* (1) 16(2), pp. 41-55.

Forsyth, A. R. (1918). *Theory of functions of a complex variable*, 2 Vols. Cambridge. Repr. New York 1965.

Fourier, J. B. J. (1888-90). *Oeuvres*, 2 Vols. G. Darboux (ed.). Paris.

___ (1798). Mémoire sur la statique, contentant la démonstration du principe des vitesses virtuelles, ... *J. Ecole polytech.* 2 [Cahier 5], pp. 20-60. In: *Oeuvres* 2, pp. 475-521.

___ (1822). Théorie analytique de la chaleur. In: *Oeuvres* 1.

Fox, R. (1974). The Rise and Fall of Laplacian Physics. *Hist. Stud. Phys. Sci.* 4, pp. 81-136.

Fraser, C. (1985). J. L. Lagrange's changing approach to the foundations of the calculus of variations. *Arch. Hist. Exact Sci.* 32, pp. 151-191.

Freudenthal, H. (1971). Did Cauchy plagiarize Bolzano? *Arch. His. Exact Sci.* 7, pp. 375-392.

Fuss, P. H. (1843). *Correspondance mathématique et physique de quelques géomètres du 18ème siècle*, Vol. I. St. Petersburg.

Galois, E. (1962). *Écrits et mémoires mathématiques d'Évariste Galois.* R. Bourgne and J.-P. Azra (eds.). Paris.

___ (1830). Démonstration d'un théorème d'analyse. In: *Écrits*, pp. 382-385.

Gauss, C. F. (1863-1933). *Werke*, 12 Vols. Leipzig.

___ (1799). Demonstratio nova theorematis omnem functionem algebraicam rationalem integram unius variabilis in factores reales primi vel secundi gradus resolvi posse. In: *Werke* 3, pp. 3-30.

___ (1811). Brief an Bessel. In: *Werke* 8, pp. 90-92.

___ (1813a). Disquisitiones generales circa seriem infinitam ... In: *Werke* 3, pp. 123-162.

___ (1813b). Theoria attractionis corporum spheroidicorum ellipticorum homogeneorum methodo nova tractata. In: *Werke* 5, pp. 3-22.

___ (1816a). Demonstratio nova et altera theorematis omnem functionem algebraicam ... In: *Werke* 3, pp. 51-56.

___ (1816b). Theorematis de risolubilitate ... In: *Werke* 3, pp. 57-64.

___ (1825). Allgemeine Auflösung der Aufgabe: die Teile einer gegebenen Fläche abzubilden ... In: *Werke* 4, pp. 189-216.

___ (1827). Disquisitiones generales circa superficies curvas. In: *Werke* 4, pp. 217-258.

___ (1831). Anzeige zu Theoria residuorum ... [1832]. In: *Werke* 2, pp. 169-178.

___ (1832). Theoria residuorum biquadraticorum, Commentatio secunda. In: *Werke* 2, pp. 93-148.

___ (1839). Allgemeine Lehrsätze in Beziehung auf die im verkehrten Verhältnisse wirkenden Anziehungs - und Abstossungs Kräfte. In: *Werke* 5, pp. 194-242.

___ (1880). *Briefwechsel zwischen Gauss und Bessel.* Leipzig.

___ (1981). Mathematisches Tagebuch 1796-1814. In: Ostwald's Klassiker, H. Wussing (ed.), 256. Leipzig.

Geppert, H. (1933). Über Gauss' Arbeiten zur Mechanik und Potentialtheorie. In: Gauss, *Werke* 10(2), Abh. 7, p. 61.

Gerver, J. (1970). The Differentiability of the Riemann Function at Certain Rational Multiples of π. *Amer. J. Math.* 92, pp. 33-55.

Gigli, D. (1924-7). Numeri complessi a due e più unità. In: F. Enriques (ed.), *Questioni riguardanti le matematiche elementari*, 2 Vols. Bologna. Vol. I, pp. 133-270.

Goldstine, H. H. (1980). *A History of the Calculus of Variations from the 17th through the 19th Century*. New York.

Goursat, E. (1884). Démonstration du théorème de Cauchy. *Acta Math.* **4**, pp. 197-200.

____ (1900). Sur la définition generale des fonctions analytiques, d'après Cauchy. *Trans. Amer. Math. Soc.* **1**, pp. 14-16.

Grabiner, J. V. (1978). The Origins of Cauchy's Theory of the Derivative. *Historia Math.* **5**, pp. 379-409.

____ (1981). *The Origins of Cauchy's Rigorous Calculus*. Cambridge, Mass.

Grattan-Guinness, I. (1970a). Bolzano, Cauchy and the 'New Analysis' of the Early Nineteenth Century. *Arch. Hist. Exact Sci.* **6**, pp. 372-400.

____ (1970b). *The Development of the Foundations of Mathematical Analysis from Euler to Riemann*. Cambridge, Mass.

____ (1972). (with J. R. Ravetz). *Joseph Fourier, 1768-1830*. Cambridge, Mass.

____ (1975). Preliminary Notes on the Historical Significance of Quantification ... *Historia Math.* **2**, pp. 475-488.

____ (1980). (ed.) *From the Calculus to Set Theory, 1630-1910; An Introductory History*. London.

____ (1981). Mathematical Physics in France, 1800-1840. In: Dauben (1981a), pp. 95-138.

____ (1983). Euler's Mathematics in French Science, 1795-1815. In: *Leohard Euler, 1707-1783*, pp. 395-408. Basel.

Gray, J. J. (1984). Fuchs and the theory of differential equations. *Bull. Am. Math. Soc.* **10**, pp. 1-26.

Green, G. (1871). *Mathematical Papers*. N. Ferrers (ed.). London. Repr. 1970.

____ (1828). An Essay on the Application of Mathematical Analysis to Theories of Electricity and Magnetism. In: *Math. Papers*, pp. 3-115.

____ (1833). On the determination of the exterior and interior attraction of ellipsoids of variable density. In: *Math. Papers*, pp. 185-222.

Hadamard, J. (1968). *Oeuvres*, 2 Vols. Paris.

____ (1892). Essai sur l'etudé des fonctions données par leur développements de Taylor. In: *Oeuvres* 1, pp. 7-92.

____ (1909). La géometrie de situation et son rôle en mathématiques. In: *Oeuvres* 2, pp. 805-827.

____ (1923). *Lectures on Cauchy's problem*. New Haven.

____ (1926). (with S. Mandelbrojt). *La série de Taylor et son prolongement analytique*. Paris.

Hamilton, W. R. (1931-67). *The Mathematical Papers of Sir William Rowland Hamilton*, 3 Vols. H. Halberstam and R. E. Ingram (eds.). Cambridge.

___ (1834-5). On a general method in dynamics. *Phil. Trans. Math. Soc. London*, 1834, pp. 247-308; 1835, pp. 95-144. In: *Math. Papers 2*, pp. 103-211.

___ (1837). Theory of conjugate functions, or algebraic couples; with a preliminary and elementary essay on algebra as the science of pure time. *Trans. Irish Acad.* 17, pp. 293-422. In: *Math. Papers 3*, pp. 3-96.

Hankel, H. (1867). *Theorie der complexen Zahlensysteme* ... Leipzig.

___ (1870). Untersuchungen über die unendlich oft oszillierenden und unstetigen Functionen. Gratulationsprogramm der Tübinger Universität. Repr. in: *Math. Annalen* 20 (1882), pp. 63-112.

___ (1871). Grenze. *All. Enzykl. der Wiss. und Künste*. Sect. 1, Teil 19, pp. 185-211. Leipzig.

Hankins, T. L. (1970). *Jean D'Alembert: Science and the Enlightenment*. Oxford.

Hardy, G. H. (1966-79). *Collected Papers*. 7 Vols. Cambridge.

___ (1910). Orders of Infinity: The Infinitärkalkül of Paul Du Bois Reymond. *Cambridge Tracts in Math.* 12.

___ (1913). Oscillating Dirichlet's Integrals: An Essay on the Infinitärkalkül of Paul du Bois Reymond. In: *Collected Papers 4*, pp. 280-341.

___ (1918). Sir George Stokes and the Concept of Uniform Convergence. In: *Collected Papers*, 7, pp. 505-513.

Harnack, A. (1888). Über Cauchy's zweiten Beweis für die Konvergenz der Fourierschen Reihen ... *Math. Annalen* 32, pp. 175-202.

Hawkins, T. (1970). *Lebesgue's Theory of Integration: Its Origins and Developments*. New York.

Heine, E. (1870). Über trigonometrische Reihen. *J. reine angew. Math.* 71, pp. 353-365.

___ (1872). Die Elemente der Funktionenlehre. *J. reine angew Math.* 72, pp. 172-188.

Helly, E. (1921). Über Systeme linearer Gleichungen mit unendlich vielen Unbekannten. *Monatschr. Math. Phys.* 31, pp. 60-91.

Herivel, J. (1975). *Joseph Fourier, The Man and the Physicist*. Oxford

Hilbert, D. (1932-35). *Gesammelte Abhandlungen*. D. Hilbert et al. (eds). Berlin.

___ (1900). Über das Dirichletsche Prinzip. In: *Gesam. Abh.* 3, pp. 10-14.

___ (1904). Über das Dirichletsche Prinzip. In: *Gesam Abh.* 3, pp. 15-37.

Hille, E. (1963). *Analytic Function Theory*, 2 Vols. New York.

Hofmann, J. E. (1949). *Leibniz in Paris, 1672-1676*. München. Eng. transl. Cambridge 1974.

Jacobi, C. G. J. (1881-91). *Gesammelte Werke*, 7 Vols. Berlin.

___ (1835). De Functionibus duarum variabilium quadrupliciter periodicis, quibus theoria transcendentium Abelianarum innititur. In: *Ges. Werke* 2, pp. 23-50.

Jourdain, P. E. B. (1905). The Theory of Functions with Cauchy and Gauss. *Bibl. Math.* (3) 6, pp. 190-207.

___ (1913). The Origins of Cauchy's Conceptions of a Definite Integral and of the Continuity of a Function. *Isis* 1, pp. 661-703.

___ (1917). The Influence of Fourier's Theory of the Conduction of Heat on the Development of Pure Mathematics. *Scientia* 22, pp. 245-254.

Klein, F. (1921-23). *Gesammelte mathematische Abhandlungen*, 3 Vols. R. Fricke et.al. (ed.). Berlin.

___ (1882). Über Riemanns Theorie der algebraischen Funktionen und ihrer Integrale. In: *Gesam. math. Abh.* 3, pp. 499-573.

___ (1894). Riemann und seine Bedeutung für die Entwicklung der modernen Mathematik. In: *Gesam. math. Abh.* 3, pp. 482-497.

___ (1895). Über die Arithmetisierung der Mathematik. In: *Gesam. math. Abh.* 2, pp. 232-240.

___ (1926). *Vorlesungen über die Entwicklung der Mathematik im 19. Jahrhundert*, R. Courant and O. Neugebauer (eds.). Berlin. Repr. New York 1967.

Kline, M. (1972). *Mathematical Thought from Ancient to Modern Times.* New York.

Kneser, A. (1900). Variationsrechunung. *Enzykl. math. Wiss.* [Art. II A 8], Vol. 2, Teil 1, 1, pp. 572-625.

Knopp, K. (1947). *Theorie und Anwendung der unendlichen Reihen.* Berlin.

Kowalewski, S. (1975). Zur Theorie der partiellen Differentialgleichungen. *J. reine angew. Math.* 80, pp. 1-32.

Kronecker, L. (1895-1931). *Werke*, K. Hensel (ed.). 5 Vols. Leipzig.

___ (1887). Über den Zahlbegriff. In: *Werke* 3, pp. 249-274.

___ (1894). *Vorlesungen über die Theorie der einfachen und vielfachen Integrale.* E. Netto (ed.). Leipzig.

Krull, W. (1971). Zahlen und Grössen, Dedekind und Eudoxos. *Mitteil. Math. Sem. Giessen* 90, pp. 29-47.

Kummer, E. E. (1975). *Collected Papers*, 2 Vols. A. Weil (ed.). Berlin-Heidelberg-New York.

___ (1836). Über die hypergeometrische Reihe ... In: *Collected Papers* 2, pp. 75-166.

Lacroix, S. F. (1797-1800). *Traité du calcul différentiel et du calcul intégral*, 3 Vols. Paris. (2nd ed. Paris 1810; 3rd ed. Paris 1837).

___ (1858). *Traité élémentaire du calcul différentiel et du calcul intégral*, 6th ed. Paris.

Lagrange, J. L. (1867-1892). *Oeuvres de Lagrange*, 14 Vols. J.-A. Serret, G. Darboux, *et al.* (eds.). Paris.

___ (1759). Recherches sur la nature et la propagation du son. In: *Oeuvres* 1, pp. 39-148.

___ (1760-61). Nouvelles recherches sur la nature et la propagation du son. In: *Oeuvres* 1, pp. 151-332.

___ (1772). Sur une nouvelle espèce de calcul relatif à la différentiation des quantités variables. In: *Oeuvres* 3, pp. 439-470.

___ (1773). Sur l'équation séculaire de la lune. In: *Oeuvres* 6, pp. 35-399.

___ (1774). Sur la forme des racines imaginaires des équations. In: *Oeuvres* 3, pp. 477-516.

___ (1797). *Théorie des fonctions analytiques*, Paris (2nd ed. 1813).

___ (1797-8). Traité de la résolution des équations numériques de tous les degrés, Paris. In: *Oeuvres* 8 (2nd ed. 1808).

___ (1798). Sur le principe des vitesses virtuelles. *J. École Polytech.* 2 [Cahier 5], pp. 115-118. In: *Oeuvres* 7, pp. 317-321.

___ (1806). Lecons sur le calcul des fonctions, 2nd ed. *J. École Polytech.* 6 [Cahier 13]. In: *Oeuvres* 10.

___ (1811-15). Mécanique analytique, 2nd ed. In: *Oeuvres* 12 (1st ed. 1788).

___ (1813). Théorie des fonctions analytiques, 2nd ed. Paris. In: *Oeuvres* 9. [N.B. This 2nd ed., which is that printed in the *Oeuvres*, differs from the 1st edition of 1797 in many respects.]

Lakatos, I. (1971). History of Science and its Rational Reconstructions. In: R. C. Buck and R. S. Cohen (eds.), *Proceedings of the Philosophy of Science Association*, 1970, Boston, pp. 91-135.

___ (1976). *Proofs and Refutations*. Cambridge.

Landen, J. (1764). *The Residual Analysis*. London.

Laplace, P. S. (1878-1912). *Oeuvres*, 14 Vols. Paris.

___ (1782). Théorie des attractions des sphéroïdes et de la figure des planetes. In: *Oeuvres* 10, pp. 339-419.

___ (1787). Mémoire sur la théorie de l'anneau de Saturne. In: *Oeuvres* 11, pp. 273-292.

___ (1799-1825). Traité de mécanique céleste. In: *Oeuvres* 1-5.

___ (1784). (with Lavoisier.) Mémoire sur la chaleur. In: *Oeuvres* 10, pp. 147-200.

Laurent, P. A. (1863). Mémoire sur la théorie des imaginaires, sur l'équilibre des temperatures et sur l'équilibre d'elasticité. *J. École Polytech.* 23, pp. 75-204.

Lebesgue, H. (1972-73). *Oeuvres scientifiques*, 5 Vols. Genève.

___ (1904). *Lecons sur l'integration et la recherche des fonctions primitives*, (2nd ed. 1928). Paris.

___ (1905). Recherches sur la convergence des séries de Fourier. In: *Oeuvres scientifiques*, 3, pp. 181-210.

___ (1906). *Lecons sur les séries trigonométriques*. Paris.

Legendre, A. M. (1782). Recherches sur l'attraction des spheroides homogenes. *Mem. Sav. Etrang.* 10, pp. 411-434.

___ (1808). *Essai sur la théorie des nombres*. Paris.

___ (1811-17). *Exercices du calcul intégral*, 3 Vols. Paris.

Lindt, R. (1904). Das Prinzip der virtuellen Geschwindigkeit. *Abh. Gesch. Math. Wiss.* (4) 8, pp. 145-196.

Liouville, J. (1839-40). Sur le principe fondamental de la théorie des équations algébriques. *J. math. pures et appl.* **4**, pp. 501-507; **5**, pp. 31-34.

____ (1844). Remarques (relatives 1° à des lignes géodesiques 2° à des fonctions doublement périodiques) à l'occasion d'une note de M. Chasles. *Comptes Rendus Acad. Sci. Paris* **19**, pp. 1261-1263.

Lipschitz, R. (1864). De explicatione per series trigonometricas ... *J. reine angew. Math.* **63**, pp. 296-308.

Listing, J. B. (1847). Vorstudien zur Topologie. *Göttinger Studien*, pp. 811-875.

Lützen, J. (1983). Sturm and Liouville's work on ordinary differential equations. The emergence of Sturm-Liouville theory. *Arch. Hist. Exact Sci.* **29**, pp. 309-376.

Maclaurin, C. (1742). *Treatise on Fluxions*, 2 Vols. Edinburgh.

Manning, K. R. (1975). The Emergence of the Weierstrassian Approach to Complex Analysis. *Arch. Hist. Exact Sci.* **14**, pp. 298-383.

Markusevic, A. I. (1955). *Skizzen zur Geschichte der analytischen Funktionen.* Berlin.

____ (1978). Some topics in the history of the theory of analytic functions in the 19th century. Unpublished trans. of Russian orig. in: *Proceedings of the International Congress of Mathematicians,* Helsinki, 1978, Vol. 2.

Mathews, J. (1978). William Rowan Hamilton's paper of 1837 on the arithmetization of analysis. *Arch. Hist. Exact Sci.* **19**, pp. 177-200.

Menchoff, D. (1936). *Les conditions de monogeneité.* Paris.

Méray, C. (1872). *Nouveau précis d'analyse infinitesimale.* Paris.

Métivier, M. (1981). (with P. Costabel and P. Dugac) (eds.). *Siméon-Denis Poisson et la science de son temps.* Palaiseau.

Mittag-Leffler, G. (1884). Sur la répresentation analytique des fonctions monogènes... *Acta. Math.* **4**, pp. 1-79.

____ (1923). Weierstrass et Sonja Kowalewsky. *Acta Math.* **39**, pp. 133-198.

Moigno, F. N. M. (1840-44). *Lecons de calcul differentiel et de calcul intégral,* 2 Vols. Paris.

Monna, A. F. (1973a). *Functional Analysis in Historical Perspective.* Utrecht.

____ (1973b). Hermann Hankel. *Nieuw Archief voor Wiskunde* (3) **21**, pp. 64-87.

____ (1975). *Dirichlet's Principle. A Mathematical Comedy of Errors.* Utrecht.

____ (1979). Evolution des problèmes d'existence en analyse. Essais historiques. *Commun. math. Inst. Rijkuniv. Utrecht* **9**.

Mourey, C. V. (1828). *La vraie théorie des quantites négatives et des quantités prétendues imaginaire dediée aux amis de l'évidence.* Repr. 1861. Paris.

Neuenschwander, E. (1978a). Der Nachlass von Casorati in Pavia. *Arch. Hist. Exact Sci.* **19**, pp. 1-89.

___ (1978b). Riemann's Example of a Continuous 'Nondifferentiable' Function. *Math. Intelligencer* **1**, pp. 40-44.

___ (1978c). The Casorati-Weierstrass theorem. (Studies in the history of complex function theory I) *Historia Math.* **5**, pp. 139-166.

___ (1980). Riemann und das "Weierstrassche" Prinzip der analytischen Fortsetzung durch Potenzenreihen. *Jahresber. Deutsch. Math.-Verein* **82**, pp. 1-11.

___ (1981a). Studies in the History of Complex Function Theory II: Interactions among the French school, Riemann and Weierstrass. *Bull. Amer. Math. Soc.* **5**, pp. 87-105.

___ (1981b). Über die Wechselwirkungen zwischen der französischen Schule, Riemann und Weierstrass. Eine Übersicht mit zwei Quellenstudien. *Arch. Hist. Exact Sci.* **24**, pp. 221-255.

Neumann, C. (1865). *Das Dirichlet'sche Prinzip in seiner Anwendung auf die Riemann'schen Flächen.* Leipzig.

___ (1877). *Über das logarithmische und Newton'sche Potential.* Leipzig.

Nový, L. (1973). *Origins of Modern Algebra.* Leiden.

Ore, O. (1959). *Niels H. Abel, Mathematician Extraordinary.* New York.

Osgood, W. F. (1901). Allgemeine Theorie der analytischen Funktionen a) einer und b) mehrerer complexen Grössen. *Enzyk. math. Wiss.* [Art II B 1], Vol. II, Teil 2, pp. 1-114.

Ostrowski, A. (1927). Über den ersten und vierten Gauss'schen Beweis des Fundamentalsatzes der Algebra. In: Gauss, *Werke,* 10(2), Abh. 3, pp. 1-18.

Petrova, S. S. (1974). Sur l'histoire des démonstrations analytique du théorème fondamental de l'algèbre. *Historia Math.* **1**, pp. 255-261.

Pfeiffer, J. (1983). Joseph Liouville (1809-1882): ses contributions à la théorie des fonctions d'une variable complexe. *Rev. d'Hist. Sci. Appl.* **36**, pp. 209-248.

Picard, J. (1879). Sur une propriété des fonctions entières. *Comptes rendus Acad. Sci. Paris* **88**, pp. 1024-1027.

___ (1893). *Traité d'analyse,* Vol. 2. Paris.

Pierpont, J. (1928). Mathematical Rigor, Past and Present. *Bull. Amer. Math. Soc.* **34**, pp. 23-53.

Pincherle, S. (1880). Saggio di una introduzione alla teoria delle funzioni analitiche secondo i principi del prof. C. Weierstrass. *Giorn. matem.* **18**, pp. 178-154, 317-357.

Poincaré, H. (1916-56). *Oeuvres,* 11 Vols. Paris.

___ (1887). Sur le problème de la distribution électrique. In: *Oeuvres* **9**, pp. 15-17.

___ (1890). Sur les équations aux derivées partielles de la physique mathématique. In: *Oeuvres* **9**, pp. 28-113.

___ (1897). La méthode de Neumann et le problème de Dirichlet. In: *Oeuvres* **9**, pp. 202-272.

Poinsot, L. (1806). Théorie générale de l'équilibre et du mouvement des systèmes. *J. École Polytech.* 6 [Cahier 13], pp. 206-241.

Poisson, S. D. (1808). Mémoire sur la propagation de la chaleur dans les corps solides. In: Fourier, *Oeuvres* 2, pp. 213-221.

___ (1811). Second mémoire sur la distribution de élasticité à la surface des corps conducteurs. *Mém. Inst. France*, 2eme partie, pp. 163-274.

___ (1813). Remarques sur une équation qui se presente dans la théorie de l'attraction des sphéroides. *Bull. Soc. Phil.* **3**, pp. 388-392.

___ (1816). Mémoire sur la théorie des ondes. *Mem. Acad. Sci. Paris* (2) **1**, pp. 71-186.

___ (1820a). Sur les intégrales oles fonctions qui passent par l'infini et sur l'usage des imaginaires ... *J. École Polytech.* **11** [Cahier 18], pp. 295-341.

___ (1820b). Mémoire sur la Maniére d'exprimer les Fonctions par des Séries de quantités périodiques ... *J. École Polytech.* **11** [Cahier 18], pp. 417-489.

___ (1823a). Mémoire sur la Distribution de la Chaleur dans les Corps solides. *J. École Polytechn.* **12** [Cahier 19], pp. 1-144.

___ (1823b). Addition au Mémoire précédent ... *J. École Polytech.* **12** [Cahier 19], pp. 145-162.

___ (1823c). Suite du mémoire sur les intégrales définies ... *J. École Polytech.* **12**, [Cahier 19], pp. 404-509.

___ (1835). *Théorie mathématique de la chaleur.* Paris.

Pont, J.-C. (1974). *La topologie algébrique des origines à Poincaré.* Paris.

Pringsheim, A. (1898). Irrationalzahlen und Konvergenz unendlicher Prozesse. *Enzyk. math. Wiss.* [Art I A3], vol. I, Teil 1, pp. 47-146.

___ (1899). Grundlagen der allgemeinen Funktionenlehre. *Enzyk. math. Wiss.* [Art. II A 1], Vol. II, Teil 1, 1, pp. 1-53.

___ (1901). Ueber den Goursat'schen Beweis des Cauchy'schen Integralsatzes. *Trans. Amer. Math. Soc.* **2**, pp. 413-421.

Puiseux, M. V. (1850). Recherches sur les fonctions algébriques. *J. math. pures et appl.* **15**, pp. 365-480.

Remmert, R. (1983). Fundamentalsatz der Algebra. In: K. Lamotke (ed.), *Zahlen.* Berlin.

___ (1984). *Funktionenlehre I.* Berlin/New York.

Riemann, G. F. B. (1953). *The Collected Works of Bernhard Riemann*, 2nd ed. H. Weber *et al.* (eds.). With a Supplement, M. Noether & W. Wirtinger (eds.). New York.

___ (1851). Grundlagen für eine allgemeine Theorie der Funktionen einer veränderlichen complexen Grösse. In: *Works*, pp. 3-48.

___ (1857a). Beiträge zur Theorie der durch die Gauss'sche Reihe $F(\alpha,\beta,\gamma,x)$ darstellbaren Funktionen. In: *Works*, pp. 67-83.

___ (1857b). Theorie der Abelschen Funktionen. In: *Works*, pp. 88-144.

___ (1859). Über die Anzahl der Primzahlen unter einer gegebenen Grösse. In: *Works*, pp. 145-156.

___ (1861). Theorie der Funktionen einer veränderlichen complexen Grösse, besonders der Elliptischen und Abelschen. Ausgearbeitet von Herrn Ed. Schultz. Göttingen. Manuscript in the Akademie der Wissenschaften zu Berlin.

___ (1867a). Über die Darstellbarkeit einer Funktion durch eine trigonometrische Reihe. In: *Works*, pp. 227-271.

___ (1867b). Über die Hypothesen, welche der Geometrie zu Grunde liegen. In: *Works*, pp. 272-287.

___ (1869). *Partielle Differentialgleichungen und ihre Anwendung auf physikalische Fragen*, Vorlesungen, K. Hattendorff (ed.). Braunschweig (3rd ed. 1882, repr. (1938).

___ (1899). *Elliptische Funktionen*, Vorlesungen. H. Stahl (ed.). Leipzig.

Riesz, F. (1913). *Les systèmes d'équations à une infinité d'inconnus*. Paris.

Rootselaar, B. van. (1962-66). Bolzano's Theory of Real Numbers. *Arch. Hist. Exact Sci.* 2, pp. 168-180.

Ruffner, A. (1962-66). Reinterpretation of the Genesis of Newton's Law of Cooling. *Arch. Hist. Exact Sci.* 2, pp. 138-152.

Russ, S. B. (1980). A Translation of Bolzano's Paper on the Intermediate Value Theorem. *Historia Math.* 7, pp. 156-185.

Rychlik, K. (1961). La théorie des nombres réels dans un ouvrage posthume manuscrit de B. Bolzano. *Rev. d'Hist. Sci. Appl.* 14, pp. 313-327.

Sachse, A. (1880). Versuch einer Geschichte der Darstellung willkürlicher Funktionen einer Variablen durch trigonometrische Reihen. *Abh. Gesch. Math.* 3, pp. 229-276.

Saint Venant, A. Barré Comte de (1845). Mémoire sur les sommes et les différences géométriques, et sur leur usage pour simplifier la mécanique. *Comptes Rendus Acad. Sci. Paris* 21, pp. 620-625.

Schlesinger, L. (1933). Über Gauss' Arbeiten zur Funktionentheorie. In: Gauss, *Werke* 10(2), Abh. 2, pp. 222.

Scholz, E. (1980). *Geschichte des Mannigfaltigkeitsbegriffs von Riemann bis Poincaré*. Basel.

___ (1982). Herbart's Influence on Bernhard Riemann. *Historia Math.* 9, pp. 413-440.

Schwarz, H. A. (1890). *Gesammelte mathematische Abhandlungen*, 2 Vols. Berlin.

___ (1870a). Ueber einen Grenzübergang durch alternierendes Verfahren. In: *Gesam. math. Abh.* 2, pp. 133-143.

___ (1870b). Ueber die Integration der partiellen Differential-gleichung $\partial^2 u/\partial x^2 + \partial^2 u/\partial y^2 = 0$ unter vorgeschriebenen Grenz - und Unstetigkeitsbedingungen. In: *Gesam. math. Abh.* 2, pp. 144-171.

___ (1872). Zur Integration der partiellen Differentialgleichungen $\partial^2 u/\partial x^2 + \partial^2 u/\partial y^2 = 0$. In: *Gesam. math. Abh.* 2, pp. 175-210.

Sebestik, J. (1964). B. Bolzano et son memoire sur le theoreme fondamental de l'analyse. *Rev. d'Hist. Sci. Appl.* 17, pp. 129-135.

Seidel, P. L. (1847). Note über eine Eigenschaft der Reihen, welche discontinuirliche Funktionen darstellen. *Abh. bayer. Akad. Wiss.* 5, pp. 381-394. Repr. in: Ostwald's Klassiker, H. Liebmann (ed.), 116, pp. 35-45 (Leipzig 1900).

___ (1871). Über die Darstellung des Kreisbogens, des Logarithmus und des elliptischen Integrals ersten Art durch unendliche Producte. *J. reine angew. Math.* 73, pp. 273-291.

Speiser, A. (1952). Editor's Introduction. In: Euler, *Opera* (1) **25**, pp. vii-xxvi.

Springer, G. (1957). *Introduction to Riemann Surfaces*. Reading, Mass.

Stokes, G. G. (1849). On the Critical Values of the Sums of Periodic Series. *Trans. Cambridge Phil. Soc.* **8**, pp. 533-583.

Sturm, C. (1836). Mémoire sur une classe d'équations à differences partielles. *J. math. pures et appl.* 1, pp. 373-444.

Sylow, L. (1902). Les études d'Abel et ses découvertes. In: Abel 1902.

Terracini, A. (1957). Cauchy a Torino. *Rend. Sem. Mat. Torino*, pp. 159-203.

Thomson, W. (1848). Theorems with reference to the solution of certain partial differential equations. *Camb. and Dublin Math. Journal* 3, pp. 84-87. In: *Math. and Phys. Papers* 1, pp. 93-96.

Todhunter, I. and Pearson, K. (1960). *A History of the Theory of Elasticity*, 2 Vols. New York.

Tonelli, A. (1873-4). Osservazioni sulla teoria della connessione. *Atti Acc. Naz. Lincei* (2) **2**, pp. 594-602.

Tonelli, L. (1928). *Serie trigonometriche*. Bologna.

Torretti, R. (1978). *Philosophy of Geometry from Riemann to Poincaré.* Dordrecht/Boston.

Truesdell, C. A. (1954). Rational fluid mechanics, 1687-1765. In: Euler, *Opera* (2) **12**, pp. ix-cxxv.

_____ (1956). Introduction. In: Euler, *Opera* (2) **13**, pp. ix-cv.

_____ (1960). The Rational Mechanics of Flexible or Elastic Bodies, 1638-1788. In: Euler, *Opera* (2) **11**, Sect. 2.

_____ (1980). *The Tragicomical History of Thermodynamics, 1822-1854.* New York.

Tucciarone, J. (1973). The Development of the Theory of Summable Divergent Series from 1880 to 1925. *Arch. Hist. Exact Sci.* **10**, pp. 1-40.

Valson, C. A. (1868). *La vie et les travaux du Baron Cauchy*, 2 Vols. Paris.

Weierstrass, K. (1894-1927). *Mathematische Werke,* 7 Vols. Berlin.

_____ (1886). *Abhandlungen aus der Functionenlehre.* Berlin.

_____ (1841a). Darstellung einer analytischer Funktion einer complexen Veränderlichen, deren absoluten Betrag zwischen zwei gegebenen Grenzen liegt. In: *Werke* 1, pp. 117-120.

_____ (1841b). Zur Theorie der Potenzreihen. In: *Werke* 1, pp. 67-74.

_____ (1842). Definition analytischer Funktionen einer Veränderlichen vermittelst algebraischer Differentialgleichungen. In: *Werke* 1, pp. 75-85.

_____ (1854). Zur Theorie der Abel'schen Funktionen. In: *Werke* 1, pp. 133-152.

_____ (1870). Über das sogenannte Dirichlet'sche Princip. In: *Werke* 2, pp. 49-54.

_____ (1872). Über continuirliche Functionen eines reellen Arguments, ... In: *Werke* 2, pp. 71-74.

___ (1876). Zur Theorie der eindeutigen analytischen Funktionen. In: *Werke* 2, pp. 77-124.

___ (1880a). Zur Funktionenlehre. In: *Werke* 2, pp. 201-230.

___ (1880b). Über einen functiontheoretischen Satz des Herrn G. Mittag-Leffler. In: *Werke* 2, pp. 189-199.

Weil, A. (1976). *Elliptic Functions according to Eisenstein and Kronecker*. New York.

Wessel, G. (1897). *Essai sur la représentation analytique de la direction*. Copenhagen. Eng. trans. in D. E. Smith, *A Source Book in Mathematics*, pp. 55-66. New York 1929.

Weyl, H. (1913). *Die Idee der Riemannschen Fläche*. Leipzig. (3rd ed. Stuttgart 1955.) Eng. trans. *The Concept of a Riemann Surface*. Princeton 1955.

___ (1919). *Raum-Zeit-Materie*, 3rd ed. Berlin. Eng. trans. *Space, Time, Matter*. New York 1921, repr. 1952.

Whiteside, D. T. (1961). Patterns of Mathematical Thought in the Seventeenth Century. *Arch. Hist. Exact Sci.* 1, pp. 179-388.

Yushkevich, A. P. (1971). Lazare Carnot and the Competition of the Berlin Academy in 1786 on the Mathematical Theory of the Infinite. In: C. G. Gillispie (ed.), *Lazare Carnot Savant*. Princeton.

___ (1974). La notion de fonction chez Condorcet. In: R. Cohen *et al.* (eds.), *For Dirk Struik*, pp. 131-139. Dordrecht.

___ (1976). The Concept of Function up to the Middle of the 19th Century. *Arch. Hist. Exact Sci.* 16, pp. 37-85.

Zariski, O. (1926). (trans.) R. Dedekind, *Essenza e significato dei numeri*; *Continuità e numeri irrazionali*. Roma.

Zermelo, E. and Hahn, H. (1904). Weiterentwicklung der Variationsrechnung in den letzten Jahren. *Enzyk. math. Wiss.* [Art. II A 8a], Vol. II, Teil 1, 1, pp. 626-641.

INDEX